Agricultural Practices to Improve Irrigation Sustainability

Agricultural Practices to Improve Irrigation Sustainability

Guest Editors

Patrícia Palma

Alexandra Tomaz

MDPI

Basel • Beijing • Wuhan • Barcelona • Belgrade • Novi Sad • Cluj • Manchester

Guest Editors
Patrícia Palma
Center for Sci-Tech Research in Earth System and Energy (CREATE)
Polytechnic Institute of Beja
Beja
Portugal

Alexandra Tomaz
Center for Sci-Tech Research in Earth System and Energy (CREATE)
Polytechnic Institute of Beja
Beja
Portugal

Editorial Office
MDPI AG
Grosspeteranlage 5
4052 Basel, Switzerland

This is a reprint of the Special Issue, published open access by the journal *Water* (ISSN 2073-4441), freely accessible at: https://www.mdpi.com/journal/water/special_issues/irrigation_sustainability.

For citation purposes, cite each article independently as indicated on the article page online and as indicated below:

Lastname, A.A.; Lastname, B.B. Article Title. *Journal Name* **Year**, *Volume Number*, Page Range.

ISBN 978-3-7258-6654-0 (Hbk)
ISBN 978-3-7258-6655-7 (PDF)
https://doi.org/10.3390/books978-3-7258-6655-7

Cover image courtesy of Alexandra Telo da Costa Trincalhetas Tomaz

Contents

About the Editors

Patrícia Palma

Patrícia Palma is a Full Professor at the Polytechnic Institute of Beja. Currently, as a researcher, she is a member of the scientific committee of the Center for Scientific and Technological Research in Earth Systems and Energy (CREATE) and associate research at the Geobiotec Research Unit. She develops her research career guided by the principle of deepening knowledge about environmental pollution and its possible consequences for environmental health. She has participated as a coordinator or researcher in several funded projects, particularly in the development of ecotoxicological tools to assess the impact of agricultural activities on heavily modified water bodies, the impact of organic waste use on the recovery of degraded soils, the environmental risk assessment of treated agro-industrial effluents, ecotoxicological tools to assess the environmental risk of pesticides, and the use of IoT to improve laboratory analyses. An active researcher with over 70 publications, she is also a dedicated academic advisor, having supervised 4 doctoral students, 17 master's students, and 28 postdoctoral researchers. She is a member of the Society of Toxicology and Environmental Chemistry, of the Portuguese Association of Toxicology, and of the Board of Directors of the Southern Regional Committee of the Water Resources Association.

Alexandra Tomaz

Alexandra Tomaz is an Assistant Professor at the School of Agriculture of the Polytechnic University of Beja (Portugal) and an integrated researcher at the CREATE Center for Sci-Tech Research in Earth System and Energy (CREATE). Her research focuses on soil and water conservation, sustainable irrigation management, and water-use efficiency in Mediterranean agroecosystems.

She has been involved in several national and European research projects addressing sustainable agricultural practices and water management under climate uncertainty. Her scientific output includes publications in international peer-reviewed journals and active collaboration within national and international research networks. Since 2022, she has been a Guest Editor of several Special Issues of the journal *Water*, focusing on irrigation sustainability and soil and water quality in agricultural systems. Her work also places strong emphasis on knowledge transfer, including organizing scientific and technical events, co-authoring technical manuals on sustainable Mediterranean agroecosystem management, and contributing to professional and outreach publications. Through research, teaching, and science communication, she actively promotes sustainable soil and water management, particularly in Mediterranean European regions.

water

MDPI

Editorial

Agricultural Practices to Improve Irrigation Sustainability

Patrícia Palma [1,2,3,*] and Alexandra Tomaz [1,3,*]

1 Instituto Politécnico de Beja, Escola Superior Agrária de Beja, Rua Pedro Soares, 7800-295 Beja, Portugal
2 Instituto de Ciências da Terra (ICT), Universidade de Évora, 7000-671 Évora, Portugal
3 GeoBioTec, NOVA School of Science and Technology, Campus da Caparica, 2829-516 Caparica, Portugal
* Correspondence: ppalma@ipbeja.pt (P.P.); atomaz@ipbeja.pt (A.T.); Tel.: +351-284314300 (P.P. & A.T.)

1. Introduction

Water is the main limiting factor in agricultural production in regions where the annual or seasonal rainfall is insufficient for the water requirements of crops. Agriculture and water use are intricately related. Globally, around 70% of all water extracted from water supplies is used in irrigation. Irrigated agriculture plays a key role in feeding the world's population, being responsible for 40% of the global food production while taking up only 20% of the global cultivated land [1,2].

Data from FAOSTAT (2024) [3] on the evolution of world agricultural and irrigation-equipped areas from 2011 to 2021 show an increase in the amount of land used for irrigation, while the global area of agriculture has slightly decreased (Figure 1). When looking at global values regarding the area and yield of three major groups of crops (cereals, fruit, and vegetables) from 2011 to 2021, it is clear that production growth was not accomplished due to an increase in the crops' area (Figure 2) [4]. On the other hand, an increase in crop productivity in recent years has been achieved thanks to the technological evolution of agriculture, no doubt including advances in the efficiency of irrigation water use, soil, water, and biodiversity conservation practices, and smart agricultural practices. However, pressures on soil and water resources have also grown: e.g., the increasing use of chemical inputs or farm mechanization; the expansion of soil degradation due to salinization, erosion, or contamination; and the rising of polluting processes in surface or groundwater resources [5,6]. Therefore, although productivity developments have been driving growth, the environment remains under pressure and inadequate agricultural practices affect ecosystems and the services they provide [6–11].

Taking the above into consideration, it is crucial to address water scarcity due to climate change while ensuring food security, enhancing water use efficiency in crop production, and minimizing the negative environmental impacts associated with intensive agricultural practices [12–16]. This editorial provides an overview of the Special Issue "Agricultural Practices to Improve Irrigation Sustainability", which focuses on agroecological practices, advances in agronomic technology, and effective management actions to promote sustainable irrigated agriculture.

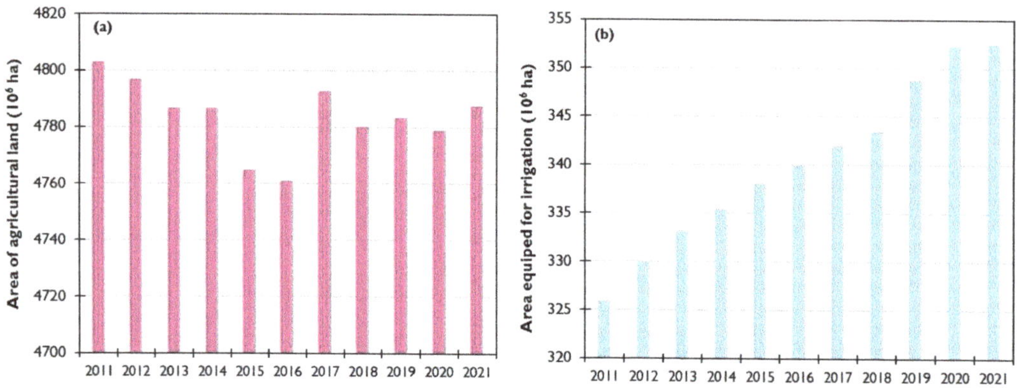

Figure 1. Evolution of (**a**) agricultural land area and (**b**) irrigation-equipped area during the period 2011–2021 (data from FAOSTAT [3]).

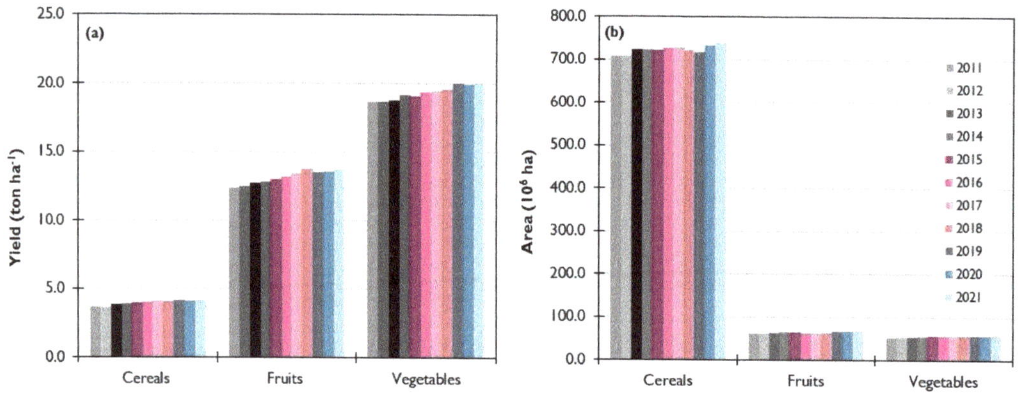

Figure 2. Evolution of average values of global cereal, fruit, and vegetable (**a**) yields and (**b**) area during the period 2011–2021 (data from FAOSTAT [4]).

2. Brief Description of the Selected Articles

The articles selected for publication, including research articles and one review, cover a wide range of topics related to irrigation sustainability, from the assessment of soil quality to the evaluation of management techniques and options, to the examination of indicators of agroecosystem sustainability.

Tomaz et al. (Contribution 1) carried out a study in irrigated farm fields in southern Portugal to evaluate the spatial and temporal variability of soil properties and their correlations with management practices, using multivariate statistical methods (factor analysis and discriminant analysis). The most influential factors and variables in temporal discrimination (sampling dates) were those related to chemical composition, with electric conductivity as the preponderant indicator. As for the spatial differentiation, the dominant factor in the surface layer (0–20 cm) was texture, and in the sub-surface layer (20–40 cm), the dominant factor was nutrient availability. The most important discriminant indicators of spatial variability were the proportion of fine sand and the available potassium, respectively, for the surface and sub-surface layers. The results showed that the multidimensional and integrated assessment of patterns of temporal and spatial variation in soil functions from agricultural practices or soil degradation processes can be valuable in improving crop productivity and soil health.

Moghbel et al. (Contribution 2) studied the well-known HYDRUS-1D numerical model, which can facilitate the exploration of management scenarios to mitigate the consequences of irrigation with poor quality water, especially soil salinization. Their research focused on calibrating the model and analyzing its parameters and the uncertainty of its outputs, using the generalized likelihood uncertainty estimation (GLUE) algorithm for simulating soil salinity in corn root zones under saline irrigation with a linear-moving sprinkle irrigation system. The results showed lower uncertainty in parameters related to water flow and solute transport compared to others and a higher level of uncertainty for the diffusion coefficient, which the authors attributed to the minor contribution of diffusion to the solute transport process in the soil compared with advection and hydrodynamic dispersion under saline water irrigation conditions. The calibrated model performed well in simulating soil water content and electrical conductivity at the corn root zone, thus providing a methodology to help manage poor-quality irrigation water and its effect on plants and soil.

A comparison between different geo-resistivity methods was carried out by Aziz et al. (Contribution 3) to evaluate the performance of mole drains in salt-affected clay soils in the Nile Delta region of Egypt. Geo-electrical surveys were conducted on three newly reclaimed farms to image the subsurface soil drainage conditions and to evaluate the efficiency of using traditional mole drain systems in these types of soils. The results showed that the proximity of buried mole drainage layers to topsoil reduced their effectivity for soil drainage and prevented deep-rooted plant growth. These results suggest that integrated models can be used to improve soil conditions and, thus, agricultural practices in these areas.

Cappelluti et al. (Contribution 4) reported a field experiment in a 5-year-old peach orchard in a Mediterranean environment to study the effect of mixed composed amendments, applied in different amounts, on the dynamics of soil water status, seeking to improve the use of rainwater and irrigation water in Mediterranean environments. The soil water balance indicators, soil water content, and relative extractable water showed that the soil storage capacity increased with the addition of amendment. Improved soil storage capacity was associated with higher values of stem water potential and stomatal conductance, while shoot and fruit growth observations were consistent with the soil water content dynamics.

Sobreiro et al. (Contribution 5) carried out a keywords-based search of peer-reviewed publications, using the following as primary keywords: irrigated olive orchards, high density/intensive/hedgerow olive orchards/groves, irrigation strategies, and soil management. Framed by the concerns about possible negative impacts of modern olive orchard production that have arisen in recent years and putting into question the trade-offs between the production benefits and the environmental costs, these authors performed a review to research the progress made regarding agronomic options that preserve ecosystem services in high-density irrigated olive orchards. They found several studies reporting that intermediate irrigation levels linked to the adoption of deficit irrigation strategies can be effective options to increase water use efficiency. Additionally, with irrigation, it is possible to implement agroecosystems with cover crops, non-tillage, and recycling of pruning residues. These practices reduce soil erosion and nutrient leaching, improve the soil organic carbon by 2 to 3 t C ha^{-1} $year^{-1}$, and increase the biodiversity of plants and animals.

Mullen and Niu (Contribution 6) developed a new methodology for comparing the cost-effectiveness of sustainable agricultural water policies during times of drought. They compared two policy options for consideration by the state of Georgia in the lower Flint River basin: namely, irrigation buyout auctions and source switching. The results of their study demonstrated, on one hand, the importance of modeling uncertainty associated with both the frequency and timing of drought and the hydrologic effects of source switching, and, on the other hand, that the cost-effectiveness of irrigation buyout auctions decreases as the frequency of drought increases.

Esenarro et al. (Contribution 7) evaluated different water management techniques in ancient hydraulic systems located in arid climate regions of Peru, Morocco, and Iran. They analyzed climatic and water supply data, as well as the structure and operation of the systems, having observed that the techniques employed in different civilizations are responses to contextual realities, offering an adaptive solution to environmental and physical challenges.

In the study by Naher et al. (Contribution 8), exploratory data analysis techniques were employed to examine historical changes in wheat and corn cropping patterns in the Texas High Plains from the perspective of geographical concentration and spatial autocorrelation, from 1978 to 2017. The results regarding the temporal changes indicated that the harvested acres of corn and wheat tended to decrease throughout the study period. Also, the total and irrigated harvested corn and wheat areas were concentrated in a smaller number of counties over time, while wheat production was mainly concentrated in the northern part of the region.

Sezen et al. (Contribution 9) assessed changes in the development parameters of *Salvia splendens* L., a commonly used plant in seasonal floriculture in urban green spaces, through the implementation of deficit irrigation practices. Their study evaluated the effect of four irrigation treatments—100% (control), 75%, 50%, and 25% of the pot's water-holding capacity—on plant parameters (number of flowers, flower stem thickness, flower diameter, flower height, leaf chlorophyll value, leaf area, and root length) at two stages of the development, cycle as well as fresh and dry weight measurements of the flowers, vegetative parts, and roots. The results revealed that, in comparison to 100% water application, *Salvia splendens* L. plants exhibited positive effects in the assessed parameters when subjected to 75% water application, except for the flower diameter parameter. Therefore, it is expected that reducing the water application by 25% when cultivating *Salvia splendens* L. can yield substantial water conservation benefits while maintaining good levels of plant development.

3. Conclusions

This Special Issue highlights the diversity and complexity of irrigation systems and the related challenges faced in different regions and contexts. It offers valuable insights into irrigation sustainability in the face of climate change and growing water demands and we expect that it will encourage more research and efforts in addressing this crucial issue. The selected papers also demonstrate the interdisciplinary and multi-scale nature of irrigation sustainability, involving different disciplines, methods, and stakeholders.

Author Contributions: Conceptualization, P.P. and A.T.; writing—original draft preparation, P.P. and A.T.; writing—review and editing, P.P. and A.T.; project administration, P.P.; funding acquisition, P.P. and A.T. All authors have read and agreed to the published version of the manuscript.

Funding: This research is co-funded by the European Union through the European Regional Development Fund, included in COMPETE 2020 (Operational Program for Competitiveness and Internationalization) through the ICT project (UIDB/04683/2020), with the reference POCI-01-0145-FEDER-007690, and through Geobiotec (UIDB/04035/2020), funded by FCT—Fundação para a Ciência e a Tecnologia, Portugal.

Conflicts of Interest: The authors declare no conflicts of interest.

List of Contributions:

1. Tomaz, A.; Martins, I.; Catarino, A.; Mourinha, C.; Dôres, J.; Fabião, M.; Boteta, L.; Coutinho, J.; Patanita, M.; Palma, P. Insights into the Spatial and Temporal Variability of Soil Attributes in Irrigated Farm Fields and Correlations with Management Practices: A Multivariate Statistical Approach. *Water* **2022**, *14*, 3216. https://doi.org/10.3390/w14203216.
2. Moghbel, F.; Mosaedi, A.; Aguilar, J.; Ghahraman, B.; Ansari, H.; Gonçalves, M.C. Bayesian Calibration and Uncertainty Assessment of HYDRUS-1D Model Using GLUE Algorithm for Simulating Corn Root Zone Salinity under Linear Move Sprinkle Irrigation System. *Water* **2022**, *14*, 4003. https://doi.org/10.3390/w14244003.
3. Aziz, A.; Berndtsson, R.; Attia, T.; Hamed, Y.; Selim, T. Noninvasive Monitoring of Subsurface Soil Conditions to Evaluate the Efficacy of Mole Drain in Heavy Clay Soils. *Water* **2023**, *15*, 110. https://doi.org/10.3390/w15010110.

4. Cappelluti, O.; Bruno, M.R.; Modugno, A.F.; Ferrara, R.M.; Gaeta, L.; De Carolis, G.; Campi, P. The Use of Mixed Composed Amendments to Improve Soil Water Content and Peach Growth (*Prunus persica* (L.) Batsch) in a Mediterranean Environment. *Water* **2023**, *15*, 1708. https://doi.org/10.3390/w15091708.
5. Sobreiro, J.; Patanita, M.I.; Patanita, M.; Tomaz, A. Sustainability of High-Density Olive Orchards: Hints for Irrigation Management and Agroecological Approaches. *Water* **2023**, *15*, 2486. https://doi.org/10.3390/w15132486.
6. Mullen, J.D.; Niu, Y. Cost-Effectiveness of Sustainable Agricultural Water Policies: Source Switching versus Irrigation Buyout Auctions in Georgia's Lower Flint River Basin. *Water* **2023**, *15*, 3381. https://doi.org/10.3390/w15193381.
7. Esenarro, D.; Vilchez, J.; Adrianzen, M.; Raymundo, V.; Gómez, A.; Cobeñas, P. Management Techniques of Ancestral Hydraulic Systems, Nasca, Peru; Marrakech, Morocco; and Tabriz, Iran in Different Civilizations with Arid Climates. *Water* **2023**, *15*, 3407. https://doi.org/10.3390/w15193407.
8. Naher, A.; Almas, L.K.; Guerrero, B.; Shaheen, S. Spatiotemporal Economic Analysis of Corn and Wheat Production in the Texas High Plains. *Water* **2023**, *15*, 3553. https://doi.org/10.3390/w15203553.
9. Sezen, I.; Yağanoğlu, S.; Akpınar Külekçi, E.; Karahan, A. Effect of Deficit Irrigation on Growth Parameters of the *Salvia splendens* L. Plant. *Water* **2023**, *15*, 4187. https://doi.org/10.3390/w15234187.

References

1. FAO. *The State of the World's Land and Water Resources for Food and Agriculture—Systems at Breaking Point*; Synthesis Report; FAO: Rome, Italy, 2021.
2. FAO. Water. Key Facts. Available online: http://www.fao.org/water/en/ (accessed on 27 September 2021).
3. FAO. Land Use. Available online: https://www.fao.org/faostat/en/#data/RL (accessed on 2 February 2024).
4. FAO. Crops and Livestock Products. Available online: https://www.fao.org/faostat/en/#data/QCL (accessed on 2 February 2024).
5. United Nations. *Global Sustainable Development Report 2023: Times of Crisis, Times of Change: Science for Accelerating Transformations to Sustainable Development*; United Nations: New York, NY, USA, 2023; p. 224.
6. OECD/FAO. *Environmental Sustainability in Agriculture 2023*; OECD: Paris, France; FAO: Rome, Italy, 2023.
7. Liu, Q.; Sun, X.; Wu, W.; Liu, Z.; Fang, G.; Yang, P. Agroecosystem Services: A Review of Concepts, Indicators, Assessment Methods and Future Research Perspectives. *Ecol. Indic.* **2022**, *142*, 109218. [CrossRef]
8. Adhikari, K.; Hartemink, A.E. Linking Soils to Ecosystem Services—A Global Review. *Geoderma* **2016**, *262*, 101–111. [CrossRef]
9. Dale, V.H.; Polasky, S. Measures of the Effects of Agricultural Practices on Ecosystem Services. *Ecol. Econ.* **2007**, *64*, 286–296. [CrossRef]
10. Alves-Ferreira, J.; Vara, M.G.; Catarino, A.; Martins, I.; Mourinha, C.; Fabião, M.; Costa, M.J.; Barbieri, M.V.; de Alda, M.L.; Palma, P. Pesticide Water Variability and Prioritization: The First Steps towards Improving Water Management Strategies in Irrigation Hydro-Agriculture Areas. *Sci. Total Environ.* **2024**, *917*, 170304. [CrossRef] [PubMed]
11. Minhas, P.S.; Ramos, T.B.; Ben-Gal, A.; Pereira, L.S. Coping with Salinity in Irrigated Agriculture: Crop Evapotranspiration and Water Management Issues. *Agric. Water Manag.* **2020**, *227*, 105832. [CrossRef]
12. Velasco-Muñoz, J.F.; Aznar-Sánchez, J.A.; Belmonte-Ureña, L.J.; López-Serrano, M.J. Advances in Water Use Efficiency in Agriculture: A Bibliometric Analysis. *Water* **2018**, *10*, 377. [CrossRef]
13. Fernández, J.E.; Alcon, F.; Diaz-Espejo, A.; Hernandez-Santana, V.; Cuevas, M.V. Water Use Indicators and Economic Analysis for On-Farm Irrigation Decision: A Case Study of a Super High Density Olive Tree Orchard. *Agric. Water Manag.* **2020**, *237*, 106074. [CrossRef]
14. Levidow, L.; Zaccaria, D.; Maia, R.; Vivas, E.; Todorovic, M.; Scardigno, A. Improving Water-Efficient Irrigation: Prospects and Difficulties of Innovative Practices. *Agric. Water Manag.* **2014**, *146*, 84–94. [CrossRef]
15. Tomaz, A.; Patanita, M.; Guerreiro, I.; Dôres, J.; Boteta, L.; Palma, J. Efficient Use of Water and Nutrients in Irrigated Cropping Systems in the Alqueva Region. *Span. J. Soil Sci.* **2018**, *8*, 12–23. [CrossRef]
16. Hoover, D.L.; Abendroth, L.J.; Browning, D.M.; Saha, A.; Snyder, K.; Wagle, P.; Witthaus, L.; Baffaut, C.; Biederman, J.A.; Bosch, D.D.; et al. Indicators of Water Use Efficiency across Diverse Agroecosystems and Spatiotemporal Scales. *Sci. Total Environ.* **2023**, *864*, 160992. [CrossRef] [PubMed]

water

MDPI

Review

Sustainability of High-Density Olive Orchards: Hints for Irrigation Management and Agroecological Approaches

Justino Sobreiro [1,2,*], Maria Isabel Patanita [1,2], Manuel Patanita [1,2] and Alexandra Tomaz [1,2]

1 Instituto Politécnico de Beja, Escola Superior Agrária de Beja, Rua Pedro Soares, 7800-295 Beja, Portugal; ipatanita@ipbeja.pt (M.I.P.); mpatanita@ipbeja.pt (M.P.); atomaz@ipbeja.pt (A.T.)
2 GeoBioTec, Nova School of Science and Technology, Campus da Caparica, 2829-516 Caparica, Portugal
* Correspondence: justino.sobreiro@ipbeja.pt

Abstract: The production of olive oil in Portugal and other countries of the Mediterranean region has greatly increased in recent years. Intensification efforts have focused on the growth of the planted area, but also on the increase of the orchards density and the implementation of irrigation systems. Concerns about possible negative impacts of modern olive orchard production have arisen in the last years, questioning the trade-offs between the production benefits and the environmental costs. Therefore, it is of great importance to review the research progress made regarding agronomic options that preserve ecosystem services in high-density irrigated olive orchards. In this literature review, a keywords-based search of academic databases was performed using, as primary keywords, irrigated olive orchards, high density/intensive/hedgerow olive orchards/groves, irrigation strategies, and soil management. Aside from 42 general databases, disseminated research, and concept-framing publications, 112 specific studies were retrieved. The olive orchards were classified as either traditional (TD) (50–200 trees ha^{-1}), medium-density (MD) (201–400 trees ha^{-1}), high-density (HD) (401–1500 trees ha^{-1}), or super-high-density (SHD) orchards (1501–2500 trees ha^{-1}). For olive crops, the crop coefficient (Kc) ranges ranges from 0.65 to 0.70, and can fall as low as 0.45 in the summer without a significant decrease in oil productivity. Several studies have reported that intermediate irrigation levels linked with the adoption of deficit irrigation strategies, like regulated deficit irrigation (RDI) or partial rootzone drying (PRD), can be effective options. With irrigation, it is possible to implement agroecosystems with cover crops, non-tillage, and recycling of pruning residues. These practices reduce the soil erosion and nutrient leaching and improve the soil organic carbon by 2 to 3 t C ha^{-1} year^{-1}. In this situation, in general, the biodiversity of plants and animals also increases. We expect that this work will provide a reference for research works and resource planning focused on the improvement of the productive and environmental performance of dense irrigated olive orchards, thereby contributing to the overall enhancement of the sustainability of these expanding agroecosystems.

Keywords: irrigated olive orchards; high density; hedgerow; irrigation strategies; soil management

1. Introduction

The world's olive cultivation area was about 10.3 Mha in 2021, yielding 23 million tons of olive fruits. From 2015 to 2021, the top European olive producers were Spain (32%), Greece (13%), Italy (10%), and Portugal (4%) [1]. Portugal is the fourth olive producer in Europe and the eighth in the world. Despite only representing 4% of the total olive cultivation area and 4% of the total production worldwide, the Portuguese olive sector is an important source of income for the country.

In Portugal, the Southern region of Alentejo is the main production province, comprising 52.4% of the total Portuguese area (377,234 ha), with a large part occupied by dense irrigated plantations (Figure 1). Irrigated orchards cover 31.7% of the total Portuguese olive tree area, 13.8% corresponding to super-high-density orchards, with over

1500 trees ha^{-1} [2] (Figure 1). The olive harvest occurs every year from early October to January. The oil content of Portuguese olives varies from 14% to 20% of the fruit fresh weight, depending on the cultivar and harvest date. Normally, early harvesting leads to lower oil content of the fruits [3–10].

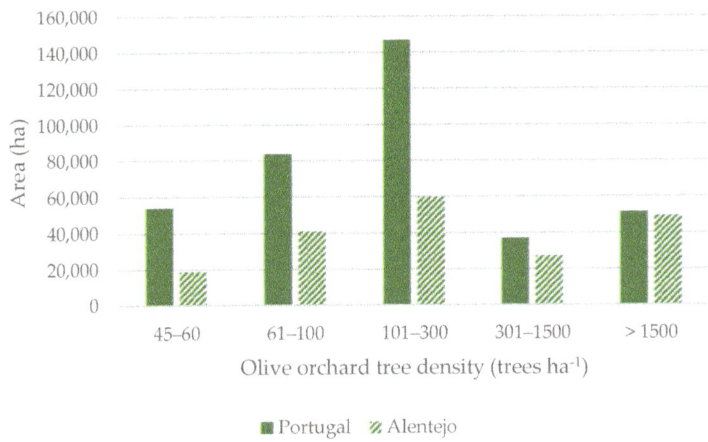

Figure 1. Comparison between olive orchard area (ha) in Portugal and the Portuguese Alentejo region by tree density classes [2].

Traditionally, olive trees have been grown in the region surrounding the Mediterranean, mainly as a rainfed crop with low productivity given the typical dry environment of this region. In recent years, the expansion of olive oil and table olive production has been achieved through both an increase in the planted area and through intensification within and beyond the Mediterranean countries by increasing the orchards' density and via the introduction of irrigation [11–13]. In fact, in the last two decades, high-density (HD; 401 to 1500 trees ha^{-1}) and super-high-density (SHD; 1501 to 2500 trees ha^{-1}) orchards, known as hedgerow olive orchards, have been developed to further reduce harvesting costs using over-the-row harvesting machines [14–16]. Because of the higher water demand of the dense canopies and the low soil volume available for each tree, irrigation is usually needed [12,17,18]. The current water scarcity in traditional olive-growing regions, like Alentejo, along with the expected increase in heat waves and droughts caused by climate change [16,19,20], imply an urgent need to reduce the use of water for irrigation of crops in these regions and to adopt measures to avoid the degradation of soil resources and biodiversity.

The adoption of management practices that maintain ecosystem services (SE), like soil and water conservation practices for the regulation of SE, or biodiversity preservation [15] for the support of SE, is a key aspect of modern agriculture [21,22]. The olive-growing sector is no exception to this premise. In fact, the increase in irrigated dense plantations of olives has led to relevant changes in the landscapes of some regions, and the risk of negative impacts of this agricultural intensification on the environment must be avoided [14,23,24].

The target of this review is to focus on the sustainability of high- and super-high-density olive orchards.

We aim to contribute towards an optimized system in terms of water management, with a focus on irrigation strategies and agroecological practices to enhance the health of these agroecosystems. It is our objective to provide: (i) a systematization of the type of olive orchards that can be found regarding their tree density; and (ii) an overall research output regarding water-saving irrigation strategies and agroecological options in irrigated dense olive orchards.

For this purpose, we applied a keywords-based search of academic databases. The primary keywords used were irrigated olive orchards, high density/intensive/hedgerow olive orchards/groves, irrigation management, and soil management. Within these, the secondary keywords used were water requirements, deficit irrigation strategies, erosion, infiltration rate, surface temperature, pesticides, herbicides, diseases, cover crops, pruning residues, organic matter, organic carbon accumulation, nitrogen accumulation, and biodiversity.

2. The Olive Orchard Mosaic

2.1. The Traditional Olive Orchards

When traveling in the Mediterranean area, one can often find olive orchards planted in the XIXth century or up to the mid-XXth century, with fewer than 50 trees ha^{-1} to a maximum of 200 trees ha^{-1}, that are still productive today. These were sometimes planted on sharp slopes or small and narrow terraces made with stone walls, as can mainly be observed in the north of Portugal, providing landscapes of great beauty.

In traditional olive orchards (TD), the management of cover crops is conducted by tillage or total herbicide coverage. Grain crops were traditionally grown within olives as primary sources of farmers' income. In these situations, the soil erosion can be quite dramatic [25–27], and at the same time, the temperature of the soil's top layer is quite high in the summer (over 40 °C). Although olive is a well-adapted species to drought conditions, the soil's exposure to direct sun and the lack of canopy shade over the tree root zone leads to water and heat stress, and can induce summer dormancy in the trees [28–30].

The farmers use few fertilizers and apply a reduced number of chemical pest and disease treatments in the olive groves. They are pruned every four years by chain saw, and the pruning residue is generally burned. The alternate bearing is very strong, with a sparse yield in the year following pruning [14]. Since these orchards are rainfed, the biodiversity of species is sometimes low due to the lack of water and cover crops [31–33].

Traditionally, the harvest is performed by hand with wood sticks, although nowadays, some growers use portable backpack shakers with or without nets covering the floor. The net production of these olive ecosystems is less than 3 t ha^{-1} of fruits. The quality of the oil produced is often affected by diseases like anthracnose (*Colletotrichum* sp.) [34] or by contamination of the fruit through direct contact with the orchard floor [35]. The overall sustainability of this traditional olive system is currently compromised due to the lack of workers and the labor price [36] (Table 1).

2.2. The Medium-Density Olive Orchards

The most common olive orchards in the Mediterranean area are those with medium density (MD; 200–400 trees ha^{-1}), which are very likely to be observed in lime soils of the southern parts of Portugal or Spain. They are rainfed or little irrigated, and the soil is kept weed-free by tillage or by partial (in the rows) or total herbicide application. Many have spontaneous cover plants, mainly in the interrows, which are used to some extent as grazing lands. In this case, animal manure provides some nutrient recycling for the ecosystem and complements the annual fertilization. The pruning is carried out in alternate years and is less intense than in the traditional orchards. The pruning residue is often burned.

The sun exposure of the soil is lower due to the improved tree shade, resulting in better development of resident herbaceous vegetation that increases insect populations, improving biodiversity, and provides more protection against soil erosion than in the TD systems.

The harvest is carried out by tree shaking using floor nets or wraps around the trees as collecting systems. These orchards have been upgraded over time by increasing plant density and providing better irrigation. This agricultural system is undergoing a fast transition to a higher-density system [37–39].

Table 1. Systematization of the most common olive orchards' agricultural systems in the Mediterranean climate and their features. Traditional (TD), medium-density (MD), high-density (HD), and super-high-density (SHD).

<table>
<tr><th>Orchard Type</th><th>Spacing Inter-row × Row (m)</th><th>Tree Density (trees ha^{-1})</th><th>Productivity (t ha^{-1})</th><th>Soil Conservation</th><th>Tree Architecture</th><th>Pruning</th><th>Irrigation and Soil Management</th><th>Harvest</th><th>Common Cultivars</th></tr>
<tr><td>Traditional (TD)</td><td>8–15 × 6–15</td><td>50–200</td><td>0.5–3</td><td>Slopes: 0 to 30%.
Strong erosion.</td><td>Trichotomic vase canopy.
Strong alternate bearing.</td><td>Every 4 years.
Chain saw.
Pruning residue is burned.</td><td>Non-irrigated.
Soil tillage, inter-row grain crops. Herbicides.</td><td>Hand branch shakers, with or without floor nets.</td><td>Galega, Verdeal, Cordovil.</td></tr>
<tr><td>Medium-density (MD)</td><td>7–8 × 3.5–6</td><td>201–400</td><td>3–6</td><td>Slopes: 0 to 15%. Some erosion.</td><td>Trichotomic vase canopy.
Alternate bearing.</td><td>Every 2 years.
Chain saw.
Pruning residue is burned.</td><td>Non-irrigated or low-irrigated.
Soil tillage, herbicides, or spontaneous weed cover, some used for animal pasture.</td><td>Trunk shaker, floor nets. Wrap around the tree collector.</td><td>Galega, Verdeal, Cordovil, Cobrançosa, Picual, Frantoio</td></tr>
<tr><td>High-density (HD)</td><td>4–7 × 1.7–3.5</td><td>401–1500</td><td>6–12</td><td rowspan="2">Slopes: 0 to 10%.
Low erosion.</td><td rowspan="2">Dichotomic vase or hedge row.
Some alternate bearing in orchards over 20 years old.</td><td rowspan="2">Every 1–2 years.
Manual shears, electric or air compressed.
Tractor disc trimmers.
Pruning residue is shredded on site.</td><td rowspan="2">Drip irrigation—250–500mm year^{-1}.
Spontaneous or sowed cover crops.
Herbicide in the tree rows or no herbicide.</td><td rowspan="2">Trunk shaker and wrap around the tree collector, or over-the-row.</td><td>Cobrançosa, Picual, Arbequina, Frantoio.</td></tr>
<tr><td>Super-high-density (SHD)</td><td>3.5–4 × 1–1.7</td><td>1501–2500</td><td>12–22</td><td>Arbequina, Arbosana, Koroneiki.</td></tr>
</table>

2.3. *The High- (HD) and Super-High-Density (SHD) Olive Orchards*

The success of the higher-density olive agricultural systems is based on water availability [12,40]. The olive tree is an evergreen species with a remarkable water control process that manages water losses, requiring less water in the summer than in the remaining period of the year [41–43]. Nevertheless, in a region with 562 mm year^{-1} of average rainfall [44], 250 mm to 500 mm year^{-1} of supplemental irrigation water are the necessary values for the trees to achieve their maximum productivity. This demand is lower when compared to the 500–800 mm year^{-1} required by other perennial species (Figure 2). Under these conditions, higher densities lead to increased productivity. The HD and SHD olive orchards are planted with 401–2500 trees ha^{-1}. Plantation is sometimes conducted in ridges of 1.0 m × 0.5 m (width × height) that are meant to prevent waterlogging and improve soil temperature in the early spring. These ridges must be made with special care; otherwise, they can prevent the natural rainfall flow and worsen the waterlogging [45].

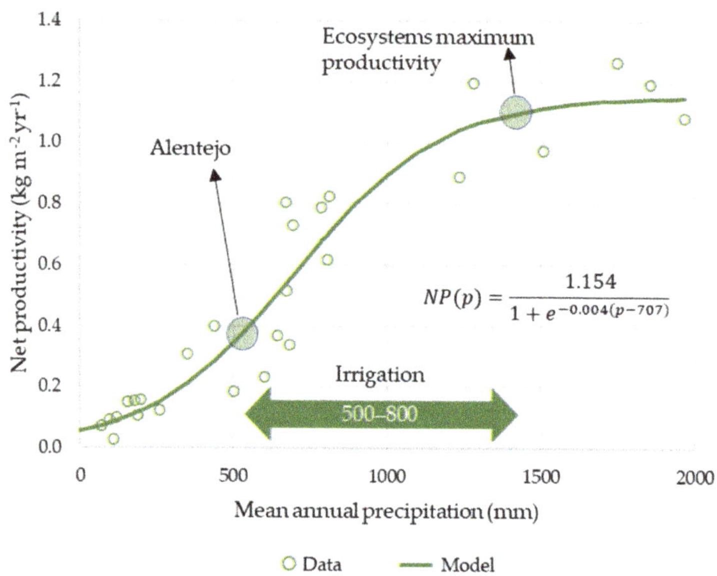

Figure 2. Biomass productivity by world ecosystems. The Mediterranean rainfall in Alentejo is signaled as well as the irrigation requirements, calculated as the difference between the ecosystem's maximum productivity and the average Alentejo rainfall. (Data from Taiz et al. [46]). NP—net productivity (kg m^{-2} year^{-1}), p—precipitation (mm).

Considering soil management, the soil is normally covered with spontaneous or sowed herbaceous vegetation to minimize soil erosion. The sowed cover species could be *Fabaceae* sp., like *Medicago sativa*, *Vicia* sp. or *Trifolium* spp., which are quite important nitrogen recyclers (Table 2). Every 2–8 t ha^{-1} of olive fruits extract 7–28 kg ha^{-1} of N, 2–8 kg ha^{-1} of P_2O_5, and 12–48 kg ha^{-1} of K_2O [47]. The cover species can provide an important contribution in the form of nitrogen balance in the cases of HD and SHD olive orchards. The spontaneous or sowed cover crops are also important refuges for beneficial insects or pollinators, which improve the general biodiversity of HD and SHD orchards [48–54]. Inter-row weed management is usually carried out by shredding 3 to 5 times a year to keep the weeds below 0.5 m in height. The shredding also recycles the pruning residues left in the topsoil of these orchards. The recycling of pruning residues is a good practice which allows the reposition of 2.9 kg t^{-1} N, 1,1 kg t^{-1} P_2O_5, and 2.9 kg t^{-1} K_2O [47], apparently without side effects related to the improvement of orchard diseases [55]. Nevertheless, soil diseases caused by *Verticilium dahliae* may occur [56].

Table 2. Seed quantity necessary to establish the cover crop and nitrogen fixed by hectare with *Fabaceae* species (Adapted from [47]).

Species	Sow Seed Quantity ($kg\ ha^{-1}$)	Nitrogen Fixed ($kg\ ha^{-1}\ yr^{-1}$)
Alfalfa (*Medicago sativa*)	10–25	114–223
Broad bean (*Vicia faba*)	150–200	160–216
Common vetch (*Vicia* sp.)	40–60	90–155
Crimson clover (*Trifolium incarnatum*)	10–20	20–64
Eggs and Bacon (*Lotus corniculatus*)	4–6	49–112
Lentil (*Lens culinaris*)	60–80	15–85
Pea (*Pisum sativum*)	70–140	37–185
Red clover (*Trifolium pratense*)	4–10	68–113
Sub clover (*Trifolium subterraneum*)	10–20	48–183
White clover (*Trifolium repens*)	8–12	165–188

If irrigation lines are directly on the soil surface, they do not allow for weed mowing in the tree lines. Therefore, weed control in the tree row normally requires herbicide application. This issue should be addressed in the near future, as the herbicide glyphosate could be banned, and other chemical solutions are currently less economical [57,58].

One advantage of HD and SHD olive orchards is the soil temperature. In the same location, the temperature of the topsoil in the summer, measured with a FLIR (Forward Looking InfraRed) device, was about 20 °C lower at the top of the cover grass when compared to bare topsoil [59,60].

Finally, HD and SHD olive orchards are more regular in yield, but do not show evidence for strong alternate crop behavior when compared with the other systems [12,14]. The cultivars in use have less vigor and, therefore, provide more regular production, at least during the first 20 years of the orchard's life [14,16,61–63].

Harvests in HD and SHD olive orchards require tractor trunk shakers with wraps around the tree collectors or over-the-row self-propelled machines. The latter can harvest up to one hectare per 1 h (12–22 t of fruits). As the fruits are never in contact with the ground, they are quite suitable for virgin or extra-virgin oil production [39]. In Portugal, the harvest is restricted to the period from sunrise to sunset in order to prevent involuntary bird losses, since these animals often use olive trees as refuges overnight [64].

3. Water Management

3.1. Water Use and Irrigation Requirements

Crop water requirements (CWR) are defined as the amounts of water needed to replace the water lost through evapotranspiration by a disease-free crop growing in large fields under no limitations regarding soil conditions, including soil water and fertility, and achieving full production potential in the given growing environment [65]. This water loss is defined as the crop evapotranspiration (ET_c) under standard conditions, given by Equation (1):

$$ET_c = ET_0 \times K_c \tag{1}$$

where ET_0 is the reference evapotranspiration of a grass-like reference crop, and K_c is the crop coefficient [66,67]. In fact, ET_0 represents an index of climatic demand, and K_c represents the influence of the specific crop characteristics [68]. The K_c in olive orchards is affected by several factors, including the canopy architecture, the fraction of ground covered by the vegetation, crop management practices, and rainfall variability [42]. In the case of olive growing under standard climatic and agronomic conditions, the Kc values recommended by FAO vary between 0.65 in the initial phase and 0.70 in the intermediate and final phases of the development cycle [67]. The monthly Kc values proposed by Pastor and Orgaz (1994) [69] vary between 0.45 in July and August and 0.65 in March and May.

To meet the reduction in the fraction of soil covered by vegetation, or the fraction of shaded area (C, in %) in an olive grove, Fereres and Castel (1981) [70] proposed that ET_c be estimated by Equation (2):

$$ET_c = ET_0 \times K_c \times K_r \tag{2}$$

where K_r should be used when the coverage fraction is less than 50% and corresponds to a reduction coefficient, obtained by Equation (3):

$$K_r = \frac{2\,C}{100} \tag{3}$$

In the case of irrigated crops, the concept of irrigation water requirement (IWR) must be considered. The IWR is the amount of water that is required to be applied to a crop to fully satisfy its specific crop water requirement whenever rainfall, soil water storage, and groundwater contributions are insufficient [68].

Olive's water requirements are a function of cultivar characteristics, management, and environmental demands. Olive trees withstand long periods of drought and can survive in very sparse plantings, even in climates with very low annual rainfall: values of 150–200 mm year^{-1} are indicated in Steduto et al. (2012) [17] and Carr (2013) [11] refers to 200–250 mm year^{-1}. However, as referenced in Section 2.3, for economic production, much higher precipitation or irrigation are required: Carr (2013) [11] states that an average annual precipitation or irrigation above 600 mm year^{-1}, in soils with good water-holding capacity, is needed for successful cultivation; Beede and Goldhamer (2005) [71] found values of around 950 mm year^{-1} for mature olive trees in clean cultivated orchards with 60% or higher shaded areas.

Olives are perennial trees that retain their canopies and use water during the entire year, but, regardless of the growing conditions affecting seasonal water use, they have different sensitivities to water deficits depending on their development stage. While water stress during the period of flower bud formation can lead to a reduced number of flowers, thereby affecting the year's yield, when it occurs during periods of shoot growth, it can affect the next year's yield, which is formed on 1-year-old shoots [45,71,72].

For olive oil production, fruit sets should be managed to maximize oil extractability and quality. Several studies have reported that intermediate irrigation levels linked with the adoption of deficit irrigation during certain stages of fruit development can increase the fruit and oil quality [13,73–76]. Additionally, the slowing of fruit development—known as the pit hardening phase—is considered as the less sensitive period of olive trees to water deficit, when it is possible to reduce or interrupt irrigation without a significant reduction in yield or in oil quality [77–80].

3.2. Irrigation Strategies

The management of irrigation in olive trees following schedules to optimize water productivity can be an effective option to balance vegetative development, yield, and fruit quality while ensuring water conservation [81–83]. These irrigation regimes include supplemental irrigation (SI) and deficit irrigation (DI) strategies. The former is used by applying irrigation in selected phenological stages and is responsible for remarkable responses even with low irrigation supplies. Its goals include achieving maximum yields and eliminating yield fluctuations caused by water deficits [83,84]. The latter are widely adopted in other drought-resistant crops, the most relevant example being grapevine (*Vitis vinifera*) [85–87], where they are commonly supported by physiologically based and soil-based monitoring tools [88–91].

Supplemental irrigation can be defined as the application of a limited amount of water to increase and stabilize crop yields when rainfall fails to provide sufficient water for plant growth [92]. Studies concerning the effect of supplemental irrigation on olive trees' productive responses involve mostly TD and MD orchards in semi-arid conditions (e.g., [76,93–95]).

Deficit irrigation strategies are based on supplying irrigation volumes lower than the irrigation crop requirements under non-limiting growing conditions, that is, below the potential ETc, allowing for water savings in regions with present or future limited water resources without compromising production [96]. Three DI strategies can be considered: (i) sustained (or continuous) deficit irrigation (SDI), (ii) regulated deficit irrigation (RDI), and (iii) partial root-zone drying (PRD). Selected research regarding the use of DI strategies in MD, HD, and SHD olive orchards is summarized in Table 3.

3.2.1. Sustained Deficit Irrigation

In SDI, the irrigation water used at any moment during the season is below the crop evapotranspiration demand. This is based on the idea of allotting the water deficit uniformly over the entire growing season [97]. Thereby, the water deficit increases progressively as the season advances due to a combination of the uniform application of a reduced amount of water and the depletion of available soil water. This allows water stress to develop slowly and for the plants to adapt to the water deficits when the soil presents significant water storage capacity [96]. One of the first and most well-known studies on the effects of SDI in olive trees was published by Goldhamer et al. (1994) [98], which tested eight irrigation rates, ranging from 232 mm (Kc = 0.16) to 1016 mm (Kc = 0.85), in mature olive trees, cv. 'Manzanillo', planted with a density of 239 trees ha^{-1}, in Madera County, California. They reported tree water stress occurring for Kc values of 0.55 or less, and a strong correlation between fruit value (USD/ha^{-1}) and applied irrigation (mm) up to 950 mm, indicating that higher amounts of irrigation water do not correspond to increased economic water productivity when a threshold value is exceeded. Grattan et al. (2006) [74], by studying the effect of different water-application treatments on oil yield in a SHD olive orchard, cv. 'Arbequina', found that oil yields can be maximized over a rather broad range of applied water, since increases in fruit yield with higher irrigation levels are offset by the reduction in the percentage of oil extracted. In a MD orchard, cv. 'Cobrançosa', in Northern Portugal, Fernandes-Silva et al. (2010) [82] reported that the oil yield increased to more than double with SDI treatment when compared to rainfed conditions. Santos et al. (2018) [99] studied the water use and productivity of the same cultivar in an orchard with 300 trees/ha^{-1}, located in Alentejo, under two deficit irrigation treatments, and found that the 70% of ETc strategy presented a higher yield and increased water use efficiency. Other studies on the SDI technique applied in medium-to-dense olive orchards usually consist of comparisons with other DI strategies, like RDI, and the reported results point to similar yield responses [78,81,100].

3.2.2. Regulated Deficit Irrigation

The RDI strategy consists of reducing or withholding irrigation water during specific periods to manipulate plants' vegetative and reproductive growth [13,101]. The less sensitive period for olive trees to water deficits is midsummer, when it is possible to reduce or interrupt irrigation without a significant yield reduction nor decreased oil quality [72,77,100] (Figure 3). However, during certain stages of the growth cycle, irrigation supplies must balance, or be close to, the crops' water needs (Figure 3). According to Fernández et al. (2013) [102], these periods are:

- From the last stages of floral development to full bloom, normally in mid-April, when water stress can affect flower fertilization.
- At the end of the first stage of fruit development, normally in June, when water stress causes reductions in fruit size.
- After the midsummer period, normally from late August to mid-September, when a marked increase in oil accumulation occurs.

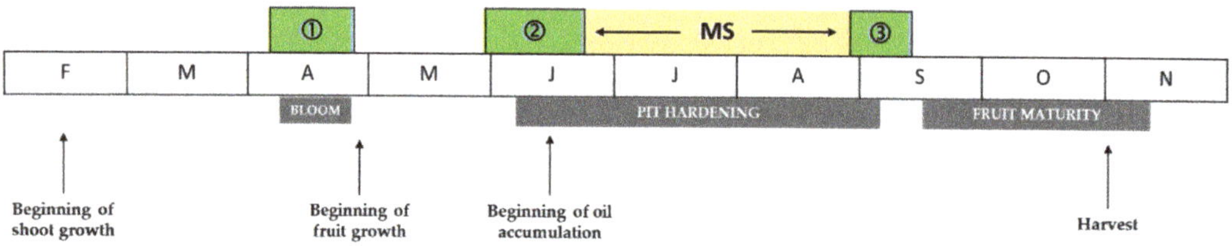

Figure 3. Olive growth cycle and periods at which olive trees are most sensitive to water stress, indicated as ①, ②, and ③. MS is the midsummer period, normally from late June to late August, during which olive trees are resistant to drought and irrigation can be reduced or withdrawn (adapted from Fernández et al. (2013) [102]).

The studies regarding RDI strategies are usually based on timing the withdrawal of or reduction in irrigation during midsummer and/or immediately before and after this period [75,81,100,102]. Usually, different percentages of reduction in irrigation are tested with the aim to understand the threshold at which the reduction in fruit yield caused by stress can be offset by the maintenance/increase in the percentage of oil extracted, as well as in the oil quality [75,102,103]. Additionally, trees' water statuses are monitored, and these measurements are used to define thresholds for irrigation scheduling. The most frequently used physiological parameters are leaf (Y) or stem water potential (Y_{Stem}), measured at different times of the day, normally at predawn or midday [80,102,104–106]; stomatal conductance (g_S); net photosynthesis (A_N); and evapotranspiration rate (E) at the leaf level [82,105,107].

3.2.3. Partial Rootzone Drying

The PRD technique requires that approximately half of the root system be maintained in a drying state while the remainder of the root system is irrigated [85,101,108,109]. The theoretical background of PRD is that irrigation of part of the root system keeps the upper part of crops in favorable water conditions, while the drought in the other part of the roots induces the formation of root chemical signals, mainly abscisic acid, which are transported to the upper parts of the plants to induce reductions in stomatal conductance and shoot growth [110–112]. The aim of PRD would then be to reduce water losses by transpiration without affecting the yield. However, as stated by Fernández et al. (2006) [107], the studies carried out to date have not always supported this hypothesis, and PRD and RDI do not differ significantly in terms of water productivity [113].

In general, studies of the effects of PRD on olive trees grown under semi-arid conditions have in common a slight PRD-induced yield reduction, although with high water productivity and no reduction in oil yield [114–116].

Table 3. Summary of selected studies on the effect of irrigation strategies on olive production in medium- to super-high-density olive orchards.

DI Strategy	Cultivar	Location	Annual Rainfall (mm)	Orchard Type	Irrigation Treatments	Main Results	Reference
SDI	Arbequina	California, USA	533 (3-year average during the experiment)	30-month commercial orchard (1709 trees ha^{-1})	7 treatments: 1–15 (28), 2–25 (33), 3–40 (55), 4–57 (75), 5–71 (93), 6–89 (117), and 7–107 (140)% ETc [(1)] (SDI in treatments 1 to 6 and 1 to 5 in the first and second year of the trial, respectively)	SDI treatments of 70–75% ETc did not reduce oil yields significantly; sustained season-long irrigation deficit of approximately 33–40% ETc maximized oil quality (chemical parameters, flavor, and stability).	Berenguer et al. (2004) [73] Grattan et al. (2006) [74]
SDI	Cobrançosa	Vilariça Valley, Portugal	520	10-year-old commercial orchard (278 trees ha^{-1})	3 treatments: R 30% ETc (SDI) FI	With SDI treatment, the oil yield increased to more than double that of rainfed conditions; 25% oil yield reduction in SDI compared to FI.	Fernandes-Silva et al. (2010) [82]
SDI	Frantoio	Venturina, Italy	635	10-year-old experimental orchard (513 trees ha^{-1})	3 treatments: FI 46–52% ETc (SDI) 2–6% ETc (SI)	The fruit yield of the SDI trees was 68% of that of FI; the fruit sets and numbers of fruits of the FI trees were similar to those of SDI trees and significantly higher than the SI trees; the oil yield of the DI treatment was 82% that of FI trees.	Caruso et al. (2013) [117]
SDI and RDI	Picual	Cordoba, Spain	602	18-year-old experimental orchard (278 trees ha^{-1})	5 treatments: FI 75% ETc and no irrigation from mid-July to mid-September (RDI) 75% ETc (SDI) Adaptation to alternate bearing habit: R during years of few or no crops and FI during heavy crop years R	Responses to deficits were similar for SDI and RDI; yield responses to FI during the bearing year and R in the nonbearing year were less favorable than those observed in SDI and RDI.	Moriana et al. (2003) [81]
SDI and RDI	Arbequina	Cordoba, Spain	502 (3-year average during the experiment)	12-year-old experimental orchard (408 trees ha^{-1})	3 treatments: FI 25% IWR (SDI) 25% IWR and no irrigation in midsummer (RDI)	RDI and SDI caused higher reductions in fresh fruit yield than oil yield due to a higher oil concentration in deficit-irrigated trees	Iniesta et al. (2009) [100]
SDI and RDI	Koroneiki	Nicosia, Cyprus	428	17-year-old commercial orchard (278 trees ha^{-1})	2 treatments: 70% ETc (SDI) 70% ETc ② → 35% ETc MS → 70% ETc ③ → 35% ETc during maturity (RDI)	No significant differences between the two irrigation treatments were found in terms of morphology, physiology, fruit yield, or oil quality; water productivity was 1.4 and 1.0 kg oil m^{-3} in SDI and RDI, respectively.	Siakou et al. (2021) [78]
RDI	Arbequina	Seville, Spain	534	4-year-old commercial orchard (1667 trees ha^{-1})	3 treatments: FI 60% IWR ② → 10% IWR MS → 30% IWR ③ (RDI1) 80% IWR ② → 20% IWR MS → 100% IWR ③ (RDI2)	RDI1 treatment showed the best balance between water saving (72%), tree vigor, and oil yield (26% reduction) when compared to FI.	Fernández et al. (2013) [102]

Table 3. *Cont.*

DI Strategy	Cultivar	Location	Annual Rainfall (mm)	Orchard Type	Irrigation Treatments	Main Results	Reference
RDI	Arbequina	Toledo, Spain	395	10-year-old commercial orchard (1250 trees ha^{-1})	4 treatments: FI 30% IWR in July and FI in the remaining growth period (RDI1) 30% IWR in August and FI in the remaining growth period (RDI2) 50% IWR in July and August and FI in the remaining growth period (RDI3)	FI trees produced more oil and fruit with higher oil percentages than RDI trees; the oil yield with RDI1 was not significantly reduced compared with FI and the oil percentage was higher; RDI1 was the most effective strategy, with 16% less water applied relative to FI.	Gómez-del-Campo (2013) [75]
RDI	Arbequina	Pencahue Valley, Chile	620	6-year-old commercial orchard (1333 tress ha^{-1})	4 treatments: FI Irrigation cut-off from fruit set until Ψstem = −3.5 MPa (RDI1) Irrigation cut-off from fruit set until Ψstem = −5.0 MPa (RDI2) irrigation cut-off from fruit set until Ψstem = −6.0 MPa (RDI3)	Fruit yield, fruit weight, and fruit diameter decreased in RDI2 and RDI3; total oil content and pulp/stone ratio were not affected by the different irrigation strategies; RDI treatments averaged 83% to 53% of applied water compared with FI.	Ahumada-Orellana et al. (2017) [104]
PRD	Picholine marocaine	Station Saada, Morocco	250	13-year-old experimental orchard (278 tress ha^{-1})	4 treatments: FI (100% ETc on both sides of the trees) 50% ETc on one side, switching every irrigation (PRD1) 50% ETc on one side, switching every two-irrigation (PRD2) 100% ETc on one side, switching every irrigation (PRD3)	Slight yield reduction (15–20%) under PRD1 and PRD2 was mainly due to a decrease in fruit number; oil percentage and oil acidity in the fruits did not show any significant differences between PRD treatments and the control; water use efficiency increased (60–70%) under PRD1 and PRD2 treatments.	Wahbi et al. (2005) [114]
PRD	Chemlali	Sfax, Tunisia	220	9-year-old experimental orchard (625 trees ha^{-1})	4 treatments: FI (100% ETc on both sides of the trees) 50% ETc on one side, switching every 15 days (PRD1) 50% ETc on one side, switching every 30 days (PRD2) R	PRD2 achieved a slight cumulative yield reduction (11%) compared to FI while applying half of the irrigation quantity; oil content showed an improvement with increasing deficits.	Ghrab et al. (2013) [115]
PRD	Arbequina, Arbosana, and Chetoui	Sidi Bouzid, Tunisia	240	11-year-old commercial orchard (1250 trees ha^{-1})	4 treatments: FI (100% ETc on both sides of the trees) 100% ETc on one side, switching every 2-weeks (PRD1) 75% ETc on one side, switching every 2-weeks (PRD2) 50% ETc on one side, switching every 2-weeks (PRD3)	Shoot length was lower under PRD irrigation treatments, mainly for Arbequina and Chetoui; reducing irrigation volumes by 25% and 50% with PRD strategy compared to the control increased oil yield and water productivity, mainly for Arbequina cultivar, without significant reductions in yield components.	Abboud et al. (2019) [116]

Notes: [(1)] Values between brackets were used in the second year of the trial. R: rainfed; SI: supplemental irrigation; SDI: sustained deficit irrigation; RDI: regulated deficit irrigation; PRD: partial rootzone drying; FI: full irrigation; IWR: irrigation water requirements (ETc—Crop evapotranspiration); MS: midsummer (late June to late August); ②—period that occurs at the end of the first phase of fruit development (normally in June) (Figure 3); ③—after the midsummer period, around 3 weeks prior to ripening, when a marked increase in oil accumulation occurs (normally from late August to mid-September) (Figure 3). Ψ_{stem}: stem water potential.

4. Agroecological Practices

4.1. Non-Tillage, Cover Crops and Herbicide Reduction

Semi-arid Mediterranean regions are among the most productive areas in the world [118]. However, the soil has a low carbon content and is susceptible to degradation [119–121]. Semi-arid soils are exposed to erosion by random and heavy precipitation, absence of herbaceous plant cover, and high rates of carbon mineralization related to high temperatures and high soil pH [45,122,123]. Intensive tillage in olive farming promotes soil organic matter degradation and general nutrient losses [124] (Table 4). Thus, tillage increases CO_2 emission at the expense of organic matter, contributing to global climate change. In irrigated olive orchards such as HD or SHD, it is possible for non-tillage practices to be implemented, fully mitigating these side effects. Normally, the organic matter in non-tillage orchards is about 0.8% or more higher than tilled ones [125]. The contribution to carbon sequestration of a non-tillage system with cover crops is 1.23 t C ha^{-1} $year^{-1}$ [126] or 1.34 t C ha^{-1} $year^{-1}$ [123] compared with bare soil. Non-tillage system avoids the propagation of soil-borne diseases such as *Verticillium dahliae*, the main soil-borne disease for this perennial species worldwide [56,127]. Preventing soil disturbance and minimizing the contact of fungus mycelia from root to root decreases the infection rate.

Herbaceous vegetation can have a positive impact on erosion reduction, especially in orchards planted on slopes [31], contributing to carbon and nitrogen sequestration and acting as a nutrient buffer. Herbaceous cover also provides shelter and food for many beneficial and pollinator insects. Nevertheless, vectors for the bacteria *Xylella fastidiosa* could also live and feed on orchard weeds. Late in the spring, as the weeds dry out, these vectors could fly from weeds to the olive canopy and infect the olive trees [128,129].

The generalized application of herbicides dramatically decreases the number of species, plants, animals, and other living organisms present in an olive orchard ecosystem [125]. For instance, the abundance and diversity of nematodes is lower in bare soils treated with herbicides, and is intermediate in non-herbicide areas [125]. Normally, tillage reduces the number of arthropod species [130,131].

The use of herbicides in the total area of an orchard increases the rainwater runoff and contributes to faster soil erosion and lower nutrient availability [132]. The use of herbicides sprayed in stripes, as in rows of trees, seems to have a lower impact on soil erosion. Weed species present on an olive orchard's floor, like *Conyza* sp., present significant challenges nowadays, as they are not effectively controlled by glyphosate spray treatment [133,134]. The eventual withdrawal of this herbicide will lead to the implementation of other non-herbicide solutions for orchard floor management [135].

4.2. Pruning Biomass Recycling

Olive orchards show a carbon accumulation rate in tree structures of 0.58 t C ha^{-1} $year^{-1}$, whereas the maximum potential rate is around 1 t C ha^{-1} $year^{-1}$ for perennial crops; 20 year-old olive orchards can have up to 11.7 t C ha^{-1} in the trees' permanent structures, and pruning residues represent an additional 2 t C ha^{-1} $year^{-1}$ [123]. The annual olive orchard carbon sequestration is higher than the amount denoted for vineyards and lower than that mentioned for other fruit trees [123].

In HD and SHD olive orchards, the pruning wood is normally shredded together with the cover weeds, and its nutrients are slowly released over time. This is a way to recycle nutrients and organic matter [136]. The presence of chopped wood pieces and weed residues on the orchard floor has four main benefits. First, it decreases the rainwater runoff speed and helps to prevent erosion [137]. Second, it promotes the rainwater infiltration rate, which is quite important in the case of heavy rain events [136]. Third, it improves machines' traction, preventing tractor or harvesters' wheels from sliding. Fourth, crossed chopped wood pieces act as a physical barrier over the floor, preventing soil compaction [138]. The last two benefits are often disregarded.

Table 4. Summary of selected studies on the effect of agroecological practices on soil factors in high-density orchards.

Soil Factor	Tillage	Pruning Residues	Herbicide	Cover Crops	Organic Farming	Main Results	Reference
Erosion	+	–	+A –P	–	+T –NT	Cover crops can reduce soil loss by more than 92% compared with tillage. The annual water runoff increased with tillage (highest runoff: tillage or full herbicide coverage; lowest runoff: cover crops and pruning residues).	Repullo–Ruibérriz de Torres et al. [139] Novara et al. [51]
Resistance to penetration	+	–	=	–	+T –NT	With cover crops, the compaction decreased at a depth of 0.3 m. Tillage reduced compaction just at the first 0.1 m of depth. Water availability improved in the soil with cover crops. However, the infiltration rate decreased.	Sastre et al. [140]
Water evaporation	–	– or =	–A =P	+	–T +NT	Cover crops increased the water consumption compared with tillage.	Novara et al. [51]
Pesticide accumulation	=	=	+	=	+Cu – Other	Total Cu in olive orchard and vineyard soils is about 5–10 times the concentration found in forest soils. Organic vs. integrated pest management: the use of fewer pesticides, but more cooper fungicides, is recommended.	Viti et al. [141] Miloš and Bensa [142]
Biodiversity	–	=	–	+	–T +NT	Tillage and herbicides decrease soil biodiversity. Tillage reduces the abundance of microarthropods.	Sánchez–Moreno et al. [124] Vignozii et al. [131]
Organic matter and carbon accumulation	–	+	–	+	–T +NT	Tillage negatively affected soil organic carbon pools in the interrow. Cover crops vs. bare soil: increase of 1.23–1.34 t C $ha^{-1}year^{-1}$. Pruning residues vs. removal: increase of 1–2 t C ha^{-1} $year^{-1}$.	Velázquez–Martí et al. [143] Repullo et al. [136]
Nitrogen accumulation	–	+	–	+	–T +NT	The N in pruned residues from a SHD orchard was 59 kg ha^{-1}. The N contained in fruits was 7 kg t^{-1}.	Zipori et al. [144]
Waterlogging	+	– or =	=	–	+T –NT	The olive trees survived if soil salinity was <4 dS m^{-1}. Wet flat land increased tree mortality due to hypoxia. Ridge plantation can prevent this.	Aragüés et al. [145]
Diseases	+	– or =	–	–	+T –NT	Tillage vs. cover crops or herbicides: verticillium wilt increased. Drip irrigation increased verticillium wilt.	Calderón et al. [127] López–Escudero and Blanco–López [146]

Notes: + Increase, − decrease, = equal. A—total coverage, P—stripes of 1 m, T—organic farming with tillage, NT—organic farming with cover crops, IPM—integrated pest management.

4.3. Adaptation of Cultivars

Due to the longevity of olive trees and the adaptation to the cultivation system, the TD and MD olive orchards present different cultivars than the HD or SHD [14]. Therefore, one can wonder whether old traditional varieties could be adapted to HD or SHD systems. According to Marino et al. [147] some old Italian cultivars can be suitable for these systems. In Portugal, the 'Cobraçosa' cv. seems to be adaptable to high-density systems. The introduction of new cultivars suitable for SH or SHD orchards has a positive impact, improving the overall genetic pool of olive orchards [14,148,149]. The use of rootstocks with low vigor makes the adoption of traditional cultivars to SH or SHD systems possible. This is a promising option for decreasing the high vigor normally associated with these cultivars. Traditional cultivars grafted on such rootstocks could live together at a high density, be adaptable to higher soil variability conditions, and present improved pathogen-resistant patterns [150].

Some authors have also referenced the negative impact of tree density on biodiversity, as in the case of bird population reduction [65,151]. Heavier machinery and increased fertilizer, pesticide, and water usage are also said to negatively impact ecosystems' biodiversity [24,152,153]. The generalized adoption of drip irrigation increases the *Verticillium dahliae* in the soil. The inoculum density in all experiments was higher in wet than in dry areas, and after 4 months of watering, the soil pathogen population increased considerably in both wet and dry areas [146]. The inoculum density remained higher in the wet soil.

5. Conclusions

The target of this work was to focus on the sustainability of high- and super-high-density olive orchards.

The increase in tree densities, the introduction of irrigation, and the development of new training systems to facilitate mechanical pruning and harvesting have contributed significantly to the intensification and expansion of olive oil and table olive production. In recent years, concerns about the potential detrimental impacts of high-density olive cultivation have emerged, bringing into question the trade-offs between production benefits and environmental costs. Water-saving irrigation practices and more sustainable soil management or other agroecological practices can mitigate the negative effects of climate change and improve the ecosystem services of dense irrigated olive cultivation.

The systematization of the various olive cultivation systems allows us to gain a better understanding of the olive orchard cultivation mosaic. The review and summary of studies and publications on deficit irrigation strategies and agroecological practices in dense olive orchards can contribute towards optimized options in terms of water, soil, and biodiversity management in order to enhance the health of these types of agroecosystems.

Author Contributions: Study conceptualization, M.I.P. and A.T.; investigation, J.S., M.I.P., M.P. and A.T.; writing—original draft preparation, J.S. and A.T.; writing—review and editing, J.S. and A.T.; funding acquisition, M.I.P., M.P. and A.T. All authors have read and agreed to the published version of the manuscript.

Funding: This work was supported by Geobiotec, funded by FCT—Fundação para a Ciência e a Tecnologia, Portugal (grant No. UIDB/04035/2020).

Data Availability Statement: No new data were created. Data sharing is not applicable to this article.

Conflicts of Interest: The authors declare no conflict of interest. The funders had no role in the design of the study; in the collection, analyses, or interpretation of data; in the writing of the manuscript; or in the decision to publish the results.

References

1. FAO. FAOSTAT Production Data: Crop and Livestock Products. Available online: https://www.fao.org/faostat/en/#data/QCL/visualize (accessed on 31 October 2022).
2. Instituto Nacional de Estatística. *Recenciamento Agrícola. Análise dos Principais Resultados: 2019*; INE: Lisboa, Portugal, 2021; ISBN 978-989-25-0562-6.

3. Yorulmaz, A.; Erinc, H.; Tekin, A. Changes in Olive and Olive Oil Characteristics during Maturation. *J. Am. Oil Chem. Soc.* **2013**, *90*, 647–658. [CrossRef]
4. Garcia, J.M.; Seller, S.; Perez-Camino, M.C. Influence of Fruit Ripening on Olive Oil Quality. *J. Agric. Food Chem.* **1996**, *44*, 3516–3520. [CrossRef]
5. Motilva, M.J.; Tovar, M.J.; Romero, M.P.; Alegre, S.; Girona, J. Influence of Regulated Deficit Irrigation Strategies Applied to Olive Trees (*Arbequina cultivar*) on Oil Yield and Oil Composition during the Fruit Ripening Period. *J. Sci. Food Agric.* **2000**, *80*, 2037–2043. [CrossRef]
6. Dag, A.; Kerem, Z.; Yogev, N.; Zipori, I.; Lavee, S.; Ben-David, E. Influence of Time of Harvest and Maturity Index on Olive Oil Yield and Quality. *Sci. Hortic.-Sci Hort-Amst.* **2011**, *127*, 358–366. [CrossRef]
7. Maaitah, M.; Al-Absi, K.; Al-Rawashdeh, A. Oil Quality and Quantity of Three Olive Cultivars as Influenced by Harvesting Date in the Middle and Southern Parts of Jordan. *Int. J. Agr. Biol.* **2009**, *1*, 266–272.
8. Dag, A.; Harlev, G.; Lavee, S.; Zipori, I.; Kerem, Z. Optimizing Olive Harvest Time under Hot Climatic Conditions of Jordan Valley, Israel. *Eur. J. Lipid Sci. Technol.* **2014**, *116*, 169–176. [CrossRef]
9. Lavee, S.; Wodner, M. The Effect of Yield, Harvest Time and Fruit Size on the Oil Content in Fruits of Irrigated Olive Trees (*Olea europaea*), Cvs. Barnea and Manzanillo. *Sci. Hortic.* **2004**, *99*, 267–277. [CrossRef]
10. Polari, J.J.; Juan, J.P.; Polari, J.J.; Lauren, M.C.; Crawford, L.M.; Crawford, L.M.; Wang, S.C. Cultivar Determines Fatty Acids and Phenolics Dynamics for Olive Fruit and Oil in Super-High-Density Orchards. *Agronomy* **2021**, *11*, 313. [CrossRef]
11. Carr, M. The Water Relations and Irrigation Requirements of Olive (*Olea europaea* L.): A Review. *Exp. Agric.* **2013**, *49*, 597–639. [CrossRef]
12. Arbizu-Milagro, J.; Castillo-Ruiz, F.J.; Tascón, A.; Peña, J.M. How Could Precision Irrigation Based on Daily Trunk Growth Improve Super High-Density Olive Orchard Irrigation Efficiency? *Agronomy* **2022**, *12*, 756. [CrossRef]
13. Martín-Gimeno, M.A.; Zahaf, A.; Badal, E.; Paz, S.; Bonet, L.; Pérez-Pérez, J.G. Effect of Progressive Irrigation Water Reductions on Super-High-Density Olive Orchards According to Different Scarcity Scenarios. *Agric. Water Manag.* **2022**, *262*, 107399. [CrossRef]
14. Lo Bianco, R.; Proietti, P.; Regni, L.; Caruso, T. Planting Systems for Modern Olive Growing: Strengths and Weaknesses. *Agriculture* **2021**, *11*, 494. [CrossRef]
15. Massenti, R.; Ioppolo, A.; Veneziani, G.; Selvaggini, R.; Servili, M.; Lo Bianco, R.; Caruso, T. Low Tree Vigor, Free Palmette Training Form, and High Planting Density Increase Olive and Oil Yield Efficiency in Dry, Sloping Areas of Mediterranean Regions. *Horticulturae* **2022**, *8*, 817. [CrossRef]
16. Camposeo, S.; Vivaldi, G.A.; Russo, G.; Melucci, F.M. Intensification in Olive Growing Reduces Global Warming Potential under Both Integrated and Organic Farming. *Sustainability* **2022**, *14*, 6389. [CrossRef]
17. Steduto, P.; Hsiao, T.; Fereres, E.; Raes, D. *Crop Yield Response to Water*; Food and Agriculture Organization of the United Nations, Ed.; FAO irrigation and drainage paper; Food and Agriculture Organization of the United Nations: Rome, Italy, 2012; ISBN 978-92-5-107274-5.
18. Connor, D. Adaptation of Olive (*Olea europaea* L.) to Water-Limited Environments. *Aust. J. Agric. Res.-Aust. J. Agr. Res.* **2005**, *56*, 1181–1189. [CrossRef]
19. Masson-Delmotte, V.P.; Zhai, H.-O.; Pörtner, D.; Roberts, J.; Skea, P.R.; Shukla, A.; Pirani, W.; Moufouma-Okia, C.; Péan, R.; Pidcock, S.; et al. *Global Warming of 1.5 °C. An IPCC Special Report on the Impacts of Global Warming of 1.5°C Above Pre-Industrial Levels and Related Global Greenhouse Gas Emission Pathways, in the Context of Strengthening the Global Response to the Threat of Climate Change, Sustainable Development, and Efforts to Eradicate Poverty*; Cambridge University Press: Cambridge, UK; New York, NY, USA, 2018; 616p. [CrossRef]
20. Branquinho, S.; Rolim, J.; Teixeira, J.L. Climate Change Adaptation Measures in the Irrigation of a Super-Intensive Olive Orchard in the South of Portugal. *Agronomy* **2021**, *11*, 1658. [CrossRef]
21. Falkenmark, M.; Finlayson, M.; Gordon, L.; Bennett, E.; Chiuta, T.M.; Coates, D.; Ghosh, N.; Gopalakrishnan, M.; De Groot, R.S.; Jacks, G.; et al. Agriculture, water, and ecosystems: Avoiding the costs of going too far. In *Water for Food Water for Life: A Comprehensive Assessment of Water Management in Agriculture*; Taylor & Francis: London, UK, 2007; pp. 233–277. [CrossRef]
22. Adhikari, K.; Hartemink, A.E. Linking Soils to Ecosystem Services—A Global Review. *Geoderma* **2016**, *262*, 101–111. [CrossRef]
23. Silveira, A.; Ferrão, J.; Muñoz-Rojas, J.; Pinto Correia, T.; Guimarães, M.H.; Schmidt, L. The Sustainability of Agricultural Intensification in the Early 21st Century: Insights from the Olive Oil Production in Alentejo (Southern Portugal). In *Changing Societies: Legacies and Challenges. Vol. iii. The Diverse Worlds of Sustainabilit*; Delicado, A., Domingos, N., de Sousa, L., Eds.; Imprensa de Ciências Sociais: Lisbon, Portugal, 2018; pp. 247–275.
24. Guerrero-Casado, J.; Carpio, A.J.; Tortosa, F.S.; Villanueva, A.J. Environmental Challenges of Intensive Woody Crops: The Case of Super High-Density Olive Groves. *Sci. Total Environ.* **2021**, *798*, 149212. [CrossRef]
25. Soriano, M.-A.; Álvarez, S.; Landa, B.B.; Gómez, J.A. Soil Properties in Organic Olive Orchards Following Different Weed Management in a Rolling Landscape of Andalusia, Spain. *Renew. Agric. Food Syst.* **2014**, *29*, 83–91. [CrossRef]
26. Beaufoy, G. Reflections from an External Evaluator on the Future of Olive Production Systems on Sloping Land. *J. Environ. Manag.* **2008**, *89*, 140–142. [CrossRef]
27. Duarte, F.; Jones, N.; Fleskens, L. Traditional Olive Orchards on Sloping Land: Sustainability or Abandonment? *J. Environ. Manag.* **2008**, *89*, 86–98. [CrossRef] [PubMed]

28. Fernández, J.-E. Understanding Olive Adaptation to Abiotic Stresses as a Tool to Increase Crop Performance. *Environ. Exp. Bot.* **2014**, *103*, 158–179. [CrossRef]
29. Haworth, M.; Marino, G.; Brunetti, C.; Killi, D.; De Carlo, A.; Centritto, M. The Impact of Heat Stress and Water Deficit on the Photosynthetic and Stomatal Physiology of Olive (*Olea europaea* L.)—A Case Study of the 2017 Heat Wave. *Plants* **2018**, *7*, 76. [CrossRef] [PubMed]
30. Brito, C.; Dinis, L.-T.; Moutinho-Pereira, J.; Correia, C.M. Drought Stress Effects and Olive Tree Acclimation under a Changing Climate. *Plants* **2019**, *8*, 232. [CrossRef]
31. Guzmán, G.; Montes-Borrego, M.; Gramaje, D.; Benítez, E.; Gómez, J.A.; Landa, B.B. Cover Crops as Bio-Tools to Keep Soil Biodiversity and Quality in Slopping Olive Orchards. In Proceedings of the 20th EGU General Assembly, Procceedings from the Conference, Viena, Austria, 4–13 April 2018; p. 9957.
32. Guzmán, G.; Boumahdi, A.; Gómez, J.A. Expansion of Olive Orchards and Their Impact on the Cultivation and Landscape through a Case Study in the Countryside of Cordoba (Spain). *Land Use Policy* **2022**, *116*, 106065. [CrossRef]
33. Gomez Calero, J.A.; Campos, M.; Guzman, G.; Castillo-Llanque, F.; Giráldez, J.V. Use of Heterogeneous Cover Crops in Olive Orchards to Soil Erosion Control and Enhancement of Biodiversity. In *The Earth Living Skin: Life and Climate Changes*; Castellaneta Marina, Italy, 2014; Available online: http://hdl.handle.net/10261/159778 (accessed on 3 July 2023).
34. Peres, F.; Talhinhas, P.; Afonso, H.; Alegre, H.; Oliveira, H.; Ferreira-Dias, S. Olive Oils from Fruits Infected with Different Anthracnose Pathogens Show Sensory Defects Earlier Than Chemical Degradation. *Agronomy* **2021**, *11*, 1041. [CrossRef]
35. Mele, M.A.; Islam, M.Z.; Kang, H.M.; Giuffrè, A.M. Pre-and Post-Harvest Factors and Their Impact on Oil Composition and Quality of Olive Fruit. *Emir. J. Food Agric.* **2018**, *592*, 592–603. [CrossRef]
36. Gomez, J.A.; Amato, M.; Celano, G.; Koubouris, G.C. Organic Olive Orchards on Sloping Land: More than a Specialty Niche Production System? *J. Environ. Manag.* **2008**, *89*, 99–109. [CrossRef]
37. Morgado, R. From Traditional to Super-Intensive: Drivers and Biodiversity Impacts of Olive Farming Intensification. Ph.D. Thesis, UL, Lisbon, Portugal, 2022. Available online: https://www.repository.utl.pt/handle/10400.5/27582 (accessed on 3 July 2023).
38. Mairech, H.; López-Bernal, Á.; Moriondo, M.; Dibari, C.; Regni, L.; Proietti, P.; Villalobos, F.J.; Testi, L. Is New Olive Farming Sustainable? A Spatial Comparison of Productive and Environmental Performances between Traditional and New Olive Orchards with the Model OliveCan. *Agric. Syst.* **2020**, *181*, 102816. [CrossRef]
39. Di Giacomo, G.; Romano, P. Evolution of the Olive Oil Industry along the Entire Production Chain and Related Waste Management. *Energies* **2022**, *15*, 465. [CrossRef]
40. Aziz, M.; Khan, M.; Anjum, N.; Sultan, M.; Shamshiri, R.R.; Ibrahim, S.M.; Balasundram, S.K.; Aleem, M. Scientific Irrigation Scheduling for Sustainable Production in Olive Groves. *Agriculture* **2022**, *12*, 564. [CrossRef]
41. Paço, T.; Paredes, P.; Pereira, L.; Silvestre, J.; Santos, F. Crop Coefficients and Transpiration of a Super Intensive Arbequina Olive Orchard Using the Dual K c Approach and the K Cb Computation with the Fraction of Ground Cover and Height. *Water* **2019**, *11*, 383. [CrossRef]
42. Paço, T.A.; Pôças, I.; Cunha, M.; Silvestre, J.C.; Santos, F.L.; Paredes, P.; Pereira, L.S. Evapotranspiration and Crop Coefficients for a Super Intensive Olive Orchard. An Application of SIMDualKc and METRIC Models Using Ground and Satellite Observations. *J. Hydrol.* **2014**, *519*, 2067–2080. [CrossRef]
43. Niinemets, Ü.; Keenan, T. Photosynthetic Responses to Stress in Mediterranean Evergreens: Mechanisms and Models. *Environ. Exp. Bot.* **2014**, *103*, 24–41. [CrossRef]
44. IPMA—Séries Longas. Available online: https://www.ipma.pt/pt/oclima/series.longas/?loc=Beja&type=raw (accessed on 21 February 2023).
45. Tombesi, A.; Tombesi, S.; Saavedra, M.M.S.; Fernández-Escobar, R.; d'Andri, R.; Lavini, A.; Ali Triki, M.; Rhouma, A.; Ksantini, A. *Production Techniques in Olive Growing*, 1st ed.; International Olive Council: Madrid, Spain, 2007; ISBN 978-84-931663-6-6.
46. Taiz, L.; Zeiger, E.; Møller, I.M.; Murphy, A. *Fisiologia e Desenvolvimento Vegetal; 6ª Edição*; ArteMed Editora Lda: Porto Alegre, Brazil, 2017; ISBN 978-85-8271-367-9.
47. Diário da Republica 2 Série, 25. Available online: https://files.dre.pt/2s/2018/02/025000000/0413204170.pdf (accessed on 19 February 2023).
48. Carpio, A.J.; Castro, J.; Tortosa, F.S. Arthropod Biodiversity in Olive Groves under Two Soil Management Systems: Presence versus Absence of Herbaceous Cover Crop. *Agric. For. Entomol.* **2019**, *21*, 58–68. [CrossRef]
49. Alcalá Herrera, R.; Ruano, F.; Gálvez Ramírez, C.; Frischie, S.; Campos, M. Attraction of Green Lacewings (Neuroptera: Chrysopidae) to Native Plants Used as Ground Cover in Woody Mediterranean Agroecosystems. *Biol. Control* **2019**, *139*, 104066. [CrossRef]
50. Vasconcelos, S.; Pina, S.; Herrera, J.M.; Silva, B.; Sousa, P.; Porto, M.; Melguizo-Ruiz, N.; Jiménez-Navarro, G.; Ferreira, S.; Moreira, F.; et al. Canopy Arthropod Declines along a Gradient of Olive Farming Intensification. *Sci. Rep.* **2022**, *12*, 17273. [CrossRef]
51. Novara, A.; Cerda, A.; Barone, E.; Gristina, L. Cover Crop Management and Water Conservation in Vineyard and Olive Orchards. *Soil Tillage Res.* **2021**, *208*, 104896. [CrossRef]
52. Sánchez-Fernández, J.; Vílchez-Vivanco, J.A.; Navarro, F.B.; Castro-Rodríguez, J. Farming System and Soil Management Affect Butterfly Diversity in Sloping Olive Groves. *Insect Conserv. Divers.* **2020**, *13*, 456–469. [CrossRef]

53. Rey, P.J.; Manzaneda, A.J.; Valera, F.; Alcántara, J.M.; Tarifa, R.; Isla, J.; Molina-Pardo, J.L.; Calvo, G.; Salido, T.; Gutiérrez, J.E.; et al. Landscape-Moderated Biodiversity Effects of Ground Herb Cover in Olive Groves: Implications for Regional Biodiversity Conservation. *Agric. Ecosyst. Environ.* **2019**, *277*, 61–73. [CrossRef]
54. Cano, Á.G.; Alejandro, H. Semi-Natural Habitats and Natural Enemies in Olive Orchards: Abundance, Function, Trophic Interactions, and Global Climate Change. Ph.D. Thesis, Universidad de Granada, Granada, Spain, 2021. Available online: http://hdl.handle.net/10481/70690 (accessed on 29 June 2023).
55. Álvarez, B.; Couanon, W.; Olivares, J.; Nigro, F. *EIP-AGRI Focus Group "Pests and Diseases of the Olive Tree" Biocontrol Agents and Cropping Practices to Control Olive Diseases*; EIP-AGRI; European Commission: Brussels, Belgium, 2019. [CrossRef]
56. Montes Osuna, N.; Mercado-Blanco, J. Verticillium Wilt of Olive and Its Control: What Did We Learn during the Last Decade? *Plants* **2020**, *9*, 735. [CrossRef] [PubMed]
57. Glyphosate: Commission Responds to European Citizens' Initiative and Announces More Transparency in Scientific Assessments. Available online: https://ec.europa.eu/commission/presscorner/detail/en/IP_17_5191 (accessed on 29 May 2023).
58. Andriukaitis, V. Pesticides in the European Union—Authorization and Use. Available online: https://ec.europa.eu/commission/presscorner/api/files/attachment/855260/Pesticides_factsheet.pdf (accessed on 29 May 2023).
59. Caruso, G.; Palai, G.; Tozzini, L.; Gucci, R. Using Visible and Thermal Images by an Unmanned Aerial Vehicle to Monitor the Plant Water Status, Canopy Growth and Yield of Olive Trees (Cvs. Frantoio and Leccino) under Different Irrigation Regimes. *Agronomy* **2022**, *12*, 1904. [CrossRef]
60. Taguas, E.V.; Marín-Moreno, V.; Díez, C.M.; Mateos, L.; Barranco, D.; Mesas-Carrascosa, F.-J.; Pérez, R.; García-Ferrer, A.; Quero, J.L. Opportunities of Super High-Density Olive Orchard to Improve Soil Quality: Management Guidelines for Application of Pruning Residues. *J. Environ. Manag.* **2021**, *293*, 112785. [CrossRef] [PubMed]
61. Pastor, M.; García-Vila, M.; Soriano, A.; Vega, V.; Fereres, E. Productivity of Olive Orchards in Response to Tree Density. *J. Hortic. Sci. Biotechnol.* **2007**, *82*, 555–562. [CrossRef]
62. Gomez-del-Campo, M.; Connor, D.J.; Trentacoste, E.R. Long-Term Effect of Intra-Row Spacing on Growth and Productivity of Super-High Density Hedgerow Olive Orchards (Cv. Arbequina). *Front. Plant Sci.* **2017**, *8*, 1790. [CrossRef] [PubMed]
63. Díez, C.M.; Moral, J.; Cabello, D.; Morello, P.; Rallo, L.; Barranco, D. Cultivar and Tree Density As Key Factors in the Long-Term Performance of Super High-Density Olive Orchards. *Front. Plant Sci.* **2016**, *7*, 1226. [CrossRef]
64. Morgado, R.; Santana, J.; Porto, M.; Sánchez-Oliver, J.S.; Reino, L.; Herrera, J.M.; Rego, F.; Beja, P.; Moreira, F. A Mediterranean Silent Spring? The Effects of Olive Farming Intensification on Breeding Bird Communities. *Agric. Ecosyst. Environ.* **2020**, *288*, 106694. [CrossRef]
65. Jiang, X.; He, L. Investigation of Effective Irrigation Strategies for High-Density Apple Orchards in Pennsylvania. *Agronomy* **2021**, *11*, 732. [CrossRef]
66. Doorenbos, J.; Pruitt, W.O. *Guidelines for Predicting Crop Water Requirements*; Irrigation and Drainage Paper; Food and Agriculture Organization of the United Nations: Rome, Italy, 1977.
67. Allen, R.G.; Pereira, L.S.; Raes, D.; Smith, M. *Crop Evapotranspiration—Guidelines for Computing Crop Water Requirements*; Irrigation and Drainage Paper; Food and Agriculture Organization of the United Nations: Rome, Italy, 1998; ISBN 92-5-104219-5.
68. Pereira, L.S.; Alves, I. Crop Water Requirements. In *Encyclopedia of Soils in the Environment*; Hillel, D., Ed.; Elsevier: Oxford, UK, 2005; pp. 322–334. ISBN 978-0-12-348530-4.
69. Pastor, M.; Orgaz, F. Riego deficitario del olivar: Los programas de recorte de riego en olivar. *Agricultura* **1994**, *746*, 768–776.
70. Fereres, E.; Castel, J.R. *Drip Irrigation Management*; Publication Leaflet; Division of Agricultural Sciences, University of California: Richmond, CA, USA, 1981.
71. Beede, R.H.; Goldhamer, D.A. Olive Irrigation Management. In *Olive Production Manual*; Sibbett, G.S., Ferguson, L., Eds.; University of California Publication; UCANR Publications: Oakland, CA, USA, 2005; pp. 61–69. ISBN 978-1-879906-15-0.
72. Lightle, D.; Connel, J. *Drought Tip: Drought Strategies for Table and Oil Olive Production*; ANR Public: Delhi, India, 2018; Volume 5.
73. Berenguer, M.J.; Grattan, S.; Connell, J.H.; Polito, V.S.; Vossen, P. Irrigation Management to Optimize Olive Growth, Production and Sensorial Oil Quality. *Acta Hortic.* **2004**, *664*, 79–85. [CrossRef]
74. Grattan, S.; Berenguer, M.J.; Connell, J.H.; Polito, V.S.; Vossen, P. Olive Oil Production as Influenced by Different Quantities of Applied Water. *Agric. Water Manag.* **2006**, *85*, 133–140. [CrossRef]
75. Gómez-del-Campo, M. Summer Deficit-Irrigation Strategies in a Hedgerow Olive Orchard Cv. 'Arbequina': Effect on Fruit Characteristics and Yield. *Irrig. Sci.* **2013**, *31*, 259–269. [CrossRef]
76. Freihat, N.M.; Shannag, H.K.; Alkelani, M.A. Effects of Supplementary Irrigation on Performance of 'Nabali' and 'Grossa de Spain' Olives under Semi-Arid Conditions in Jordan. *Sci. Hortic.* **2021**, *275*, 109696. [CrossRef]
77. Palese, A.M.; Nuzzo, V.; Favati, F.; Pietrafesa, A.; Celano, G.; Xiloyannis, C. Effects of Water Deficit on the Vegetative Response, Yield and Oil Quality of Olive Trees (*Olea europaea* L., Cv Coratina) Grown under Intensive Cultivation. *Sci. Hortic.* **2010**, *125*, 222–229. [CrossRef]
78. Siakou, M.; Bruggeman, A.; Eliades, M.; Zoumides, C.; Djuma, H.; Kyriacou, M.C.; Emmanouilidou, M.G.; Spyros, A.; Manolopoulou, E.; Moriana, A. Effects of Deficit Irrigation on 'Koroneiki' Olive Tree Growth, Physiology and Olive Oil Quality at Different Harvest Dates. *Agric. Water Manag.* **2021**, *258*, 107200. [CrossRef]
79. Zeleke, K.; Mailer, R.; Eberbach, P.; Wünsche, J. Oil Content and Fruit Quality of Nine Olive (*Olea europaea* L.) Varieties Affected by Irrigation and Harvest Times. *Null* **2012**, *40*, 241–252. [CrossRef]

80. Trentacoste, E.R.; Connor, D.J.; Gómez-del-Campo, M. Response of Oil Production and Quality to Hedgerow Design in Super-High-Density Olive Cv. Arbequina Orchards. *Agronomy* **2021**, *11*, 1632. [CrossRef]
81. Moriana, A.; Orgaz, F.; Pastor, M.; Fereres, E. Yield Responses of a Mature Olive Orchard to Water Deficits. *J. Am. Soc. Hortic. Sci. Jashs* **2003**, *128*, 425–431. [CrossRef]
82. Fernandes-Silva, A.A.; Ferreira, T.C.; Correia, C.M.; Malheiro, A.C.; Villalobos, F.J. Influence of Different Irrigation Regimes on Crop Yield and Water Use Efficiency of Olive. *Plant Soil* **2010**, *333*, 35–47. [CrossRef]
83. Fernández, J.E.; Diaz-Espejo, A.; Romero, R.; Hernandez-Santana, V.; García, J.M.; Padilla-Díaz, C.M.; Cuevas, M.V. Chapter 9—Precision Irrigation in Olive (*Olea europaea* L.) Tree Orchards. In *Water Scarcity and Sustainable Agriculture in Semiarid Environment*; García Tejero, I.F., Durán Zuazo, V.H., Eds.; Academic Press: Cambridge, MA, USA, 2018; pp. 179–217; ISBN 978-0-12-813164-0.
84. Debaeke, P.; Aboudrare, A. Adaptation of Crop Management to Water-Limited Environments. *Eur. J. Agron.* **2004**, *21*, 433–446. [CrossRef]
85. Loveys, B.R.; Dry, P.R.; Stoll, M.; Mccarthy, M. Using Plant Physiology to Improve the Water Efficiency of Horticultural Crops. *Acta Hortic.* **2000**, *537*, 187–197. [CrossRef]
86. Mccarthy, M.; Loveys, B.R.; Dry, P.R.; Stoll, M. Regulated Deficit Irrigation and Partial Rootzone Drying as Irrigation Management Techniques for Grapevines. *Deficit Irrig. Pract.* **2000**, *22*, 79–87.
87. Flexas, J.; Galmés, J.; Gallé, A.; Gulías, J.; Pou, A.; Ribas-Carbo, M.; Tomàs, M.; Medrano, H. Improving Water Use Efficiency in Grapevines: Potential Physiological Targets for Biotechnological Improvement. *Aust. J. Grape Wine Res.* **2010**, *16*, 106–121. [CrossRef]
88. Cifre, J.; Bota, J.; Escalona, J.M.; Medrano, H.; Flexas, J. Physiological Tools for Irrigation Scheduling in Grapevine (*Vitis vinifera* L.): An Open Gate to Improve Water-Use Efficiency? *Agric. Ecosyst. Environ.* **2005**, *106*, 159–170. [CrossRef]
89. Tomaz, A.; Coleto Martinez, J.M.; Arruda Pacheco, C. Yield and Quality Responses of 'Aragonez' Grapevines under Deficit Irrigation and Different Soil Management Practices in a Mediterranean Climate. *Ciência Téc. Vitiv.* **2015**, *30*, 9–20. [CrossRef]
90. Tomaz, A.; Pacheco, C.A.; Coleto Martinez, J.M. Influence of Cover Cropping on Water Uptake Dynamics in an Irrigated Mediterranean Vineyard. *Irrig. Drain.* **2017**, *66*, 387–395. [CrossRef]
91. Tomaz, A.; Coleto Martínez, J.; Arruda Pacheco, C. Effects of Cover Crops and Irrigation on 'Tempranillo' Grapevine and Berry Physiology: An Experiment under the Mediterranean Conditions of Southern Portugal. *OENO One* **2021**, *55*, 191–208. [CrossRef]
92. Oweis, T. *Supplemental Irrigation: A Highly Efficient Water-Use Practice*; International Center for Agricultural Research in the Dry Areas (ICARDA): Aleppo, Syria, 1997; ISBN 92-9127-070-9.
93. Attalla, A.; Abdel-Sattar, M.; Mahrous, A.; Abdel-Azeez, A. Olive Trees Productivity in Response to Supplemental Irrigation under North-Western Coastal Conditions in Egypt. *Am. Eurasian J. Agric. Environ. Sci.* **2011**, *11*, 609–615.
94. Draie, R.; Alhaj-Rabie, W.; Al-Mahmoud, A. Effect of Supplementary Spring and Summer Irrigation on the Growth and Productivity of the Olive Tree (Sourani Variety) in the Region of Al-Rouj Plain. *Int. Res. J. Innov. Eng. Technol.* **2020**, *4*, 45–55.
95. Lodolini, E.M.; Polverigiani, S.; Ali, S.; Mutawea, M.; Qutub, M.; Pierini, F.; Neri, D. Effect of Complementary Irrigation on Yield Components and Alternate Bearing of a Traditional Olive Orchard in Semi-Arid Conditions. *Span. J. Agric. Res* **2016**, *14*, e1203. [CrossRef]
96. Fereres, E.; Soriano, M.A. Deficit Irrigation for Reducing Agricultural Water Use. *J. Exp. Bot.* **2006**, *58*, 147–159. [CrossRef]
97. Galindo, A.; Collado-González, J.; Griñán, I.; Corell, M.; Centeno, A.; Martín-Palomo, M.J.; Girón, I.F.; Rodríguez, P.; Cruz, Z.N.; Memmi, H.; et al. Deficit Irrigation and Emerging Fruit Crops as a Strategy to Save Water in Mediterranean Semiarid Agrosystems. *Agric. Water Manag.* **2018**, *202*, 311–324. [CrossRef]
98. Goldhamer, D.A.; Dunai, J.; Ferguson, L.F. Irrigation Requirements of Olive Trees and Responses to Sustained Deficit Irrigation. *Acta Hortic. Int. Soc. Hortic. Sci. (ISHS)* **1994**, *356*, 172–175. [CrossRef]
99. Santos, F.L. Olive Water Use, Crop Coefficient, Yield, and Water Productivity under Two Deficit Irrigation Strategies. *Agronomy* **2018**, *8*, 89. [CrossRef]
100. Iniesta, F.; Testi, L.; Orgaz, F.; Villalobos, F.J. The Effects of Regulated and Continuous Deficit Irrigation on the Water Use, Growth and Yield of Olive Trees. *Eur. J. Agron.* **2009**, *30*, 258–265. [CrossRef]
101. Dry, P.R.; Loveys, B.R.; Mccarthy, M.G.; Stoll, M. Strategic Irrigation Management in Australian Vineyards. *OENO One* **2001**, *35*, 129–139. [CrossRef]
102. Fernández, J.E.; Perez-Martin, A.; Torres-Ruiz, J.M.; Cuevas, M.V.; Rodriguez-Dominguez, C.M.; Elsayed-Farag, S.; Morales-Sillero, A.; García, J.M.; Hernandez-Santana, V.; Diaz-Espejo, A. A Regulated Deficit Irrigation Strategy for Hedgerow Olive Orchards with High Plant Density. *Plant Soil* **2013**, *372*, 279–295. [CrossRef]
103. Vossen, P.M.; Berenguer, M.J.; Grattan, S.R.; Connell, J.H.; Polito, V.S. The Influence of Different Levels of Irrigation on the Chemical and Sensory Properties of Olive Oil. *Acta Hortic. Int. Soc. Hortic. Sci. (ISHS)* **2008**, *791*, 439–444. [CrossRef]
104. Ahumada-Orellana, L.E.; Ortega-Farías, S.; Searles, P.S.; Retamales, J.B. Yield and Water Productivity Responses to Irrigation Cut-off Strategies after Fruit Set Using Stem Water Potential Thresholds in a Super-High Density Olive Orchard. *Front. Plant Sci.* **2017**, *8*, 1280. [CrossRef]
105. Marino, G.; Caruso, T.; Ferguson, L.; Marra, F.P. Gas Exchanges and Stem Water Potential Define Stress Thresholds for Efficient Irrigation Management in Olive (*Olea europea* L.). *Water* **2018**, *10*, 342. [CrossRef]

106. Ben-Gal, A.; Ron, Y.; Yermiyahu, U.; Zipori, I.; Naoum, S.; Dag, A. Evaluation of Regulated Deficit Irrigation Strategies for Oil Olives: A Case Study for Two Modern Israeli Cultivars. *Agric. Water Manag.* **2021**, *245*, 106577. [CrossRef]
107. Fernández, J.E.; Díaz-Espejo, A.; Infante, J.M.; Durán, P.; Palomo, M.J.; Chamorro, V.; Girón, I.F.; Villagarcía, L. Water Relations and Gas Exchange in Olive Trees under Regulated Deficit Irrigation and Partial Rootzone Drying. *Plant Soil* **2006**, *284*, 273–291. [CrossRef]
108. Davies, W.J.; Wilkinson, S.; Loveys, B. Stomatal Control by Chemical Signalling and the Exploitation of This Mechanism to Increase Water Use Efficiency in Agriculture. *New Phytol.* **2002**, *153*, 449–460. [CrossRef] [PubMed]
109. Collins, M.; Fuentes, S.; Barlow, E. Partial Rootzone Drying and Deficit Irrigation Increase Stomatal Sensitivity to Vapour Pressure Deficit in Anisohydric Grapevines. *Funct. Plant Biol.* **2010**, *37*, 128–138. [CrossRef]
110. Dodd, I.C.; Theobald, J.C.; Bacon, M.A.; Davies, W.J. Alternation of Wet and Dry Sides during Partial Rootzone Drying Irrigation Alters Root-to-Shoot Signalling of Abscisic Acid. *Funct. Plant Biol.* **2006**, *33*, 1081–1089. [CrossRef]
111. Jovanovic, Z.; Stikic, R. Partial Root-Zone Drying Technique: From Water Saving to the Improvement of a Fruit Quality. *Front. Sustain. Food Syst.* **2018**, *1*, 3. [CrossRef]
112. Chaves, M.M.; Pereira, J.S.; Maroco, J.; Rodrigues, M.L.; Ricardo, C.P.P.; Osório, M.L.; Carvalho, I.; Faria, T.; Pinheiro, C. How Plants Cope with Water Stress in the Field? Photosynthesis and Growth. *Ann. Bot.* **2002**, *89*, 907–916. [CrossRef]
113. Adu, M.O.; Yawson, D.O.; Armah, F.A.; Asare, P.A.; Frimpong, K.A. Meta-Analysis of Crop Yields of Full, Deficit, and Partial Root-Zone Drying Irrigation. *Agric. Water Manag.* **2018**, *197*, 79–90. [CrossRef]
114. Wahbi, S.; Wakrim, R.; Aganchich, B.; Tahi, H.; Serraj, R. Effects of Partial Rootzone Drying (PRD) on Adult Olive Tree (*Olea europaea*) in Field Conditions under Arid Climate: I. Physiological and Agronomic Responses. *Agric. Ecosyst. Environ.* **2005**, *106*, 289–301. [CrossRef]
115. Ghrab, M.; Gargouri, K.; Bentaher, H.; Chartzoulakis, K.; Ayadi, M.; Mimoun, M.B.; Masmoudi, M.M.; Mechlia, N.B.; Psarras, G. Water Relations and Yield of Olive Tree (Cv. Chemlali) in Response to Partial Root-Zone Drying (PRD) Irrigation Technique and Salinity under Arid Climate. *Agric. Water Manag.* **2013**, *123*, 1–11. [CrossRef]
116. Abboud, S.; Dbara, S.; Abidi, W.; Braham, M. Differential Agro-Physiological Responses Induced by Partial Root-Zone Drying Irrigation in Olive Cultivars Grown in Semi-Arid Conditions. *Environ. Exp. Bot.* **2019**, *167*, 103863. [CrossRef]
117. Caruso, G.; Rapoport, H.F.; Gucci, R. Long-Term Evaluation of Yield Components of Young Olive Trees during the Onset of Fruit Production under Different Irrigation Regimes. *Irrig. Sci.* **2013**, *31*, 37–47. [CrossRef]
118. Rundel, P.W.; Montenegro Rizzardini, G.; Jaksic, F.M. *Landscape Disturbance and Biodiversity in Mediterranean-Type Ecosystems*; Springer: Berlin/Heidelberg, Germany, 2011; ISBN 978-3-642-08416-4.
119. Freibauer, A.; Rounsevell, M.D.A.; Smith, P.; Verhagen, J. Carbon Sequestration in the Agricultural Soils of Europe. *Geoderma* **2004**, *122*, 1–23. [CrossRef]
120. Chiti, T.; Gardin, L.; Perugini, L.; Quaratino, R.; Vaccari, F.P.; Miglietta, F.; Valentini, R. Soil Organic Carbon Stock Assessment for the Different Cropland Land Uses in Italy. *Biol. Fertil. Soils* **2012**, *48*, 9–17. [CrossRef]
121. Rodríguez Martín, J.A.; Álvaro-Fuentes, J.; Gonzalo, J.; Gil, C.; Ramos-Miras, J.J.; Grau Corbí, J.M.; Boluda, R. Assessment of the Soil Organic Carbon Stock in Spain. *Geoderma* **2016**, *264*, 117–125. [CrossRef]
122. Pleguezuelo, C.R.R.; Zuazo, V.H.D.; Martínez, J.R.F.; Peinado, F.J.M.; Martín, F.M.; Tejero, I.F.G. Organic Olive Farming in Andalusia, Spain. A Review. *Agron. Sustain. Dev.* **2018**, *38*, 20. [CrossRef]
123. Petersson, T.; Perugini, L.; Chiriacò, M.V. *D2 Report: Quality and Quantity of Data Available for Each Identified Crop/Livestock Carbon Farming Practice*; European Commission Life 20 PRE IT/017; European Commission: Bruxells, Belgium, 2022.
124. Sánchez-Moreno, S.; Castro, J.; Alonso-Prados, E.; Alonso-Prados, J.L.; García-Baudín, J.M.; Talavera, M.; Durán-Zuazo, V.H. Tillage and Herbicide Decrease Soil Biodiversity in Olive Orchards. *Agron. Sustain. Dev.* **2015**, *35*, 691–700. [CrossRef]
125. Parras-Alcántara, L.; Lozano-García, B. Conventional Tillage versus Organic Farming in Relation to Soil Organic Carbon Stock in Olive Groves in Mediterranean Rangelands (Southern Spain). *Solid Earth* **2014**, *5*, 299–311. [CrossRef]
126. Márquez-García, F.; González-Sánchez, E.J.; Castro-García, S.; Ordóñez-Fernández, R. Improvement of Soil Carbon Sink by Cover Crops in Olive Orchards under Semiarid Conditions. Influence of the Type of Soil and Weed. *Span. J. Agric. Res.* **2013**, *11*, 335. [CrossRef]
127. Calderón, R.; Navas-Cortés, J.A.; Zarco-Tejada, P.J. Early Detection and Quantification of Verticillium Wilt in Olive Using Hyperspectral and Thermal Imagery over Large Areas. *Remote Sens.* **2015**, *7*, 5584–5610. [CrossRef]
128. Morente, M.; Cornara, D.; Plaza, M.; Durán, J.M.; Capiscol, C.; Trillo, R.; Ruiz, M.; Ruz, C.; Sanjuan, S.; Pereira, J.A.; et al. Distribution and Relative Abundance of Insect Vectors of Xylella Fastidiosa in Olive Groves of the Iberian Peninsula. *Insects* **2018**, *9*, 175. [CrossRef] [PubMed]
129. Morelli, M.; García-Madero, J.M.; Jos, Á.; Saldarelli, P.; Dongiovanni, C.; Kovacova, M.; Saponari, M.; Baños Arjona, A.; Hackl, E.; Webb, S.; et al. Xylella Fastidiosa in Olive: A Review of Control Attempts and Current Management. *Microorganisms* **2021**, *9*, 1771. [CrossRef] [PubMed]
130. Simoni, S.; Caruso, G.; Vignozzi, N.; Gucci, R.; Valboa, G.; Pellegrini, S.; Palai, G.; Goggioli, D.; Gagnarli, E. Effect of Long-Term Soil Management Practices on Tree Growth, Yield and Soil Biodiversity in a High-Density Olive Agro-Ecosystem. *Agronomy* **2021**, *11*, 1036. [CrossRef]

131. Vignozzi, N.; Agnelli, A.E.; Brandi, G.; Gagnarli, E.; Goggioli, D.; Lagomarsino, A.; Pellegrini, S.; Simoncini, S.; Simoni, S.; Valboa, G.; et al. Soil Ecosystem Functions in a High-Density Olive Orchard Managed by Different Soil Conservation Practices. *Appl. Soil Ecol.* **2019**, *134*, 64–76. [CrossRef]
132. Marañón-Jiménez, S.; Serrano-Ortíz, P.; Peñuelas, J.; Meijide, A.; Chamizo, S.; López-Ballesteros, A.; Vicente-Vicente, J.L.; Fernández-Ondoño, E. Effects of Herbaceous Covers and Mineral Fertilizers on the Nutrient Stocks and Fluxes in a Mediterranean Olive Grove. *Eur. J. Agron.* **2022**, *140*, 126597. [CrossRef]
133. Calha, I. Avoadinha-peluda—Conyza bonariensis resistente ao glifosato. In *Boletim Técnico*; Instituto Nacional de Investigação Agrária: Oeiras, Portugal, 2011; UIPP-BT/09; pp. 1–2.
134. Sansom, M.; Saborido, A.A.; Dubois, M. Control of Conyza Spp. with Glyphosate-a Review of the Situation in Europe. *Plant Prot. Sci.* **2013**, *49*, 44–53. [CrossRef]
135. Assirelli, A.; Ciaccia, C.; Giorgi, V.; Zucchini, M.; Neri, D.; Lodolini, E.M. An Alternative Tool for Intra-Row Weed Control in a High-Density Olive Orchard. *Agronomy* **2022**, *12*, 605. [CrossRef]
136. Repullo, M.; Carbonell- Bojollo, R.; Hidalgo, J.; Rodríguez-Lizana, A.; Ordóñez-Fernández, R. Using Olive Pruning Residues to Cover Soil and Improve Fertility. *Soil Tillage Res.* **2012**, *124*, 36–46. [CrossRef]
137. Moreno-García, M.; Repullo-Ruibérriz de Torres, M.A.; Carbonell-Bojollo, R.M.; Ordóñez-Fernández, R. Management of Pruning Residues for Soil Protection in Olive Orchards. *Land Degrad. Dev.* **2018**, *29*, 2975–2984. [CrossRef]
138. Flamand, I. *Olive Farmers' Compliance to Soil-Erosion control Policies in the Protected Designation of Origin Estepa*; Wageningen University: Wageningen, The Netherlands, 2020.
139. Repullo-Ruibérriz de Torres, M.A.; Ordóñez-Fernández, R.; Giráldez, J.V.; Márquez-García, J.; Laguna, A.; Carbonell-Bojollo, R. Efficiency of Four Different Seeded Plants and Native Vegetation as Cover Crops in the Control of Soil and Carbon Losses by Water Erosion in Olive Orchards. *Land Degrad. Dev.* **2018**, *29*, 2278–2290. [CrossRef]
140. Sastre, B.; Marques, M.J.; García-Díaz, A.; Bienes, R. Three Years of Management with Cover Crops Protecting Sloping Olive Groves Soils, Carbon and Water Effects on Gypsiferous Soil. *Catena* **2018**, *171*, 115–124. [CrossRef]
141. Viti, C.; Quaranta, D.; De Philippis, R.; Corti, G.; Agnelli, A.; Cuniglio, R.; Giovannetti, L. Characterizing Cultivable Soil Microbial Communities from Copper Fungicide-Amended Olive Orchard and Vineyard Soils. *World J. Microbiol. Biotechnol.* **2008**, *24*, 309–318. [CrossRef]
142. Miloš, B.; Bensa, A. The Copper Content in Soil of Olive Orchards from Dalmatia, Croatia. *Eurasian Soil Sci.* **2021**, *54*, 865–874. [CrossRef]
143. Velázquez-Martí, B.; Fernández-González, E.; López-Cortés, I.; Salazar-Hernández, D.M. Quantification of the Residual Biomass Obtained from Pruning of Trees in Mediterranean Olive Groves. *Biomass Bioenergy* **2011**, *35*, 3208–3217. [CrossRef]
144. Zipori, I.; Erel, R.; Yermiyahu, U.; Ben-Gal, A.; Dag, A. Sustainable Management of Olive Orchard Nutrition: A Review. *Agriculture* **2020**, *10*, 11. [CrossRef]
145. Aragüés, R.; Puy, J.; Isidoro, D. Vegetative Growth Response of Young Olive Trees (*Olea europaea* L., Cv. Arbequina) to Soil Salinity and Waterlogging. *Plant Soil* **2004**, *258*, 69–80. [CrossRef]
146. López-Escudero, F.J.; Blanco-López, M.A. Effects of Drip Irrigation on Population of Verticillium Dahliae in Olive Orchards. *J. Phytopathol.* **2005**, *153*, 238–239. [CrossRef]
147. Marino, G.; Macaluso, L.; Grilo, F.; Marra, F.P.; Caruso, T. Toward the Valorization of Olive (*Olea europaea* Var. Europaea L.) Biodiversity: Horticultural Performance of Seven Sicilian Cultivars in a Hedgerow Planting System. *Sci. Hortic.* **2019**, *256*, 108583. [CrossRef]
148. Camposeo, S.; Vivaldi, G.A.; Montemurro, C.; Fanelli, V.; Cunill Canal, M. Lecciana, a New Low-Vigour Olive Cultivar Suitable for Super High Density Orchards and for Nutraceutical EVOO Production. *Agronomy* **2021**, *11*, 2154. [CrossRef]
149. Rallo, L.; Barranco, D.; Castro-García, S.; Connor, D.J.; Gómez Del Campo, M.; Rallo, P. High-Density Olive Plantations. In *Horticultural Reviews Volume 41*; Janick, J., Ed.; John Wiley & Sons, Inc.: Hoboken, NJ, USA, 2014; pp. 303–384. ISBN 978-1-118-70741-8.
150. Torres-Sánchez, J.; de la Rosa, R.; León, L.; Jiménez-Brenes, F.M.; Kharrat, A.; López-Granados, F. Quantification of Dwarfing Effect of Different Rootstocks in 'Picual' Olive Cultivar Using UAV-Photogrammetry. *Precis. Agric* **2022**, *23*, 178–193. [CrossRef]
151. Morgado, R.; Ribeiro, P.F.; Santos, J.L.; Rego, F.; Beja, P.; Moreira, F. Drivers of Irrigated Olive Grove Expansion in Mediterranean Landscapes and Associated Biodiversity Impacts. *Landsc. Urban Plan.* **2022**, *225*, 104429. [CrossRef]
152. Pérez-Ruiz, M.; Rallo, P.; Jiménez, M.R.; Garrido-Izard, M.; Suárez, M.P.; Casanova, L.; Valero, C.; Martínez-Guanter, J.; Morales-Sillero, A. Evaluation of Over-The-Row Harvester Damage in a Super-High-Density Olive Orchard Using On-Board Sensing Techniques. *Sensors* **2018**, *18*, 1242. [CrossRef] [PubMed]
153. Martínez-Guanter, J.; Garrido-Izard, M.; Agüera, J.; Valero, C.; Pérez-Ruiz, M. Over-the-Row Harvester Damage Evaluation in Super-High-Density Olive Orchard by on-Board Sensing Techniques. *Adv. Anim. Biosci.* **2017**, *8*, 487–491. [CrossRef]

water

MDPI

Article

Insights into the Spatial and Temporal Variability of Soil Attributes in Irrigated Farm Fields and Correlations with Management Practices: A Multivariate Statistical Approach

Alexandra Tomaz [1,2,*], Inês Martins [1], Adriana Catarino [1], Clarisse Mourinha [1], José Dôres [1], Marta Fabião [3], Luís Boteta [3], João Coutinho [4], Manuel Patanita [1,2] and Patrícia Palma [1,5]

1 Instituto Politécnico de Beja, Escola Superior Agrária de Beja, Rua Pedro Soares, 7800-295 Beja, Portugal
2 GeoBioTec, Campus da Caparica, NOVA School of Science and Technology, 2829-516 Caparica, Portugal
3 Centro Operativo e de Tecnologia de Regadio, Quinta da Saúde, Apartado 354, 7800-999 Beja, Portugal
4 Centro de Química, Universidade de Trás-os-Montes e Alto Douro, Quinta de Prados, 5000-801 Vila Real, Portugal
5 Instituto de Ciências da Terra (ICT), Universidade de Évora, 7000-671 Évora, Portugal
* Correspondence: atomaz@ipbeja.pt; Tel.: +351-284-314-300

Abstract: The evaluation of the spatial and temporal variability of soil properties can be valuable to improve crop productivity and soil health. A study of soil properties was carried out in southern Portugal, in three farm fields with irrigated annual crops (layers 0–20 cm and 20–40 cm), over three years. Factor Analysis (FA) and Discriminant Analysis (DA) were used to analyze the data. With FA, the observed variables were grouped into a smaller number of latent variables related to soil attributes. Discriminant Analysis was used to classify and identify the most dominant attributes and indicators for the time and space variability of soil parameters. The FA performed for the surface layer included factors related to texture, water and nutrient retention capacity, chemical composition, and soil fertility. In the sub-surface layer, the factor structure was similar, with four factors related to texture, chemical composition, nutrient availability, and soil fertility. The most influential factors and variables in temporal discrimination (sampling dates) in both layers were those related to chemical composition, with electric conductivity as the preponderant indicator. As for the spatial differentiation (fields), the dominant factor in the surface layer was texture, and in the sub-surface layer, nutrient availability. The most important discriminant indicators of spatial variability were fine sand proportion and available potassium, respectively, for the surface and sub-surface layers. The results obtained showed potential for the multidimensional and integrated assessment of patterns of temporal and spatial variation of soil functions from agricultural practices or soil degradation processes.

Keywords: irrigation; soil indicators; factor analysis; discriminant analysis; agronomical practices; soil health

1. Introduction

Soil is a fundamental component of the biosphere, on which essential ecosystem services depend, such as the production of food and fiber or the maintenance of environmental quality and biodiversity [1]. Soil quality is a critical component of sustainable agriculture and can be defined as "the capacity of a soil to function both within its ecosystem boundaries and with the environmental external to that ecosystem" [2]. More recently, the soil health concept has also been used, being characterized by the continuum of some properties that reflects multiple decisions regarding land use and management practices [3]. Non-suitable land use contributes to soil degradation, negatively affecting the environment, plant productivity, and human health.

Irrigated agriculture plays a key role in providing food security, accounting for 20% of the total cultivated land and contributing to 40% of the total food produced worldwide [4]. However, the modification of the soil matrix by irrigation could affect soil health due to changes in soil properties [5]. Regardless of the physical or chemical soil degradation processes that can result from water erosion or soil salinization, the irrigation, even with appropriate intensity and suitable water quality, may cause: (i) soil leaching, leading to a gradual decline in soil fertility in semi-arid, semi-humid, and humid tropical zones [6]; (ii) nutrient imbalances, resulting from high fertilization rates in more intensively irrigated farming systems [7]; (iii) mineral weathering acceleration, change in soil structure, and raising of water tables [8]; (iv) increased clay illuviation, mineral weathering, and rate of pedogenic activity due to long-term irrigation in semi-arid conditions [9].

The knowledge of the relationships between crop management practices and soil properties can provide a better understanding of spatial and temporal variability influencing land productivity and the environmental impacts resulting from these practices [10]. The complexity of the processes involved in the soil-ecosystem relationships requires the collection of data from various spatial and temporally dynamic properties, leading to large sets of measurements whose understanding implies the use of statistical tools capable of analyzing the variables involved simultaneously. Multivariate statistical methods are useful tools for the analysis of complex datasets, allowing the detection of similarities between variables and the identification of patterns [11], and are suitable for application in various fields of agricultural research [12]. Research involving the application of these methods on data from in situ collections of soil samples have focused mainly on: (i) finding discriminant soil properties, using Principal Components Analysis (PCA) and Discriminant Function Analysis (DA) [13], Cluster Analysis (CA) and PCA [14], or Factor Analysis (FA) and DA [15]; (ii) relating crop yields and soil variables using FA [16] or PCA and Multiple Regression Analysis (MRA) [10,17]; (iii) determining soil quality or soil health indicators through the use of FA and DA [18] or CA and PCA [19]. These studies were mostly controlled trials in non-irrigated soils, involving established experimental designs or well-defined sampling grids. However, it is useful to understand if the space-time variability of soil properties and attributes using on-farm research, in non-controlled conditions, follows similar trends, and this is somewhat influenced by the spatial and temporal dimensions of the natural soil variability, irrigation volumes and schedules, different crops or cropping systems, and of farmers' agronomic management options.

Taking all this into consideration, in this work, the temporal and spatial variability of several soil properties in irrigated farm fields was evaluated using FA and DA. With FA, the observed variables in a data matrix are grouped into fewer variables, the so-called factors, of latent (unobserved) common characteristics, which translate soil attributes. DA is used to distinguish the dominant attributes and variables for spatial and temporal discrimination. This integrated and multidimensional approach in processing agricultural soil data can be applied to the development of soil quality/health indicators or to assess patterns of environmental change caused by agronomic management practices in irrigated agriculture.

2. Materials and Methods

2.1. Study Area

A short-term on-farm study was carried out over three years (2018 to 2020) in three fields of annual crops irrigated by center-pivot, named Pivot 3 (P3), Pivot 4 (P4), and Pivot 5 (P5). In P3, with an area of 13.1 ha, the crop succession throughout the three years was sunflower (*Helianthus annus* L.), maize (*Zea mays* L.), and sunflower; in P4, 15.0 ha, the crops were sunflower, arrowleaf clover (*Trifolium vesiculosum* Savi) for seed production, and onion (*Allium cepa* L.); the crops in P5, with 10.3 ha, were maize, sunflower, and maize. The fields were located in the Brinches-Enxoé hydro-agricultural area (HAA) with 5061 ha (Figure 1), which is one of the 21 areas of the Alqueva irrigation plan, part of the Multipurpose Development of Alqueva (EFMA—Empreendimento de Fins Múltiplos de Alqueva), centered in the Alqueva reservoir, Guadiana River Basin, southern Portugal.

The HAA of Brinches-Enxoé has both pressurized and gravity conveyance networks, with origins in the Laje reservoir and in the Montinhos reservoir, respectively.

The climate in the region is temperate with hot and dry summers (Mediterranean), with an annual precipitation and average mean monthly temperature of, respectively, 558 mm and 16.9 °C (long-term means for the 1981–2010 period, [20]). During the three years of study (2018–2020), data from an automatic meteorological station (37.96833° N; 7.55083° W) located in the HAA, showed that the annual precipitation was 603 mm, 343 mm, and 615 mm, respectively. The mean temperature was 16.7 °C, 17.3 °C, and 17.8 °C, respectively, in 2018, 2019, and 2020 [21]. Predominant soils in P3 and P5 are Calcaric Cambissols and Chromic Vertissols, while in P4 soils are mainly Pelic Vertisols and Calcaric Vertisols [22].

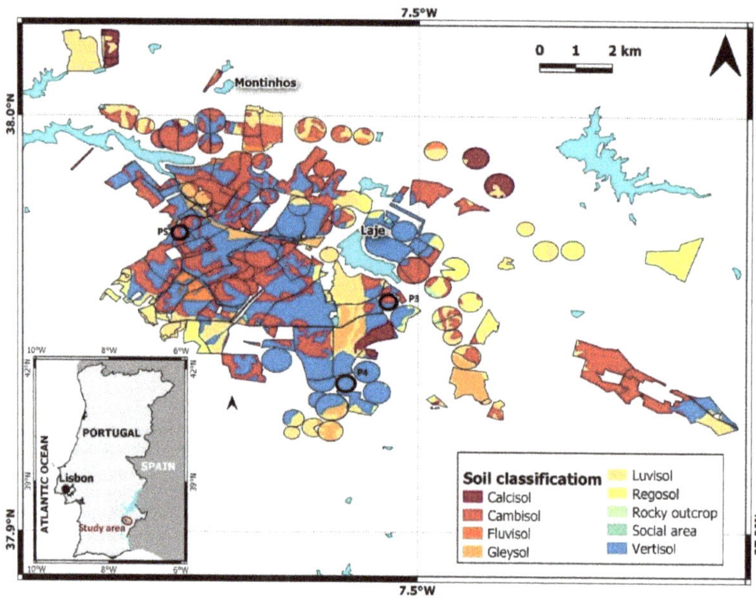

Figure 1. Hydro-agricultural area (HAA) of Brinches-Enxoé, prevalent reference soil groups, and location of the farm plots under study (P3, P4, and P5, marked with bold black circles). The bottom-left inset shows the location of the HAA in southern Portugal (adapted from [23]).

Management practices data were provided by the farmers and are described in Table 1. The soils of the studied fields were conventionally tilled. The fertilizers used were primarily formulations of Nitrogen (NO_3^-, NH_4^+, urea), Phosphorus (P_2O_5), and Potassium (K_2O), applied at sowing. Water-soluble and liquid fertilizers, applied over the crops' cycle through irrigation water, were mostly nitrogen fertilizers but also included other fertilizers containing formulations of sulfur (SO_3), iron (Fe chelates), or calcium (CaO). Foliar applications, containing boron (B), zinc (Zn), magnesium (MgO), and/or manganese (Mn), were employed in some of the crops. The pesticides used were mainly herbicides, but some insecticides were also applied in the sunflower and maize crops.

Table 1. Main crop management data in each year (2018 to 2020) and field (P3, P4, and P5) (information provided by the farmers).

Year	Field	Crop	Sowing (dd/mm)	Seasonal Irrigation Water (m^3 ha^{-1})	First Irrigation (dd/mm)	Last Irrigation (dd/mm)	Fertilizer N (kg N ha^{-1})	Fertilizer P (kg P_2O_5 ha^{-1})	Fertilizer K (kg K_2O ha^{-1})	Other Fertilizers (kg ha^{-1})	Pesticides (Active Substance)	Harvest (dd/mm)	Yield (kg ha^{-1})
2018	P3	Sunflower	18/04	2517	19/04	01/08	127	34	-	16 SO_3; 0.2 B	pre-emergence herbicide (pendimethalin); insecticide (deltamethrin)	27/08	3470
	P4	Sunflower	27/04	4606	28/04	26/08	109	40	12	16 SO_3	pre-emergence herbicide (pendimethalin)	18/09	4156
	P5	Maize	18/07	4800	18/07	04/10	202	144	216	27 SO_3	post-emergence herbicide (foramsulfuron + isoxadifen-ethyl)	17/01 [1]	5500
2019	P3	Maize	13/06	7500	[2]	[2]	253	-	-	-	post-emergence herbicide (mesotrione + S-metolachlor + terbuthylazine); insecticide (lambda-cyhalothrin)	17/11	11000
	P4	Clover	03/01	1510	18/04	24/06	-	88	-	0.2 SO3; 0.2 B; 0.1 MgO	-	18/09	1703
	P5	Sunflower	16/05	3570	20/05	30/08	81	19	20	7 SO_3	pre-emergence herbicides (pendimethalin, glyphosate)	15/09	3257
2020	P3	Sunflower	09/03	5420	[2]	[2]	[2]	[2]	[2]	[2]	[2]	13/08	8660
	P4	Onion	11/01	3210	11/01	24/08	113	1	0.1	210 SO_3; 65 CaO; 2 Zn; 1 Mn; 0.1 MgO	pre- and post-emergence herbicides (aclonifen, clethodime)	21/08	26,848
	P5	Maize	15/06	5160	15/06	15/09	82	59	121	5 SO_3; 3 Zn	pre- and post-emergence herbicides (glyphosate, MCPA, 2,4-D + florasulam, mesotrione + S-metolachlor + terbuthylazine); insecticides (lambda-cyhalothrin, chlorantraniliprole)	15/10	9182

[1] Delayed harvest due to the occurrence of a long rainy period following the physiological maturity stage; [2] No data available (the plot was leased in 2019 and 2020)—no application.

2.2. Sampling and Data Collection

Soil sampling was carried out on 5 dates (T1 to T5), over the 3 years under study: T1 and T2, in 2018; T3 and T4, in 2019; and T5 in 2020. T1 and T3 took place in spring (March/April); T2, T4, and T5 were carried out at the end of the irrigation period (September/October). The sampling methodology consisted of collecting a composite sample using an open-end soil probe, of every 5-ha area, attending also to surface heterogeneity whenever variations in soil color or slope were observed. Samples were collected in two layers, 0–20 cm and 20–40 cm, obtained from a mixture of approximately 5 sub-samples collected at randomly selected points following a zig-zag trajectory [24]. Samples were then air-dried and sieved with a 2 mm mesh for analysis of the following physical–chemical properties in the <2 mm fraction: particle size distribution (coarse sand (C. Sand), fine sand (F. Sand), Silt and Clay; g kg^{-1}), following ISO 11277:2020 (ISO, 2020); cation exchange capacity (CEC), exchangeable calcium (Ca), magnesium (Mg), potassium (K), sodium (Na), and aluminum (Al) (cmol (+) kg^{-1}), following ISO 11260:2018 (ISO, 2018); pH (H_2O 1:2.5 (p/v)); electrical conductivity of the saturated soil extract (EC; dS m^{-1}) (H_2O 1:2 (p/v)); soil organic matter (SOM; g kg^{-1}), following the Walkley-Black method [25]; total nitrogen (N; %), determined by the Kjeldahl method [26]; available P (mg P_2O_5 kg^{-1}) and K (mg K_2O kg^{-1}), determined by the Egner–Rhiem method [27]. Particle size distribution, exchange cations, and CEC were obtained only in the first sampling (T1) for the initial characterization of the fields. The resulting data matrix consisted of 11 observed variables of 226 composite soil samples.

2.3. Statistical Analysis

Data analyses were performed with Statistica 7 [28] and were conducted separately for each layer. Matrices of Spearman's correlation coefficients were computed for a preliminary examination of the relationships between the observed soil variables. The FA, using standardized data ((raw value-mean)/standard deviation), was performed to explore the data structure and to reduce the observed variables correlated with each other to a smaller number of independent variables, named factors [11]. Factors were extracted by the PCA method, and the factor loading matrix was subjected to varimax rotation to produce a factor structure that was simpler to interpret [29]. Factors were retained when presenting eigenvalues > 1, a contribution for the proportion of variance > 10%, and at least two observed variables contributing to absolute factor loadings > 0.50 [10,29,30]. Following the methodology reported in Shukla et al. (2006) [18], the scores of the factor models obtained were used to conduct a DA to differentiate sampling dates and fields based on soil attributes, determining which factor, date, or site most contributed to this differentiation. Factors presenting the lowest significant partial Wilks' Lambda were selected as the dominant discriminant factors [31]. To analyze in detail how each factor may have influenced the temporal and spatial discrimination, a Canonical Analysis (CCA) was carried out, and canonical roots, or discriminant functions, were obtained. The standardized coefficients of the first discriminant function, that is, the one that explains the largest proportion of the model's variance [31], were selected to verify the weight of each factor in the temporal and spatial discrimination. Finally, having found the preponderant factor for spatial and temporal variability in each soil layer, subsequent Discriminant and Canonical analyses were performed with the original standardized variables, which were highly correlated with each factor, thereby, distinguishing which variable had the greatest influence on the spatial and temporal discrimination of soil attributes and exploring its relationship with the agronomical practices.

3. Results and Discussion

3.1. Soil Physical–Chemical Properties

Overall, soils in the studied sites were rich in clay with medium to fine textures (Tables 2 and 3). In P3, the soils presented, in both layers, a clay-loam texture; in P4, textures varied from clay-loam to silty-clay; in P5, texture was mainly silty-clay, but there was some spatial variability, with textures also varying from loam to clay-loam and silty-clay-loam. The CEC was very high in every site, in accordance with the clayey nature of these soils and the richness in smectites, characteristic of Vertisols [24,32,33].

Table 2. Average values (± standard error) of particle size fractions (coarse (C) sand, fine (F) sand, Silt, and Clay) and Cation Exchange Capacity (CEC) in each site ((P3 (n = 9), P4 (n = 9), P5 (n = 12)), in the 0–20 cm layer.

Site	C. Sand (g kg^{-1})	F. Sand (g kg^{-1})	Silt (g kg^{-1})	Clay (g kg^{-1})	CEC (cmol $_{(+)}$ kg^{-1})
P3	197.6 (±10.0)	230.7 (±6.0)	192.1 (±15.5)	379.6 (±1.8)	56.5 (±0.3)
P4	160.5 (±11.1)	159.2 (±6.5)	248.3 (±1.4)	432.0 (±16.1)	57.0 (±0.8)
P5	163.7 (±9.8)	177.5 (±14.2)	287.6 (±19.1)	371.2 (±25.9)	53.6 (±2.1)

P3—Pivot 3; P4—Pivot 4; P5—Pivot 5; C. Sand—Coarse sand; F. Sand—Fine sand; CEC—Cation Exchange Capacity.

Table 3. Average values (± standard error) of particle size fractions (coarse (C) sand, fine (F) sand, Silt, and Clay) and Cation Exchange Capacity (CEC) in each site (P3 (n = 9), P4 (n = 9), P5 (n = 12)), in the 20–40 cm layer.

Site	C. Sand (g kg^{-1})	F. Sand (g kg^{-1})	Silt (g kg^{-1})	Clay (g kg^{-1})	CEC (cmol $_{(+)}$ kg^{-1})
P3	192.2 (±11.00)	232.4 (±5.6)	192.7 (±15.7)	375.7 (±2.3)	58.1 (±0.6)
P4	164.1 (±19.08)	167.2 (±6.4)	246.7 (±9.5)	422.0 (±17.0)	57.9 (±1.2)
P5	176.0 (±9.82)	173.3 (±12.6)	243.8 (±5.8)	406.9 (±17.4)	53.9 (±1.4)

P3—Pivot 3; P4—Pivot 4; P5—Pivot 5; C. Sand—Coarse sand; F. Sand—Fine sand; CEC—Cation Exchange Capacity.

The measured soil chemical properties evaluated throughout the study showed a non-saline condition of the soils, in both layers, at every date (Tables 4 and 5). Soils in the superficial layer (0–20 cm) presented a slight to medium alkaline reaction due to their calcareous nature, with average pH values in the range of 7.8–8.5 [32]. The alkaline nature of these soils is associated with deficiencies in nutrients such as Zn, Cu, Fe, Mn, and B, which are especially common in irrigated crops [34]; hence, it is common to apply these nutrients, either in the form of chelates (in the case of iron) or by using foliar applications of Zn, Mn, and B, practices that were adopted by the farmers, as reported in Section 2.1. The studied soils presented low levels of SOM, a feature of regions in arid or semi-arid climates, and reduced values of total N, in accordance with their low organic content, which increases the fertilization rates required to meet the crops' extraction. Available P presented high to very high levels, especially in P5 at T1. The same occurs regarding the levels of available K. The values at different dates point to a decreasing trend in P_2O_5 and K_2O between the beginning (T1, T3, T5) and the end (T2, T4) of the crops' cycle/irrigation season, which could be a result of the nutrient's extraction by the plants or, in the case of P, of the possible formation of insoluble calcium-phosphate compounds (P fixation), a common occurrence in soils with abundant Ca^{2+} or Mg^{2+} [34]. However, the levels of these nutrients remaining in the soil at the end of the crops' cycle was still high or very high. The fixation of P added by fertilization results in low uptake during the year of application and, normally, only approximately 10% to 20% of applied P is used by the plants during the first year. Therefore, the repeated application of P by fertilization leads to soils becoming sufficiently high in this nutrient [32]. This buildup of the available P in soils to levels beyond the amounts required by the crops means that these high amounts of the element are prone to leaching and can contribute to eutrophication processes [15].

Table 4. Average values (± standard error) of soil chemical parameters in each site (P3 (n = 9), P4 (n = 9), P5 (n = 12)), in the 0–20 cm layer, during the study (T1, T2, T3, T4, and T5 sampling dates).

Date	Site	pH	EC (dS m^{-1})	SOM (g kg^{-1})	N (%)	P (mg P_2O_5 kg^{-1})	K (mg K_2O kg^{-1})
T1	P3	8.39 (±0.03)	0.16 (±0.00)	15.8 (±0.5)	0.08 (±0.00)	248.61 (±31.32)	115.07 (±9.21)
	P4	8.28 (±0.06)	0.13 (±0.00)	12.8 (±0.4)	0.06 (±0.00)	221.34 (±19.07)	206.59 (±14.06)
	P5	8.02 (±0.03)	0.18 (±0.01)	11.4 (±0.6)	0.08 (±0.00)	418.63 (±88.63)	236.78 (±14.02)
T2	P3	8.34 (±0.03)	0.35 (±0.01)	12.2 (±0.6)	0.09 (±0.00)	124.46 (±17.60)	106.57 (±7.30)
	P4	7.89 (±0.03)	0.35 (±0.02)	19.6 (±2.4)	0.07 (±0.00)	143.64 (±10.22)	149.26 (±4.68)
	P5	8.45 (±0.02)	0.24 (±0.01)	19.1 (±1.8)	0.08 (±0.01)	326.53 (±32.45)	367.04 (±16.31)
T3	P3	8.40 (±0.01)	0.28 (±0.02)	10.6 (±0.8)	0.09 (±0.00)	145.24 (±12.73)	102.62 (±13.52)
	P4	[1]					
	P5	[1]					
T4	P3	8.08 (±0.02)	0.29 (±0.01)	10.0 (±0.4)	0.10 (±0.00)	148.57 (±14.68)	94.98 (±6.93)
	P4	8.25 (±0.02)	0.26 (±0.01)	6.9 (±0.4)	0.08 (±0.00)	115.81 (±6.99)	145.16 (±5.96)
	P5	8.06 (±0.04)	0.30 (±0.01)	16.0 (±0.6)	0.10 (±0.00)	239.86 (±21.76)	226.06 (±14.34)
T5	P3	7.92 (±0.01)	0.58 (±0.01)	15.0 (±0.2)	0.10 (±0.00)	180.82 (±18.98)	147.75 (±18.55)
	P4	7.80 (±0.01)	0.46 (±0.02)	10.4 (±0.2)	0.09 (±0.01)	106.32 (±6.64)	261.43 (±48.24)
	P5	[1]					

[1] No data available (samples were not collected). P3—Pivot 3; P4—Pivot 4; P5—Pivot 5. EC—Electrical Conductivity; SOM—Soil Organic Matter; N—Nitrogen; P—Phosphorus; K—Potassium.

Table 5. Average values (± standard error) of soil chemical parameters in each site (P3 (n = 9), P4 (n = 9), P5 (n = 12)), in the 20–40 cm layer, during the study (T1, T2, T3, T4, and T5 sampling dates).

Date	Site	pH	EC (dS m^{-1})	SOM (g kg^{-1})	N (%)	P (mg P_2O_5 kg^{-1})	K (mg K_2O kg^{-1})
T1	P3	8.49 (±0.02)	0.15 (±0.00)	16.9 (±1.4)	0.07 (±0.00)	144.85 (±7.02)	94.82 (±7.05)
	P4	8.19 (±0.05)	0.16 (±0.00)	8.2 (±0.2)	0.05 (±0.00)	131.13 (±4.44)	139.03 (±3.62)
	P5	8.09 (±0.07)	0.14 (±0.01)	7.9 (±0.6)	0.06 (±0.00)	258.99 (±67.76)	150.23 (±11.43)
T2	P3	8.40 (±0.03)	0.30 (±0.02)	11.2 (±0.4)	0.08 (±0.00)	106.20 (±15.77)	96.99 (±6.14)
	P4	7.97 (±0.02)	0.29 (±0.02)	18.5 (±3.1)	0.07 (±0.00)	83.74 (±1.71)	139.38 (±3.09)
	P5	8.49 (±0.02)	0.30 (±0.02)	16.8 (±1.5)	0.07 (±0.00)	197.73 (±18.97)	183.27 (±9.77)
T3	P3	8.35 (±0.02)	0.29 (±0.01)	10.2 (±0.9)	0.08 (±0.01)	148.44 (±12.28)	121.46 (±19.84)
	P4	[1]					
	P5	[1]					
T4	P3	8.10 (±0.01)	0.32 (±0.01)	9.8 (±0.8)	0.10 (±0.00)	118.42 (±11.96)	86.27 (±7.98)
	P4	[1]					
	P5	7.98 (±0.01)	0.37 (±0.01)	15.1 (±0.8)	0.09 (±0.00)	192.95 (±16.97)	195.35 (±16.83)
T5	P3	8.11 (±0.01)	0.28 (±0.00)	13.3 (±0.4)	0.09 (±0.00)	133.02 (±14.65)	73.48 (±9.30)
	P4	7.83 (±0.05)	0.41 (±0.04)	9.0 (±0.4)	0.09 (±0.01)	74.64 (±4.27)	164.78 (±11.08)
	P5	[1]					

[1] No data available (samples were not collected). P3—Pivot 3; P4—Pivot 4; P5—Pivot 5. EC—Electrical Conductivity; SOM—Soil Organic Matter; N—Nitrogen; P—Phosphorus; K—Potassium.

The maintenance of high solution potassium concentrations is determined by the K buffer capacity of the soil, which is high in the case of fine-textured soils containing abundant vermiculite and ilite clay minerals, with large amounts of interlayer K. In these soils, with high K-fixing capacity, much of the K applied by fertilization would be lost to fixation but, in the absence of easily supplied fertilizer K, a significant portion of K required by plants comes from the interlayer K, which is indicative of the beneficial role of the fixed K [35].

From the 0–20 cm to the sub-superficial layer (20–40 cm), the values of available P and K decrease, showing that both the application of fertilizers and the nutrient uptake by plants occurs mainly in the surface layer, where roots are more active (Table 5).

3.2. Correlation and Data Structure

3.2.1. Layer 0–20 cm

The Spearman correlation coefficients between the 11 soil physical–chemical parameters, measured in the 0–20 cm layer, can be observed in Table 6. Significant moderate positive correlations (>0.40) were found between EC and N, SOM and P, P and K, K and silt, clay and CEC. Significant negative correlations <−0.40 were observed between pH and EC, EC and P, K and sand fractions.

Table 6. Spearman correlation matrix of soil physical–chemical parameters in the 0–20 cm layer.

	pH	EC	SOM	N	P	K	C. Sand	F. Sand	Silt	Clay	CEC
pH	1.000										
EC	**−0.487**	1.000									
SOM	0.093	0.011	1.000								
N	**−0.270**	**0.591**	0.150	1.000							
P	**0.273**	**−0.461**	**0.426**	0.021	1.000						
K	−0.108	−0.124	**0.348**	0.020	**0.538**	1.000					
C. Sand	0.068	0.050	−0.075	0.047	−0.162	**−0.449**	1.000				
F. Sand	0.103	0.108	−0.025	0.178	−0.119	**−0.441**	**0.920**	1.000			
Silt	−0.078	−0.117	**0.198**	−0.030	0.119	**0.447**	**−0.556**	**−0.396**	1.000		
Clay	−0.123	0.086	−0.180	−0.086	0.088	0.155	**−0.554**	**−0.564**	**−0.197**	1.000	
CEC	−0.170	0.164	**−0.245**	0.043	**−0.196**	−0.151	−0.033	−0.097	**−0.302**	**0.434**	1.000

Bold values are significant at $p < 0.05$. EC—Electrical Conductivity; SOM—Soil Organic Matter; N—Nitrogen; P—Phosphorus; K—Potassium; C. Sand—Coarse sand; F. Sand—Fine sand; CEC—Cation Exchange Capacity.

The FA performed for the 0–20 cm layer allowed for the extraction of four factors, accounting for 75.04% of the total variance in the dataset (Table 7).

Table 7. Factor loadings, eigenvalues, percentage of total variance, and accumulated variance, in a four-factor model of 11 observed soil variables in the 0–20 cm layer.

	Factor 1	Factor 2	Factor 3	Factor 4
pH	0.215	0.043	**−0.719**	−0.186
EC	0.093	0.111	**0.893**	0.070
SOM	−0.061	−0.195	0.100	**−0.542**
N	0.198	0.035	**0.753**	−0.251
P	0.015	0.068	−0.257	**−0.799**
K	−0.323	−0.092	0.073	**−0.741**
C. Sand	**0.897**	−0.259	0.004	0.152
F. Sand	**0.943**	−0.256	0.053	0.100
Silt	**−0.793**	−0.527	−0.042	−0.169
Clay	−0.413	**0.861**	0.001	0.008
CEC	0.008	**0.830**	0.088	0.183
Eigenvalues	2.988	2.224	1.837	1.206
% Total variance	27.16	20.22	16.70	10.97
%. Accumulated variance	27.16	47.38	64.08	75.04

Bold values correspond to the higher factor loadings (>0.50) of the variables in each factor. EC—Electrical Conductivity; SOM—Soil Organic Matter; N—Nitrogen; P—Phosphorus; K—Potassium; C. Sand—Coarse sand; F. Sand—Fine sand; CEC—Cation Exchange Capacity.

Factor 1 accounted for 27.16% of the total variance and presented factor loadings with absolute values > 0.50 for C. Sand (0.897), F. Sand (0.943), and Silt (−0.793), indicating it was a factor translating the attribute *texture*. Factor 2 (20.22% of total variance) presented high positive loadings of the variables Clay (0.861) and CEC (0.830), thereby being a factor that characterized *water and nutrients retention capacity*. Factor 3 described 16.70% of the models' variance and was highly correlated with pH (−0.719), EC (0.893), and N (0.753), therefore representing an attribute of *chemical composition*. Factor 4, accounting for 10.97% of total variance, was related to *fertility*, because it was highly correlated with SOM (−0.542), available P (−0.799), and available K (−0.741).

3.2.2. Layer 20–40 cm

Correlation coefficients between chemical parameters in the sub-surface layer were lower than the ones obtained for the superficial layer, probably resulting from an attenuation of the influence of agronomical practices, such as tillage or fertilization, with increasing depth (Table 8). Moderate absolute correlations (>0.40) were found only between EC and N and K, F. Sand, and Silt. Regarding physical–chemical characteristics, high positive correlations were found between coarser (fine and coarse sand) and smaller separates (silt and clay). A correlation coefficient of −0.517 was found between C. Sand and CEC.

Table 8. Spearman correlation matrix of soil physical–chemical parameters in the 20–40 cm layer.

	pH	EC	SOM	N	P	K	C. Sand	F. Sand	Silt	Clay	CEC
pH	1.000										
EC	**−0.301**	1.000									
SOM	**0.234**	0.124	1.000								
N	−0.172	**0.672**	**0.326**	1.000							
P	**0.397**	−0.111	**0.224**	−0.042	1.000						
K	**−0.232**	**0.209**	0.183	−0.036	**0.302**	1.000					
C. Sand	0.019	0.092	0.135	0.054	0.082	−0.073	1.000				
F. Sand	**0.266**	0.001	0.062	0.176	−0.013	**−0.541**	**0.712**	1.000			
Silt	−0.181	−0.064	−0.065	−0.118	−0.105	**0.405**	**−0.868**	**−0.820**	1.000		
Clay	−0.130	−0.022	−0.120	−0.157	0.103	**0.286**	**−0.800**	**−0.894**	**0.691**	1.000	
CEC	−0.026	−0.147	−0.178	−0.023	−0.134	**−0.190**	**−0.517**	−0.121	**0.278**	**0.289**	1.000

Bold values are significant at $p < 0.05$. EC—Electrical Conductivity; SOM—Soil Organic Matter; N—Nitrogen; P—Phosphorus; K—Potassium; C. Sand—Coarse sand; F. Sand—Fine sand; CEC—Cation Exchange Capacity.

In the 20–40 cm layer, the FA retained four factors responsible for 75.72% of the total variance, with a structure similar to the model found for the surface layer (Table 9). Factor 1 accounted for 32.85% of total variance and presented absolute loads > 0.50 of the particle size variables—C. Sand (0.929), F. Sand (0.925), Silt (−0.869), and Clay (−0.869)—therefore, it was a proxy of soil *texture*. Factor 2, representing 17.75% of total variance of the model, presented high positive loads of EC (0.863) and N (0.894), hence being related to *chemical composition*. Factor 3 (14.65% of total variance) was highly positively correlated with CEC (0.758) and negatively correlated with available K (−0.695); therefore, we considered it as essentially representative of a *nutrient availability* attribute. Factor 4 (10.47% of total variance) showed high factor loadings of pH (0.752), SOM (0.562), and available P (0.594) and was thereby named a *fertility* attribute.

Table 9. Factor loadings, eigenvalues, percentage of total variance, and accumulated variance, in a four-factor model of 11 observed soil variables in the 20–40 cm layer.

	Factor 1	Factor 2	Factor 3	Factor 4
pH	0.195	−0.288	0.241	**0.752**
EC	0.038	**0.863**	−0.148	−0.200
SOM	0.063	0.363	−0.092	**0.562**
N	0.032	**0.894**	0.124	0.066
P	−0.141	−0.211	−0.443	**0.594**
K	−0.464	0.233	**−0.695**	0.228
C. Sand	**0.929**	0.024	−0.275	0.005
F. Sand	**0.925**	0.041	0.237	0.085
Silt	**−0.869**	−0.007	−0.099	−0.050
Clay	**−0.869**	−0.052	0.127	−0.034
CEC	−0.302	0.043	**0.758**	0.094
Eigenvalues	3.614	1.953	1.611	1.152
% Total variance	32.85	17.75	14.65	10.47
%. Accumulated variance	32.85	50.60	65.25	75.72

Bold values correspond to the higher factor loadings (>0.50) of the variables in each factor. EC—Electrical Conductivity; SOM—Soil Organic Matter; N—Nitrogen; P—Phosphorus; K—Potassium; C. Sand—Coarse sand; F. Sand—Fine sand; CEC—Cation Exchange Capacity.

3.3. Discriminant Factors and Variables

3.3.1. Layer 0–20 cm

The DA performed with the factor scores of cases obtained for the 0–20 cm layer indicated that Factor 3, with the lowest significant partial Wilks' Lambda of 0.597 ($p < 0.001$) was the most influent factor for temporal variability (sampling dates). Therefore, the differences in soil surface samples over time were predominantly affected by variations in the chemical composition, namely, by pH, EC, and N. Factor 1, showing the lowest significant partial Wilks' Lambda of 0.648, was the dominant factor for spatial variability (fields) ($p < 0.001$). Therefore, it was the texture attribute that most contributed to the spatial differentiation in the soil surface.

The first discriminant functions (roots) for factors (Y) considering time (sampling dates) and space (crop fields), and their respective factor coefficients obtained with the CCA for the surface layer confirmed the previous DA: higher absolute roots were obtained for Factor 3 and Factor 1, respectively, in temporal and spatial discrimination ((Equations (1) and (2)). The discriminant functions found accounted for 71.03% and 85.52% of the total variance ((respectively, Equations (1) and (2)):

Temporal variability:

$$Y_T = -0.252\ (\text{Factor } 1) - 0.175\ (\text{Factor } 2) + 0.973\ (\text{Factor } 3) - 0.170\ (\text{Factor } 4), \quad (1)$$

Spatial variability:

$$Y_S = -1.941\ (\text{Factor } 1) - 0.563\ (\text{Factor } 2) + 0.062\ (\text{Factor } 3) - 0.054\ (\text{Factor } 4), \quad (2)$$

The DA and CCA performed with the variables highly correlated with the dominant factors in the temporal and spatial variability allowed us to find the following first discriminant functions for the variables (Y′), accounting, respectively, for 86.50% and 92.89% of the total variance (Equations (3) and (4), respectively):

Temporal variability:

$$Y'_T = 0.130\ (\text{pH}) + 1.220\ (\text{EC}) - 0.462\ (\text{N}), \quad (3)$$

Spatial variability:

$$Y'_S = -0.469\ (\text{C. Sand}) + 1.886\ (\text{F. Sand}) - 0.368\ (\text{Silt}), \quad (4)$$

Considering the standardized coefficients of each variable, the differences over time in soil surface were predominantly identified by EC, available N, and pH. These results may be indicative of seasonal changes in soil salinity due to irrigation water quality degradation [23,36], along with the influence of fertilization and fertigation, whose management deserves attention in order to avoid salinization or N losses to the environment through leaching, ammonia (NH_3) volatilization, or bacterial denitrification of nitrate (NO_3-N) [37]. Fine sand proportion was the variable that mostly influenced the spatial variation, followed by Coarse sand and Silt proportions.

3.3.2. Layer 20–40 cm

In the case of the sub-surface layer, the DA revealed a preponderance of Factor 2 (chemical composition) and Factor 3 (nutrient availability) in temporal (partial Wilks' Lambda of 0.358; $p < 0.001$) and spatial (partial Wilks' Lambda of 0.532; $p < 0.001$) discrimination, respectively. The discriminant functions considering temporal and spatial variation (Y_T and Y_S, respectively) accounted for 79.39% and 97.92% of total variance and presented the coefficients presented in Equations (5) and (6), respectively:

Temporal variability:

$$Y_T = 0.284\ (\text{Factor } 1) - 1.012\ (\text{Factor } 2) - 0.319\ (\text{Factor } 3) + 0.081\ (\text{Factor } 4), \quad (5)$$

Spatial variability:

$$Y_S = 0.719\ (\text{Factor 1}) + 0.284\ (\text{Factor 2}) + 0.964\ (\text{Factor 3}) + 0.554\ (\text{Factor 4}), \quad (6)$$

The coefficients of each factor confirm that the chemical composition attribute (Factor 3 and Factor 2, in each layer, respectively) is the one that had the most influence on temporal differentiation, denoting the influence of crop management, e.g., fertilization and irrigation, or, indirectly, of the soil organic content, through the variable N total [18,38].

While texture (Factor 1) was predominant in the differentiation between fields, in the case of the surface layer, in the sub-surface layer this differentiation was mostly influenced by the nutrient availability (Factor 3). These results point to the importance of the soil particle distribution, the presence of clay and its close relationship with CEC, and the high to very high levels of available K found throughout the study, for the spatial variability.

The first discriminant functions for the variables highly correlated with the identified factors (Y'), accounting, respectively, for 84.63% and 80.96% of the total variance, are presented in Equations (7) and (8):

Temporal variability:

$$Y'_T = -0.741\ (\text{EC}) - 0.502\ (\text{N}), \quad (7)$$

Spatial variability:

$$Y'_S = -0.904\ (\text{K}) + 0.607\ (\text{CEC}), \quad (8)$$

In both layers, EC is the soil parameter that most contributes to the variability that occurs over time, followed by N total. As usual, nitrogen fertilization rates were higher than any of the other applied nutrients in every field and year, except for 2020, where 350 kg ha^{-1} of ammonium sulfate (20.5% N + 60% SO_3) was applied in the onion crop, significantly increasing the amount of sulfur applied compared to any other nutrient in the P4 field. Moreover, N is employed not only at sowing, but also in top dressing fertilizations, and in soluble formulations applied with irrigation during the crop development. Therefore, the influence of N in temporal discrimination was expected, and specific consideration should be taken regarding N fertilization management. The importance of EC is linked not only with irrigation water quality, but also with the correlation between this parameter and many of the main physicochemical properties of the soil, such as proportion of clay, soil water content, soil bulk density, and soil organic content, which are subject to spatial and temporal changes in agricultural soils. In fact, this correlation is the basis for the use of soil apparent electrical conductivity measured by expeditious methods in distinguishing agronomic management zones and productivity maps, presently, with widespread application in precision agriculture [39–41].

4. Conclusions

The integrated knowledge of the temporal and spatial variation of soil properties can improve crop productivity and prevent soil degradation. Given the multidimensional and multivariate nature of soil sampling data, the use of multivariate statistical analysis tools may help to better understand the dynamics of variation in time and space and the relationships between agronomic parameters and soil functions. The Factor Analysis performed with data collected in three fields of irrigated annual crops during the period 2018–2020, in two soil layers (0–20 cm and 20–40 cm) allowed us to obtain two models of four factors for each layer. The model obtained for the surface layer included factors related to the attributes texture, water and nutrient retention capacity, chemical composition, and soil fertility. In the sub-surface layer, the model of factor structure was similar. The most influential factor in temporal discrimination (sampling dates) was related to chemical composition, indicating the influence of crop management practices such as fertilization and irrigation. As for the spatial differentiation (fields), the dominant factor in the surface layer was texture, and in the sub-surface layer, nutrient availability. The variable most influent in temporal discrimination in both the surface and sub-surface layers was EC. The most preponderant spatial discriminant variables were fine sand proportion and available potassium in the 0–20 cm layer and 20–40 cm layer, respectively. Further studies on this subject should consider the integration of biological and ecotoxicological indicators as soil health indicators to attend to the intensification in the use of fertilizers and plant protection products in irrigated agriculture, thereby understanding better the implications of larger inputs of these products on the sustainability of irrigated soils.

Author Contributions: Conceptualization, A.T. and P.P.; methodology, A.T. and P.P.; validation, A.T., J.D., M.F., L.B., M.P., and P.P.; investigation, A.T., I.M., A.C., C.M., J.D., M.F., L.B., J.C., M.P. and P.P.; writing—original draft preparation, A.T.; writing—review and editing, A.T., J.C. and P.P.; supervision, P.P.; project administration, P.P.; funding acquisition, A.T., J.C., M.P. and P.P. All authors have read and agreed to the published version of the manuscript.

Funding: This research is co-funded by the European Union through the European Regional Development Fund, included in the COMPETE 2020 (Operational Program Competitiveness and Internationalization) through the ICT project (UIDB/04683/2020), with the reference POCI-01-0145-FEDER-007690, through Geobiotec (UIDB/04035/2020) and Centro de Química (UIBD/00616/2020), funded by FCT—Fundação para a Ciência e a Tecnologia, Portugal, and through the FitoFarmGest Operational Group (PDR2020-101-030926).

Conflicts of Interest: The authors declare no conflict of interest.

References

1. Adhikari, K.; Hartemink, A.E. Linking Soils to Ecosystem Services—A Global Review. *Geoderma* **2016**, *262*, 101–111. [CrossRef]
2. Larson, W.E.; Pierce, F.J. The Dynamics of Soil Quality as a Measure of Sustainable Management. In *Defining Soil Quality for a Sustainable Environment*; Doran, J.W., Coleman, D.C., Bezdicek, D.F., Stewart, B.A., Eds.; SSSA Special Publication Number: Wisconsin, WI, USA, 1994; pp. 37–51.
3. Karlen, D.L.; Veum, K.S.; Sudduth, K.A.; Obrycki, J.F.; Nunes, M.R. Soil Health Assessment: Past Accomplishments, Current Activities, and Future Opportunities. *Soil Tillage Res.* **2019**, *195*, 104365. [CrossRef]
4. FAO Water. Key Facts. Available online: http://www.fao.org/water/en/ (accessed on 27 September 2021).
5. Adejumobi, M.A.; Awe, G.O.; Abegunrin, T.P.; Oyetunji, O.M.; Kareem, T.S. Effect of Irrigation on Soil Health: A Case Study of the Ikere Irrigation Project in Oyo State, Southwest Nigeria. *Environ. Monit. Assess.* **2016**, *188*, 696. [CrossRef] [PubMed]
6. Nikolskii, Y.; Aidarov, I.P.; Landeros-Sanchez, C.; Pchyolkin, V.V. Impact of Long-Term Freshwater Irrigation on Soil Fertility. *Irrig. Drain.* **2019**, *68*, 993–1001. [CrossRef]
7. Bendra, B.; Fetouani, S.; Laffray, X.; Vanclooster, M.; Sbaa, M.; Aleya, L. Effects of Irrigation on Soil Physico-Chemistry: A Case Study of the Triffa Plain (Morocco). *Irrig. Drain.* **2012**, *61*, 507. [CrossRef]
8. Murray, R.S.; Grant, C.D. *The impact of irrigation on soil structure. Land & Water Australia*; Land & Water Australia: Camberra, Australia, 2007; p. 31. Available online: https://library.dbca.wa.gov.au/static/FullTextFiles/070521.pdf (accessed on 27 September 2021).
9. Presley, D.R.; Ransom, M.D.; Kluitenberg, G.J.; Finnell, P.R. Effects of Thirty Years of Irrigation on the Genesis and Morphology of Two Semiarid Soils in Kansas. *Soil Sci. Soc. Am. J.* **2004**, *68*, 1916–1926. [CrossRef]
10. Jagadamma, S.; Lal, R.; Hoeft, R.G.; Nafziger, E.D.; Adee, E.A. Nitrogen Fertilization and Cropping System Impacts on Soil Properties and Their Relationship to Crop Yield in the Central Corn Belt, USA. *Soil Tillage Res.* **2008**, *98*, 120–129. [CrossRef]
11. Härdle, W.K.; Simar, L. *Applied Multivariate Statistical Analysis*, 4th ed.; Springer: Heidelberg, NY, USA; Dordrecht, The Netherlands; London, UK, 2015; ISBN 978-3-662-45171-7.
12. Yeater, K.; Villamil, M. Multivariate Methods for Agricultural Research. In *Applied Statistics in Agricultural, Biological, and Environmental Sciences*; Glaz, B., Yeater, K.M., Eds.; American Society of Agronomy, Inc.; Soil Science Society of America, Inc.; Crop Science Society of America, Inc.: Madison, WI, USA, 2018; ISBN 978-0-89118-359-4. [CrossRef]
13. Quiroga, A.R.; Buschiazzo, D.E.; Peinemann, N. Management Discriminant Properties in Semiarid Soils. *Soil Sci.* **1998**, *163*, 591–597. [CrossRef]
14. Sena, M.M.; Frighetto, R.T.S.; Valarini, P.J.; Tokeshi, H.; Poppi, R.J. Discrimination of Management Effects on Soil Parameters by Using Principal Component Analysis: A Multivariate Analysis Case Study. *Soil Tillage Res.* **2002**, *67*, 171–181. [CrossRef]
15. Preston, W.; Nascimento, C.; Silva, Y.; Silva, D.; Ferreira, H. Soil Fertility Changes in Vineyards of a Semiarid Region in Brazil. *J. Soil Sci. Plant Nutr.* **2017**, *17*, 672–685. [CrossRef]
16. Mallarino, A.; Oyarzabal, E.; Hinz, P. Interpreting Within-Field Relationships Between Crop Yields and Soil and Plant Variables Using Factor Analysis. *Precis. Agric.* **1999**, *1*, 15–25. [CrossRef]
17. Juhos, K.; Szabó, S.; Ladányi, M. Influence of Soil Properties on Crop Yield: A Multivariate Statistical Approach. *Int. Agrophys.* **2015**, *29*, 433–440. [CrossRef]

18. Shukla, M.K.; Lal, R.; Ebinger, M. Determining Soil Quality Indicators by Factor Analysis. *Soil Tillage Res.* **2006**, *87*, 194–204. [CrossRef]
19. Firdous, S.; Begum, S.; Yasmin, A. Assessment of Soil Quality Parameters Using Multivariate Analysis in the Rawal Lake Watershed. *Environ. Monit. Assess.* **2016**, *188*, 533. [CrossRef]
20. IPMA Climate normal—1981–2010—Beja. Available online: https://www.ipma.pt/en/oclima/normais.clima/1981-2010/002/ (accessed on 6 July 2022).
21. COTR SAGRA—Sistema Agrometeorológico Para a Gestão Da Rega No Alentejo (Agrometeorological System for Irrigation Management in Alentejo). Available online: http://www.cotr.pt/servicos/sagranet.php (accessed on 15 February 2022).
22. *IUSS Working Group WRB World Reference Base for Soil Resources 2014*; Update 2015; FAO: Rome, Italy, 2014.
23. Tomaz, A.; Costa, M.J.; Coutinho, J.; Dôres, J.; Catarino, A.; Martins, I.; Mourinha, C.; Guerreiro, I.; Pereira, M.M.; Fabião, M.; et al. Applying Risk Indices to Assess and Manage Soil Salinization and Sodification in Crop Fields within a Mediterranean Hydro-Agricultural Area. *Water* **2021**, *13*, 3070. [CrossRef]
24. Varennes, A.D. *Produtividade Dos Solos e Ambiente (Productivity of Soils and Environment)*; Escolar Editora: Lisboa, Portugal, 2003.
25. Walkley, A.; Black, L.A. An Examination of the Dgtjareff Method for Determining Soil Organic Matter, and a Proposed Modification of the Chromic Acid Titration Method. *Soil Sci.* **1934**, *37*, 29–38. [CrossRef]
26. Kjeldahl, J. Neue Methode zur Bestimmung des Stickstoffs in organischen Körpern (New method for the determination of nitrogen in organic substances). *Z. Für Anal. Chem.* **1883**, *22*, 366–383. [CrossRef]
27. Egner, H.; Riehm, H.; Domingo, W.R. Untersuchhungen Uber Die Chemische Boden: Analyse Als Grundlage Fur Die Beurteilung Der Nahrstoffzustandes Der Boden. II. Chemique Extractions, Methoden Zur Phosphor, Und Kalium-Bestimmung (Investiga-Tions on the Chemical Soil Analysis as a Basis for Assess-Ing the Soil Nutrient Status II: Chemical Extractionmethods for Phosphorus and Potassium Determination). *K. Lantbr. Ann.* **1960**, *26*, 199–215.
28. StatSoft, Inc. *STATISTICA (Data Analysis Software System)*; StatSoft, Inc.: Tulsa, OK, USA, 2004. Available online: https://www.tibco.com/products/tibco-statistica (accessed on 15 February 2022).
29. Tomaz, A.; Palma, J.F.; Ramos, T.; Costa, M.N.; Rosa, E.; Santos, M.; Boteta, L.; Dôres, J.; Patanita, M. Yield, Technological Quality and Water Footprints of Wheat under Mediterranean Climate Conditions: A Field Experiment to Evaluate the Effects of Irrigation and Nitrogen Fertilization Strategies. *Agric. Water Manag.* **2021**, *258*, 107214. [CrossRef]
30. Lee, K.-M.; Herrman, T.J.; Lingenfelser, J.; Jackson, D.S. Classification and Prediction of Maize Hardness-Associated Properties Using Multivariate Statistical Analyses. *J. Cereal Sci.* **2005**, *41*, 85–93. [CrossRef]
31. TIBCO TIBCO Statistica®User's Guide, Version 14.0. Available online: https://docs.tibco.com/pub/stat/14.0.0/doc/html/UsersGuide/GUID-058F49FC-F4EF-4341-96FB-A785C2FA76E9-homepage.html (accessed on 22 April 2020).
32. Foth, H.D. *Fundamentals of Soil Science*, 8th ed.; John Wiley & Sons, Inc.: New York, NY, USA, 1990; ISBN 0-471-52279-1.
33. Parfitt, R.; Giltrap, D.; Whitton, J. Contribution of Organic Matter and Clay Minerals to the Cation Exchange Capacity of Soil. *Commun. Soil Sci. Plant Anal.* **1995**, *26*, 1343–1355. [CrossRef]
34. Weil, R.R.; Brady, N.C. *The Nature and Properties of Soils*, 15th ed.; Pearson: Columbus, OH, USA, 2016; ISBN 978-0-13-325448-8.
35. Chatterjee, D.; Datta, S.C.; Manjaiah, K.M. Fractions, Uptake and Fixation Capacity of Phosphorus and Potassium in Three Contrasting Soil Orders. *J. Soil Sci. Plant Nutr.* **2014**, *14*, 640–656. [CrossRef]
36. Tomaz, A.; Palma, P.; Fialho, S.; Lima, A.; Alvarenga, P.; Potes, M.; Costa, M.; Salgado, R. Risk Assessment of Irrigation-Related Soil Salinization and Sodification in Mediterranean Areas. *Water* **2020**, *12*, 3569. [CrossRef]
37. Sapkota, T.B.; Singh, L.K.; Yadav, A.K.; Khatri-Chhetri, A.; Jat, H.S.; Sharma, P.C.; Jat, M.L.; Stirling, C.M. Identifying Optimum Rates of Fertilizer Nitrogen Application to Maximize Economic Return and Minimize Nitrous Oxide Emission from Rice–Wheat Systems in the Indo-Gangetic Plains of India. *Arch. Agron. Soil Sci.* **2020**, *66*, 2039–2054. [CrossRef]
38. Simfukwe, P.; Hill, P.W.; Emmett, B.A.; Jones, D.L. Identification and Predictability of Soil Quality Indicators from Conventional Soil and Vegetation Classifications. *PLoS ONE* **2021**, *16*, e0248665. [CrossRef] [PubMed]
39. Corwin, D.L.; Lesch, S.M. Apparent Soil Electrical Conductivity Measurements in Agriculture. *Comput. Electron. Agric.* **2005**, *46*, 11–43. [CrossRef]
40. Fortes, R.; Millán, S.; Prieto, M.H.; Campillo, C. A Methodology Based on Apparent Electrical Conductivity and Guided Soil Samples to Improve Irrigation Zoning. *Precis. Agric.* **2015**, *16*, 441–454. [CrossRef]
41. Moral, F.J.; Terrón, J.M.; Silva, J.R.M. da Delineation of Management Zones Using Mobile Measurements of Soil Apparent Electrical Conductivity and Multivariate Geostatistical Techniques. *Soil Tillage Res.* **2010**, *106*, 335–343. [CrossRef]

water MDPI

Article

Bayesian Calibration and Uncertainty Assessment of HYDRUS-1D Model Using GLUE Algorithm for Simulating Corn Root Zone Salinity under Linear Move Sprinkle Irrigation System

Farzam Moghbel [1,2,3], Abolfazl Mosaedi [1,*], Jonathan Aguilar [2,3,*], Bijan Ghahraman [1], Hossein Ansari [1] and Maria C. Gonçalves [4]

1 Department of Water Science and Engineering, Faculty of Agriculture, Ferdowsi University of Mashhad, Mashhad 9177948978, Iran
2 Southwest Research–Extension Center, Kansas State University, 4500 E. Mary St., Garden City, KS 67846, USA
3 Biological and Agricultural Engineering Department, Kansas State University, 1016 Seaton Hall 920N. Martin Luther King Jr. Drive, Manhattan, KS 66506, USA
4 Instituto Nacional de Investigação Agrária e Veterinária (INIAV), Av. República, 2780-157 Oeiras, Portugal
* Correspondence: mosaedi@um.ac.ir (A.M.); jaguilar@ksu.edu (J.A.); Tel.: +1-(620)-275-9164 (J.A.)

Abstract: Soil salinization is one of the significant concerns regarding irrigation with saline waters as an alternative resource for limited freshwater resources in arid and semi-arid regions. Thus, the investigation of proper management methods to control soil salinity for irrigation with saline waters is inevitable. The HYDRUS-1D model is a well-known numerical model that can facilitate the exploration of management scenarios to mitigate the consequences of irrigation with saline waters, especially soil salinization. However, before using the model as a decision support system, it is crucial to calibrate the model and analyze the model's parameters and outputs' uncertainty. Therefore, the generalized likelihood uncertainty estimation (GLUE) algorithm was implemented for the HYDRUS-1D model in the R environment to calibrate the model and assess the uncertainty aspects for simulating soil salinity of corn root zone under saline irrigation with linear move sprinkle irrigation system. The results of the study have detected a lower level of uncertainty in the α, n, and θ_s (saturated soil water content) parameters of water flow simulations, dispersivity (λ), and adsorption isotherm coefficient (K_d) parameters of solute transport simulations comparing to the other parameters. A higher level of uncertainty was found for the diffusion coefficient as its corresponding posterior distribution was not considerably changed from its prior distribution. The reason for this phenomenon could be the minor contribution of diffusion to the solute transport process in the soil compared with advection and hydrodynamic dispersion under saline water irrigation conditions. Predictive uncertainty results revealed a lower level of uncertainty in the model outputs for the initial growth stages of corn. The analysis of the predictive uncertainty band also declared that the uncertainty in the model parameters was the predominant source of uncertainty in the model outputs. In addition, the excellent performance of the calibrated model based on 50% quantiles of the posterior distributions of the model parameters was observed in terms of simulating soil water content (SWC) and electrical conductivity of soil water (ECsw) at the corn root zone. The ranges of NRMSE for SWC and ECsw simulations at different soil depths were 0.003 to 0.01 and 0.09 to 0.11, respectively. The results of this study have demonstrated the authenticity of the GLUE algorithm to seek uncertainty aspects and calibration of the HYDRUS-1D model to simulate the soil salinity at the corn root zone at field scale under a linear move irrigation system.

Keywords: Bayesian; calibration; corn; GLUE; HYDRUS-1D; irrigation; root water uptake; salinity; solute transport; uncertainty

1. Introduction

The limited availability of high-quality water for agricultural purposes is a critical issue in arid and semi-arid regions. However, frequently significant quantities of water with different salinity levels exist in these areas that could be alternative resources for dealing with this problem [1,2]. Groundwater, agricultural till water or drainage water, and municipal or industrial wastewater are the main sources of saline waters [3–7]. In the last three decades, irrigation with saline water has been extended in semi-arid regions [8]. Expansion of saline irrigation practices without proper management could increase the risk of degrading soil quality and consequently losing agricultural lands in the long term [9]. About 77 million hectares of fields have been impacted by salinization worldwide [10]. Crops yield and yield components reductions as subsequences of irrigation with saline waters have been reported multiple times [11–13]. Every year about 1 to 2% of irrigated agricultural lands are reduced due to high soil salinity [10]. Therefore, appropriate analysis of crop root zone salinity for irrigation with saline waters are critical to explore proper management methods to control and alleviate salinization. Field experiments are usually time and money-consuming, and adapting the results to other locations with different agronomic and irrigation managements is complicated. Hence, using agro-hydrological models could be a reasonable answer to this issue. To date, several agro-hydrological models have been introduced in the literature to simulate the soil water and salinity dynamics under different climate, irrigation, and agronomic conditions. These models are categorized as steady-state and transient based on their developing approach. For instance, the TETrans [14], SaltMod [15], SALEACH [16], and WSBM [17] are among the models which have been developed so far to simulate soil water content and salinity for different cultivations.

TETrans is a transient model that simulates changes in the solute and water content at the crop root zone through a series of events within a finite discrete soil depth. These events include infiltration of irrigation water, draining soil profile to field capacity, root water uptake through transpiration, and evaporation through the soil surface. The assumptions have been made to pursue the development of this model, which might not be realistic. It has been assumed that each process or event occurs in sequential order within given soil depths. The soil is homogenously distributed over discrete depths. Depletion of soil water content through evapotranspiration does not go below the crop's threshold to water stress, and dispersion is mostly negligible [18]. The model uses these sequential components to calculate the salt balance for simulating solute transport at the crop root zone [14]. Thus, the assumptions are also reflected in solute transport process calculations.

The SaltMod model is another transient model that predicts the salinity of soil water, drainage water, and groundwater. Furthermore, the model computes water table depth and the drainage water quantity in irrigated agricultural fields [15]. The model computations are mainly based on the water model, salt balance model, and seasonal agronomic practices. The SaltMod has been known to be more reliable for its seasonal rather than daily outputs; because it implements the seasonal water balance for the region of interest, not daily bases [19]. However, this approach reduces the number of the model's inputs while neglecting in-season soil water and salinity dynamics, which could substantially affect the model's predictions.

Shahrokhnia and Wu, 2021, developed the web-based SALEACH model based on a steady-state approach to reproduce crop root zone and drainage water salinity to estimate leaching requirement and irrigation depth [16]. The models that simulate soil salinity changes based on steady approaches assume that irrigation water flows continuously downward at a constant rate, and the evapotranspiration rate is constant during the growing season. Also, soil-soluble salt concentrations are constant at any point [20]. However, comparing the assumptions with observational data reveals that these assumptions are unrealistic.

Liu et al., 2022, have introduced the WSBM model that predicts the general trend of soil salinization in the long-term aspect [17]. The model was developed based on water and salt balance at the crop root zone and groundwater. The model provides preliminary information to explore well-canal irrigation water quality strategies' effects on the soil. Therefore, the model tends to provide a general perspective rather than accurate simulations to explore the effects of irrigation water management strategies on agricultural lands. However, among the agro-hydrological models, the HYDRUS-1D has been known as a comprehensive model. Only some simplifications have been implemented for the model development. The HYDRUS-1D simulations of soil water flow and solute transport processes are closer to reality than the majority of the agro-hydrological models.

The HYDRUS-1D model is a distinguished numerical model that simulates soil water flow and solute transport [21], and its reliability has been proved multiple times in the literature. Liu et al., 2022 have illustrated the reliability of the HYDRUS-1D model through calibration and validation process to simulate soil water content and salt movement in 300 cm soil profiles at an irrigated cropland and unirrigated grassland [22]. Noshadi et al., 2020 have found the HYDRUS-1D model very accurate in simulating soil water content and salt dynamic under wheat cultivation for irrigation with waters with different levels of salinity [23]. Kanzari et al., 2018 concluded that the HYDRUS-1D model duplicated soil water and salinity dynamics of soil in a semi-arid region [24]. A study conducted in southwest Queensland, Australia, showed that the HYDRUS-1D model successfully simulates salt leaching in amended profiles [25]. Ramous et al., 2011 indicated that robust outputs of the HYDRUS-1D model regarding the reduction of maize water and nutrient uptake under osmotic stress were observed in Alvalade and Mitra, Portugal [26]. Askri et al., 2014, have investigated the interaction effects of waterlogging, salinity, and water shortage on root water uptake of date palms in an oasis by using the HYDRUS-1D model. The model was calibrated and validated using sap flow density, soil hydraulic characteristics, and applied irrigation data. Their results have demonstrated the acceptable performance of the HYDRUS-1D in simulating water uptake of the palm tree [27]. Skaggs et al., 2006, have studied the performance of the HYDRUS-1D model to imitate the root water uptake and drainage of lysimeter observational data under forage crops (alfalfa and tall wheatgrass) cultivations irrigated with synthetic drainage waters. The researchers have found good agreement between observational data and model simulation for wide ranges of irrigation water salinities (2.5 to 28 dS/m) [28]. Ali et al., 2021 have inspected the capabilities of the HYDRUS-1D model in quantifying the soil hydraulic conductivity reduction under marginal quality water irrigation. The model outputs were compared with leaching column observational data. The study's results emphasized that the HYDRUS-1D standard hydraulic conductivity reduction scaling factor needs adjustments. They recommended the non-linear approach as an alternative for determining the hydraulic conductivity scaling factor to enhance model water flow and solute transport outputs [29]. The performance of the HYDRUS-1D model to simulate water balance and salinity build-up at rice root zone cultivated in micro-lysimeters have been tested by Phogat et al., 2010. The rice crop was irrigated with waters with different salinity levels (0.4 to 10 dS/m). The statistical analysis showed close agreement between model outputs and observational data [30]. Moreover, this study has shown the reliability of the HYDRUS-1D outputs to be used for calculations of rice water productivity.

To use the HYDRUS-1D model, the proper calibration of the model is vital. Simultaneously the parameters and outputs of the model are subject to uncertainty, and it would be beneficial if they could be quantified. One of the promising approaches for calibration and uncertainty analysis of the models is exploiting Bayesian statistics concepts. The Bayesian statistics concepts use the prior knowledge (prior distribution) about the model parameters and combine this information with observational data (likelihood function) to calculate the uncertainty of the model by achieving posterior distribution of the parameters [31].

Several algorithms have been developed based on Bayesian statistics concepts to quantify the posterior distributions of parameters. Vrugt et al., 2003, introduced the Shuffled Complex Evolution Metropolis (SCEM-UA) algorithm for uncertainty analysis and optimizing parameters of the hydrological models [32]. This SCEM-UA was developed to obtain the posterior distribution of the hydrologic model parameters. This algorithm constantly modifies the proposal distribution by combining the strength of the Metropolis algorithm, random search, competitive evolution, and complex shuffling to enhance the trajectory of the algorithm for converging toward the posterior distribution of parameters. Ter Braak, 2006, developed the algorithm known as Differential Evolution Markov Chain (DE-MC), which is the Markov Chain Monte Carlo (MCMC) version of the genetic algorithm Differential Evolution [33]. DE-MC algorithm runs multiple chains in parallel. The algorithm updates the chains based on the difference between two random parameter vectors, and the Metropolis ratio controls the selection of these vectors. The superiority of the DE-MC algorithm over the conventional MCMC algorithm is the speed of calculation and convergence of the algorithm. Vrugt et al., 2009, introduced the MCMC-based algorithm known as Differential Evolution Adaptive Metropolis (DREAM) [34]. The DREAM algorithm significantly improved the efficiency of the MCMC simulation. This algorithm runs multiple chains in parallel and adjusts the scale and orientation of the proposal distribution. The DREAM algorithm has been improved multiple times so far, and several versions entitled DREAM (D), DREAM (ZS), and DREAM (Kzs) have been introduced [35–37]. However, the complexity of implementing these algorithms for agro-hydrological models would lead the studies to seek alternative algorithms that are less intricate and robust enough to explore uncertainty analysis. One well-known algorithm that fulfills these characteristics is Generalized Likelihood Uncertainty Estimation.

The Generalized Likelihood Uncertainty Estimation (GLUE) is a Bayesian theorem-based algorithm that uses prior information about the parameters and measurement data to estimate uncertainty in model parameters [38]. Estimating the posterior distribution of the parameters is attained by sampling an enormous number of parameters from the prior distribution by Monte Carlo simulations [39]. Then, the generated parameter sets' performances are evaluated by the likelihood function. The uncertainty estimated by GLUE considers all sources of uncertainty, including input, structure, and parameter uncertainties. The calculated value by the likelihood regarding the parameter sets reflects all sources of error with the model [40]. In the recent decade, the GLUE algorithm has received attention in agro-hydrological studies. Li et al., 2018 have proved that the calibrated DSSAT-CERES based on the GLUE method could accurately reproduce leaf area index, above-ground biomass, grain yield, and above-ground nitrogen of winter wheat in the Beijing area [41]. Sun et al., 2016 successfully calibrated the parameters of the RZWQM-DSSAT (RZWQM2) model for simulating crop growth and transporting water and nitrogen in a wheat-maize cropping system using the GLUE method with some adjustments in its likelihood function [42]. Their uncertainty analysis results declared that the parameters related to soil-saturated hydraulic conductivity, nitrogen nitrification and denitrification, and urea hydrolysis were the most effective ones in simulating crop yield components. Sheng et al., 2019 compared two Bayesian theorem-based algorithms of GLUE and DREAM for estimating parameters associated with cultivars in the APSIM-maize model. They found similar performance between GLUE and DREAM algorithms [43]. Uncertainty assessment of the SWAP model for simulating soil water content (SWC) at a field scale using the GLUE method has been explored by Shafiei et al., 2014 [40]. Their results revealed that the predictive uncertainty in simulating SWC was low, which proved the good performance of the SWAP model in dry regions at the field scale.

To date, few studies have focused on Bayesian calibration and uncertainty analysis of HYDRUS-1D, explicitly using the GLUE method for simulating soil salt dynamics under sprinkler irrigation systems at field scale. Thus, the main objective of this research was to calibrate and assess the uncertainty of the HYDRUS-1D model for simulating transient conditions of salts in the corn root zone in western Kansas based on the GLUE method.

2. Materials and Methods

2.1. Site Description

Field experiments were conducted at Kansas State University, Southwest Research-Extension Center, near Garden City, Kansas. The geographical coordinates of the experimental field are 38°01′20.87″ N, 100°49′26.95″ W, and the altitude of the area is 887 m above sea level. According to the long-term weather data of Garden City, KS, the total annual precipitation and evapotranspiration are 477 and 1810 mm, respectively. The well-drained Ulysses silt loam was the predominant soil type [44]. The physical characteristics of the soil (fine-silty, mixed, mesic Aridic Haplustoll) are presented in Table 1. In addition, the soil pH was 8.3, and the experimental area has 170 days frost-free period [44,45]. The electrical conductivity of irrigation water was 1.2 dS/m and its sodium adsorption ratio (SAR) was 2.6 $(meq/L)^{0.5}$. Therefore, the irrigation water chemical characteristics were categorized as C1-S1 in the USSL classification (Table 2). A four-span linear move irrigation system (model 8000, Valmont Corp., Valley, NE, USA) was used to implement irrigation for the aims of the study, and the plots' sizes of the study were 13.7 m × 27.4 m.

Table 1. The physical characteristics of the soil at the experimental field near Garden City, KS.

Depth (cm)	Soil Texture	Sand (%)	Silt (%)	Clay (%)	Wilting Point (%)	Field Capacity (%)	Saturation (%)	Bulk Density ($g.cm^{-2}$)
0–240	Silt loam	18.5	55.5	26	15	33	45	1.38

Table 2. The chemical characteristics of irrigation water.

	Units	Value
EC	dS/m	1.2
SAR	$(meq/L)^{0.5}$	2.64
Na^+	mg/L	120
Ca^{2+}	mg/L	58
Mg^{2+}	mg/L	95
SO_4^{2-}	mg/L	200
PO_4^{3-}	mg/L	6.3
NO_3^-	mg/L	118
K^+	mg/L	6.1

2.2. Data Collection and Management

The conventional corn was planted on 11 May 2016, with 84,000 seeds/ha seeding rate, and harvested on 6 October 2016. The corn plant spacing was 20 cm, and the row spacing was 70 cm. The irrigation events started on 1 April 2016 and terminated on 15 September 2016. The irrigation frequency was determined according to soil water content readings based on 50% soil water depletion. The TDRs (CS655 manufactured by Campbell Scientific, Logan, UT, USA) with 12 cm rods were installed at 16, 46, and 76 cm of soil depth, respectively, to monitor the volumetric soil water content (SWC) and soil salinity (ECsw) [46,47] at the same time during and out of the growing season. The sensors were installed for three different plots with the same installation depths. The SWC and ECsw were collected every 30 min using the Campbell Scientific data loggers. The observational data were collected for 235 days (11 May 2016 to 31 December 2016). The agronomic practices, including weed management and fertilizer application, were based on the recommended procedures for southwest Kansas. The initial ECsw was 2.88, 2.89, and 5.09 dS/m for 16, 46, and 76 cm soil depth, respectively. Daily meteorological data, including minimum and maximum temperature, relative humidity, solar radiation, wind speed, and precipitation, were obtained from the nearby tower of the K-State MESONET network. The FAO-Penman Monteith method was used to calculate daily evapotranspiration [48]. Daily crop evapotranspiration and precipitation during and out of the growing season are depicted in Figure 1. The 25.4 mm irrigation depth was constant for each irrigation event.

In addition, crop mapping was done multiple times during the growing season to measure the leaf area index (LAI) and record crop growth stages. All of these data were used as input for atmospheric boundary conditions and observational data for calibration purposes. In addition, the SWC and ECsw data were measured during and out of the growing season to comprehend salts and SWC dynamics in the field condition. The initial soil profile water content and salinity at different depths are presented in Table 3.

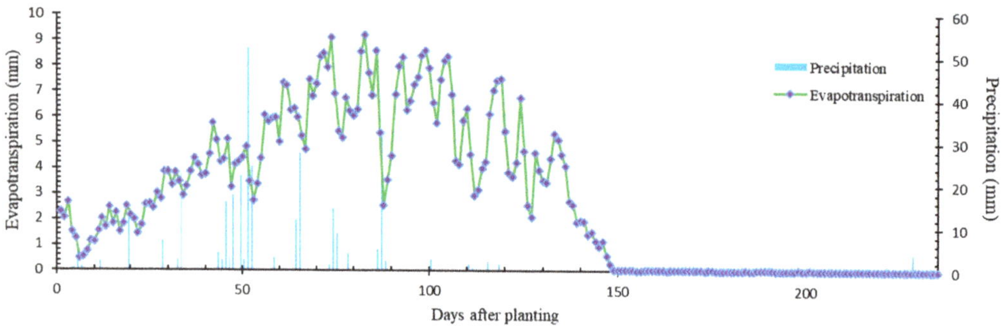

Figure 1. Time series of daily crop evapotranspiration and precipitation (15 May 2016 to 31 December 2016).

Table 3. The initial soil water content and soil salinity.

Soil Depth (cm)	Soil Water Content ($cm^3.cm^{-3}$)	ECsw (dS/m)
0–30	0.289	2.842
30–60	0.247	2.87
60–120	0.208	5.043
120–200	0.20	5.0

ECsw = electrical conductivity of soil water.

2.3. HYDRUS-1D Model

The simulations were done from the planting date to the beginning of the next growing season.

2.3.1. Water Flow Modeling

To simulate the water flow in the soil, the Richards' Equation (1) was solved numerically in HYDRUS-1D using Galerkin finite element scheme [21]:

$$\frac{\partial\theta}{\partial t} = \frac{\partial}{\partial z}\left[K(h)\left(\frac{\partial h}{\partial z} - 1\right)\right] - S(z,t) \tag{1}$$

where θ is soil volumetric water content (L^3L^{-3}), t is time (T), z is vertical space coordinate (L), K(h) is unsaturated hydraulic conductivity function (LT^{-1}), h is soil pressure head (L), and S(z,t) is sink term ($L^3L^{-3}T^{-1}$) representing the unit volume of water removed by the crop from a unit of volume of soil per time unit.

The HYDRUS-1D model utilized the following equation to represent the relations between θ and h [49]:

$$\theta(h) = \begin{cases} \frac{\theta_s - \theta_r}{1+\left(-\alpha(h)^n\right)^{1-\frac{1}{n}}} + \theta_r & h < 0 \\ \theta_s & h > 0 \end{cases} \tag{2}$$

where θ_s is saturated soil volumetric water content (L^3L^{-3}), θ_r is residual soil volumetric water content (L^3L^{-3}), and α and n are empirical parameters.

The model used the following unsaturated soil hydraulic conductivity function [49]:

$$K(h) = \begin{cases} K_s Se^l \left[1 - \left(1 - S_e^{\frac{n}{n-1}}\right)^{1-\frac{1}{n}}\right]^2 & h < 0 \\ K_s & h > 0 \end{cases} \quad (3)$$

where K_s is saturated hydraulic conductivity (LT^{-1}), l is an empirical parameter known as tortuosity parameter (dimensionless), S_e is relative soil effective saturation

$$(S_e = \theta - \theta_r/\theta_s - \theta_r)$$

2.3.2. Root Water Uptake

The sink term (S(z,t)) in Richards' equation which is known as the root water uptake term as well, was calculated in the following equation (Equation (4)) to account for water and salinity stress in the multiplicative form:

$$S(z,t) = \alpha(h,\pi)\beta(z)T_p \quad (4)$$

where the $\alpha(h, \pi)$ is water and salinity stress reduction function (dimensionless), $\beta(z)$ is normalized root density distribution (L^{-1}), T_p is potential transpiration rate ($L^3L^{-2}.T^{-1}$)

The α function that accounts for the combination of water and salinity stresses has been built from the multiplication of the S-shape water stress function and threshold and slope function for salinity stress as follows [50]:

$$\alpha(h,\pi) = \frac{1}{1+\left(\frac{h}{h_{50}}\right)^p}[1+b(\pi - a)] \quad (5)$$

where h is soil pressure head (L), h_{50} is the pressure head at which water uptake is reduced by 50% and negligible osmotic (salinity) stress exists, and p is an empirical parameter. The π is soil salinity (dS/m), b is the slope of root water uptake reduction, and a is the root water uptake threshold to soil salinity (dS/m).

The normalized root distribution (Equation (6)) as a function of root depth is as follows:

$$\beta(z) = \begin{cases} 1, & Z = 0 \\ -0.007Z + 1, & 0 < Z < 150 \text{ cm} \\ 0, & Z > 150 \text{ cm} \end{cases} \quad (6)$$

where z is root depth (L).

To calculate the potential transpiration (T_p), initially, the potential evapotranspiration (ET_p) was calculated by the FAO Penman-Monteith equation [48]. For the next step, the potential evaporation was computed using the following equation [51]:

$$Ep = ET_p e^{-kLAI} \quad (7)$$

where k is crop canopy radiation attenuation coefficient (dimensionless), which is usually taken as 0.4, and LAI is leaf area index. Then the potential transpiration ($L\ T^{-1}$) was estimated by:

$$T_p = ET_p - E_p \quad (8)$$

2.3.3. Solute Transport

The advection-dispersion equation (ADE) was solved numerically by HYDRUS-1D using the Galerkin finite elements scheme for space weighting and the Crank-Nicholson scheme for time weighting [52]:

$$\frac{\partial c}{\partial t} + \frac{\rho_0}{\theta}\frac{\partial s}{\partial t} = D_e\frac{\partial^2 c}{\partial^2 z} - \frac{q_w}{\theta}\frac{\partial c}{\partial z} \tag{9}$$

where c is liquid phase (dissolved) solute concentration (ML^{-3}), s is solid phase concentration (MM^{-1}), t is time (T), θ is volumetric soil water content (L^3L^{-3}), D_e is effective dispersion coefficient (L^2T^{-1}), q_w is soil water flux ($L^3L^{-2}T^{-1}$), z is vertical coordinates (L).

The effective dispersion coefficient accounts for diffusion and hydrodynamic dispersion coefficients [52]:

$$D_e = D_l^s + D_{lh} \tag{10}$$

$$D_{lh} = \lambda\frac{q_w}{\theta} \tag{11}$$

$$D_l^s = \frac{\theta^{\frac{7}{3}}}{\theta_s^2}D_l^w \tag{12}$$

where $D_l{}^s$ is soil effective diffusion coefficient (L^2T^{-1}), D_{lh} is coefficient of hydrodynamic dispersion (L^2T^{-1}), λ is known as dispersivity [L], θ_s is saturated volumetric water content, and $D_l{}^w$ is diffusion coefficient in free water (L^2T^{-1}).

In this research, adsorption of salts was assumed to be the linear equilibrium (instantaneous):

$$s = K_d c \tag{13}$$

where K_d is known as the distribution coefficient or adsorption isotherm coefficient (L^3M^{-1}).

2.4. The HYDRUS-1D Model Setup

The soil salinity of the corn root zone irrigated with saline water was simulated using the HYDRUS-1D model. The first 200 cm layer of the soil profile was considered for simulations. One soil material and layer were considered because the soil profile at the experimental field had a relatively uniform texture to 240 cm deep. The simulations were done on a daily basis for 235 days (from 11 May 2016, to 31 December 2016). The Van Genuchten-Mualem (Equations (2) and (3)) hydraulic model was chosen to perform the water flow simulation and describe the unsaturated soil hydraulic properties. The initial arbitrary values of the water flow parameters were obtained from the HYDRUS-1D library for the silt loam soil. The atmospheric boundary condition with the surface layer was chosen as the study was conducted in the field. The free drainage boundary condition was used for the bottom of the soil profile. The linear equilibrium adsorption was used to imitate the adsorption process under saline irrigation conditions. The upper and bottom boundary conditions for the Advection-Dispersion equation were concentration flux and zero concentration, respectively. The root water uptake simulations were done by applying the S-Shape model as a reduction function for water stress, and simultaneously the slope and threshold salinity stress reduction function was considered in a multiplicative approach.

2.5. Uncertainty Assessment and GLUE Method

The uncertainty in water management has been illustrated as the degree of confidence (probability) in the decision-making process using simulation tools [53]. To pursue the uncertainty analysis, the GLUE method was implemented for the HYDRUS-1D model to simulate the dynamic of salts under linear move irrigation. It was assumed that the results of this study would assess the degree of confidence in using the HYDRUS-1D to explore salinity management scenarios, primarily through the leaching application process. Therefore, the following sets of parameters were subjected to uncertainty analysis:

Water flow simulation parameters: [θ_r, θ_s, α, n, K_s, l]
Solute transport parameters: [λ, $D^l{}_w$, K_d]
Root water uptake = [a, b, h_{50}, P1]

The existing values of parameters in the slope-threshold method (salinity stress) in the literature [54,55] are very general. Thus, they need to be determined for a specific location with its particular soil and weather characteristics. Because of this reason, the a and b were subjected to calibration and uncertainty analysis.

To successfully perform uncertainty analysis using the GLUE method, a subjective threshold for likelihood values should be assigned to find behavioral parameter sets. Those parameter sets found behavioral were used for uncertainty analysis. Furthermore, another prevalent issue in HYDRUS-1D simulations is the non-convergence of the model outputs for some parameter sets. On that account, all of those parameter sets that resulted in model non-convergence were eliminated during computations. The last elimination procedure was called the screening operation.

The following steps were used to apply the GLUE algorithm [41]:

1. The prior distributions of parameters were identified based on the HYDRUS-1D model library and existing values in the literature (Table 4). The priors were considered uniformly distributed.

Table 4. The prior distribution of the parameters used in the HYDRUS-1D.

Parameters	Units	Range	Mean	SD	CV
θ_r	-	0.05–0.08	0.065	0.0086	0.1332
θ_s	-	0.3–0.5	0.4	0.0577	0.1443
α	1/cm	0.001–0.2	0.1005	0.0574	0.5716
K_s	cm/days	5–40	22.5	10.1036	0.4490
n	-	1–3	2	0.5773	0.2886
l	-	0.1–1	0.55	0.2598	0.4723
λ	cm^2/day	5–30	17.5	7.2168	0.4123
$D_l{}^w$	cm^2/day	1–2	1.5	0.2886	0.1924
K_d	cm^3/g	0.1–1	0.55	0.2598	0.4723
a	dS/m	2–5	3.5	0.7500	0.2142
b	%	4–8	4.5	1.4433	0.3207
h_{50}	cm	−5000–−800	−2900	1212.43	−0.418
P1	-	1.5–3	2.25	0.433	0.19

SD = Standard deviation, CV = Coefficient of variation.

2. The parameters' ranges were randomly sampled n times based on the Monte Carlo approach in RStudio 1.41717 environment.
3. The HYDRUS-1D model was run in the RStudio environment for each parameter set already sampled.
4. The likelihood values were calculated using inverse error variance as the likelihood function:

$$L(\theta_j|O) = \left(\frac{\sum_{i=1}^{n}(O_i - y_i(\theta_j))^2}{n-2}\right)^{-1} \quad (14)$$

where θ_j is *j*th parameter set, O_i is *i*th observation, y is the model output n is the number of observations.

5. The threshold for likelihood values for behavioral parameter sets was specified. In this study, 10% of successful parameters after screening operation were used for uncertainty analysis.
6. The probability of each parameter set was computed using the following equation:

$$p(\theta_j) = \frac{L(\theta_j|O)}{\sum_{J=1}^{n} L(\theta_j|O)} \quad (15)$$

where $p(\theta_j)$ is the probability (likelihood weight) of the *j*th parameter set, n is the number of parameters' sets, and $L(\theta_j | O)$ is the value of the likelihood of the *j*th parameter set.

7. The posterior distributions of the parameters and statistics were constructed. The empirical posterior distributions of parameters were achieved by pairs of parameters' sets (θ_j) and their corresponding probabilities. Then, by using the following equations, the mean and variance of the parameters were calculated:

$$\mu_{post} = \sum_{j=1}^{n} p(\theta_i)\theta_i \quad (16)$$

$$Var_{post} = \sum_{j=1}^{n} p(\theta_i)(\theta_i - \mu_{post})^2 \quad (17)$$

where μ_{post} and Var_{post} are the mean and variance of the posterior distribution of the parameters.

8. For the final step, the simulated values of soil water salinity by the HYDRUS-1D model were sorted based on the corresponding probabilities to create a cumulative distribution function of model outputs (predictive uncertainty). Then 95% confidence intervals for model outputs were retrieved [40].

The HYDRUS-1D model was run 126,000 times in the R environment to implement the GLUE algorithm and achieve this study's goals. After the screening procedure and applying threshold values for exploring behavioral parameters sets, the 1000 sets of parameters remained for uncertainty analysis.

The 50% and 97.5% quantiles of the posterior distributions of the parameters were considered as calibrated values of the parameters for investigating the model performance in terms of simulating soil water content and ECsw. In addition, the single parameter set that resulted in the highest likelihood value was used as an alternative scenario for the calibration of the model.

2.6. Evaluation of the Model Performance

Three statistical indices were used to evaluate the model performance: root mean square error (RMSE), normalized root mean square error (NRMSE), and coefficient of determination (R^2):

$$RMSE = \sqrt{\frac{1}{n}\sum_{i=1}^{n}(M_i - S_i)^2} \quad (18)$$

$$NRMSE = \frac{RMSE}{\overline{M}} \quad (19)$$

$$R^2 = \frac{\left(\sum_{i=1}^{n}(M_i - \overline{M})(S_i - \overline{S})\right)}{\sqrt{\sum_{i=1}^{n}(M_i - \overline{M})^2 \sum_{i=1}^{n}(S_i - \overline{S})^2}} \quad (20)$$

where M_i, S_i, and $\overline{M}$ are measured value, simulated value, and the average value of measurements. The R2 values close to 1 indicate the good performance model. The NRMSE ranges of <10%, 10–20%, 20–30%, and >30% categorize the model performance as excellent, good, fair, and poor.

3. Results and Discussion

In this section, the uncertainty in the model parameters and simulated outputs were initially quantified and then presented and discussed. This information would provide interesting clues to find roots of error in simulating soil salinity during and out of corn growing season irrigated with saline water under a linear move irrigation system. Afterward, the results of the performance analysis of the model calibrated with posterior values of the parameters using the GLUE algorithm are presented and discussed.

3.1. Parameters Uncertainty

The posterior distributions of the parameters indicate different levels of uncertainty in derived (SD) parameters using the GLUE algorithm. Comparing the statistical indices, specifically standard deviations associated with prior and posterior distributions (Tables 4 and 5) of the parameters reveals that the algorithm was able to reduce the uncertainty of a certain number of parameters. By using observational ECsw data, the algorithm was able to estimate the θ_s, n, and α as 26, 22, and 8% reduction was obtained in their posterior standard deviation values compared to their priors. Likewise, the posterior SD values of λ and K_d parameters of solute transport parameters were reduced by 16 and 15% compared with their corresponding prior values. Furthermore, the reduction in SD values of posterior distributions root water uptake reduction function for salinity stresses indicated a lower level of uncertainty remaining in these parameters (a and b). Contrastingly, estimations of the rest of the parameters were accomplished with lower confidence (higher uncertainty) as there was no considerable difference between the SD values of their posteriors and priors. The histograms of the parameters' posterior distributions are presented in Figures 2–4. The x-axis of the graphs was considered equal to the prior distribution of the parameters to compare posteriors and priors. Among the water flow simulating parameters, the skewed and peaked posterior distributions have been observed for θ_s, α, and n. It indicates the lower level of uncertainty remained in these parameters after implementing the GLUE algorithm for the HYDRUS-1D model to simulate soil salinity under irrigation with saline water using a linear move irrigation system. The posteriors of θ_r, K_s, and l were slightly changed from their priors as they uniformly covered the upper and lower bounds of the prior distribution. Posterior distributions of solute transport parameters show (Figure 3) that the dispersivity (λ) and adsorption isotherm coefficient (K_d) approximately follow a normal distribution, indicating a reduction in the uncertainty of their estimations.

Table 5. Posterior distribution of the HYDRUS-1D parameters for simulating soil salinity dynamics.

Parameters	Mean	SD	CV	Quantiles				
				2.5%	25%	50%	75%	97.5%
θ_r	0.0642	0.00907	0.14127	0.0505	0.0557	0.0637	0.0716	0.0796
θ_s	0.442	0.04214	0.09533	0.3417	0.412	0.4542	0.4743	0.497
α	0.0614	0.05269	0.85814	0.0062	0.018	0.0436	0.0881	0.185
K_s	22.97	9.653	0.42024	6.563	14.72	23.7	31.16	38.736
n	1.725	0.4497	0.26069	1.1	1.337	1.69	1.992	2.702
l	0.547	0.2568	0.46946	0.121	0.3341	0.5361	0.7533	0.993
λ	15.42	6.023	0.39059	6.416	10.47	14.65	19.34	28.428
D_l^w	1.53	0.2738	0.17895	1.029	1.306	1.567	1.76	1.975
K_d	0.5744	0.2201	0.38318	0.129	0.4378	0.5898	0.7277	0.964
a	3.324	0.3926	0.11811	2.637	2.996	3.346	3.648	3.964
b	5.924	1.136	0.19176	4.122	4.938	5.993	6.82	7.873
h_{50}	−2846	1238	−0.43499	−4942	−3901	−2777	−1776	−852.265
P1	2.243	0.45	0.20062	1.523	1.828	2.24	2.642	2.952

SD = Standard deviation, CV = Coefficient of variation.

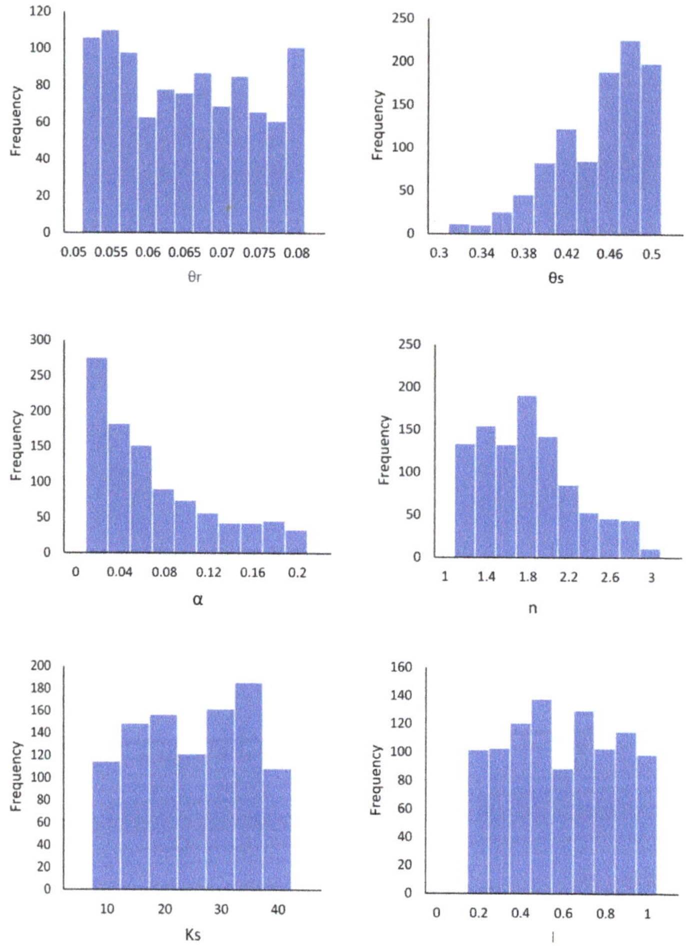

Figure 2. Posterior distributions of the water flow simulation parameters.

Nevertheless, the posterior distribution of the diffusion coefficient (D^l_w) was not noticeably different from its prior distribution. This could be because the diffusivity contribution to the solute transport procedure for saline water irrigation conditions was trivial compared with advection and hydrodynamic dispersion. In addition, the posterior distribution of the root water uptake threshold (a) was picked and concentrated around the median. However, the other parameters did not significantly change from their priors, indicating a higher level of uncertainty in these parameters after derivation by the GLUE algorithm [23]. This is presumably because of the scale of the study, as our experimental field had plots 13.7 m × 27.4 m dimensions. Studies conducted at the field scale can potentially increase the uncertainty in observational data due to several phenomena, such as uniformity distribution of irrigation applications, preferential flows as a consequence of compaction or shrinkage of dry soil before irrigation events, and soil water redistribution in the soil. On the other hand, the results showed that threshold reduction in root water uptake, which is presented generally in the literature, could be adequately derived for a specific location by the GLUE algorithm. This threshold value could be used as a guideline for leaching requirement specifications and designing irrigation systems. To compare the computed value of root water uptake threshold with the existing values in the literature [54],

The value should be divided by 2 because the HYDRUS-1D model uses the values of electrical conductivity of soil water (ECsw). The values in the literature are presented as the electrical conductivity of soil-saturated paste extract (ECe). Our results found ECsw = 3.324 dS/m or ECe = 1.662 dS/m, which was very close to the ones reported by Mass and Hoffman, 1977 (ECe = 1.7 dS/m)—[54]. The relative sensitivity of the parameters can be analyzed by comparing the CV values of the prior and posterior distribution of the parameters. The α, a, and b were the three most sensitive parameters. In contrast, the three l, h_{50}, and P1 were the least sensitive parameters, respectively, for simulating soil salinity at the corn root zone at the field scale. The scatter plots of likelihood values of behavioral parameters' sets related to water flow, solute transport, and root water uptake processes are presented in Figure 5 to explore further aspects in identifying the parameters. These plots demonstrate that the single optimum parameter set is identifiable because the parameter set's corresponding likelihood value was significantly higher in the parameters' response surface than the majority of the other parameter's values. The optimum water flow simulation parameters sets are $\theta r = 0.0637$, $\theta s = 0.4575$, $\alpha = 0.0119$, Ks = 35.71, n = 1.647, and l = 0.3797. The optimum solute transport and root water uptake parameters are $\lambda = 8.712$, $D^l_w = 1.35$, Kd = 0.83, $h_{50} = -2845$, P1 = 2.17, a = 3.63, and b = 5.913. Furthermore, based on Figure 5, ranges with higher probability or lower uncertainty in the parametric surface response of parameters can be identified only for θs and α. The results show that the ranges are: θs = [0.45–0.49], and α = [0.01–0.065].

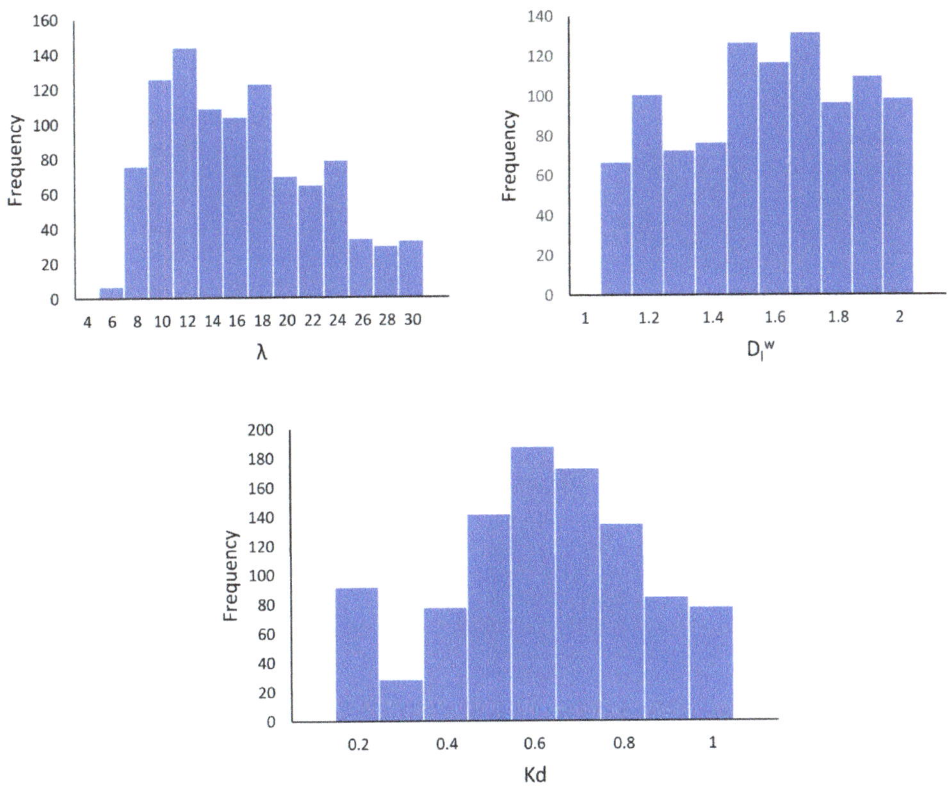

Figure 3. Posterior distributions of the solute transport parameters.

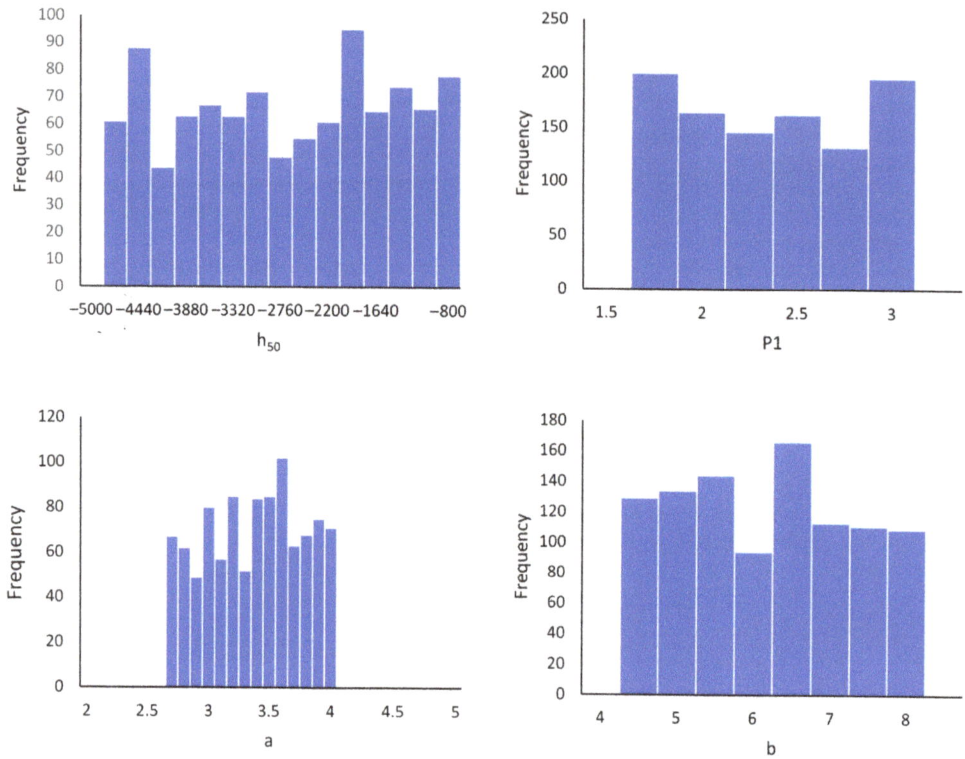

Figure 4. Posterior distributions of the root water uptake parameters.

3.2. *Predictive Uncertainty*

Analysis of the predictive uncertainty (model output uncertainty) provides additional information regarding the sources of uncertainty in model outputs and the successfulness of the GLUE algorithm in the calibration of the HYDRUS-1D model. The output uncertainty of the model based on 95% CI of behavioral parameters is depicted in Figure 6 for soil depths of 16, 46, and 76 cm. The shaded areas are the predictive uncertainty of the model, and green triangular points are observations of ECsw. The results declare that the uncertainty in the model parameters has reached the model outputs as the observational data are primarily covered in the predictive uncertainty band (shaded area). Moreover, covering the majority of the observational points by the 95% CI band show that the primary source of uncertainty in simulating ECsw of corn root zone under saline irrigation is uncertainty in the parameters' estimations. As it is expressed in Figure 6, the predictive uncertainty band varies during the growing season (The corn was harvested 128 days after the planting date), and it is approximately constant for the out of the growing season period. This is related to the absence of irrigation or significant precipitation after the growing season and the lack of root water uptake. Thus, the only remaining factor affecting the soil water and salinity balance was substantially low evaporation (Figure 1) in such a way that it could not change the uniformity of the uncertainty band out of the growing season.

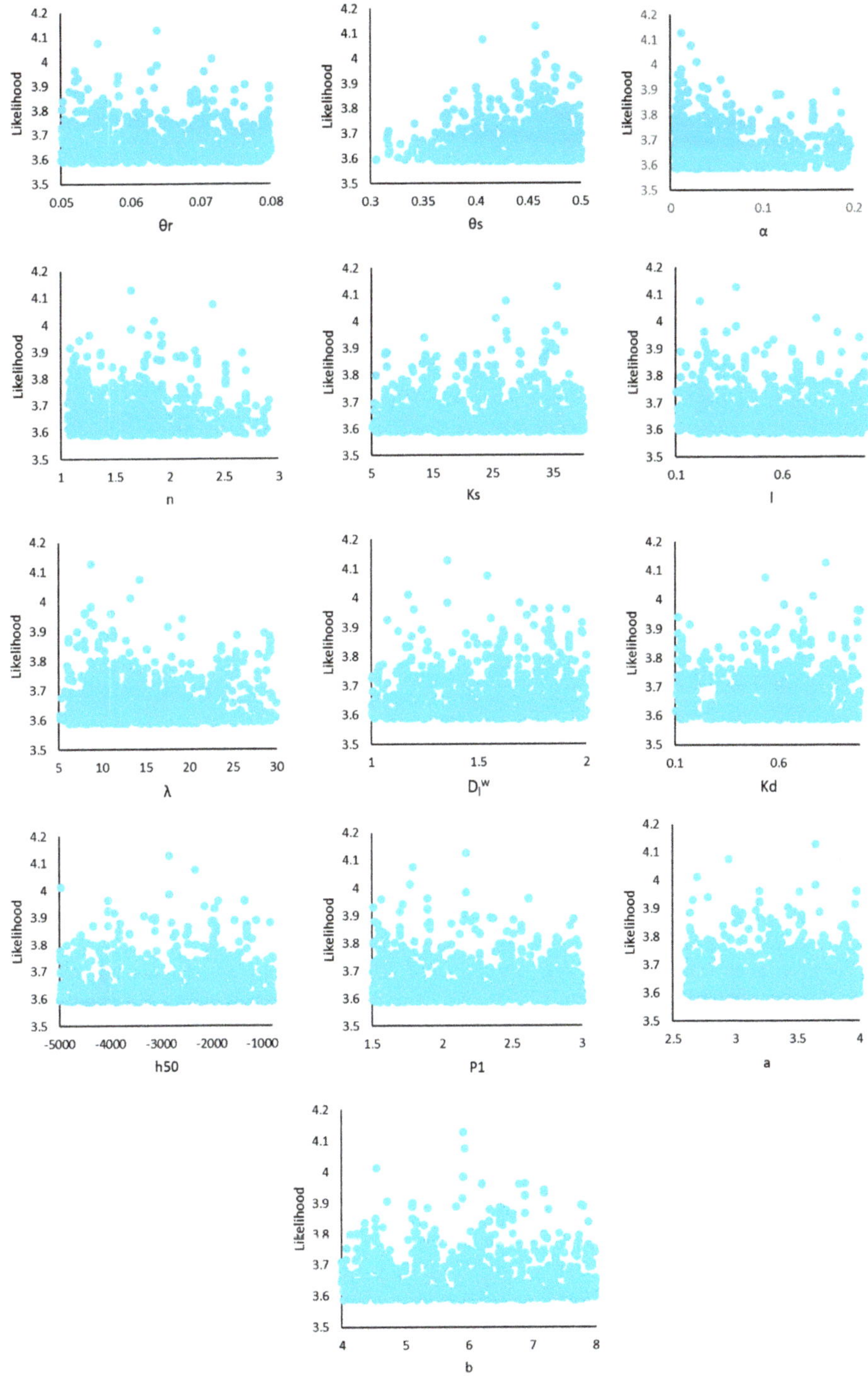

Figure 5. Scatterplots of likelihood values vs. the parameters of the HYDRUS-1D model.

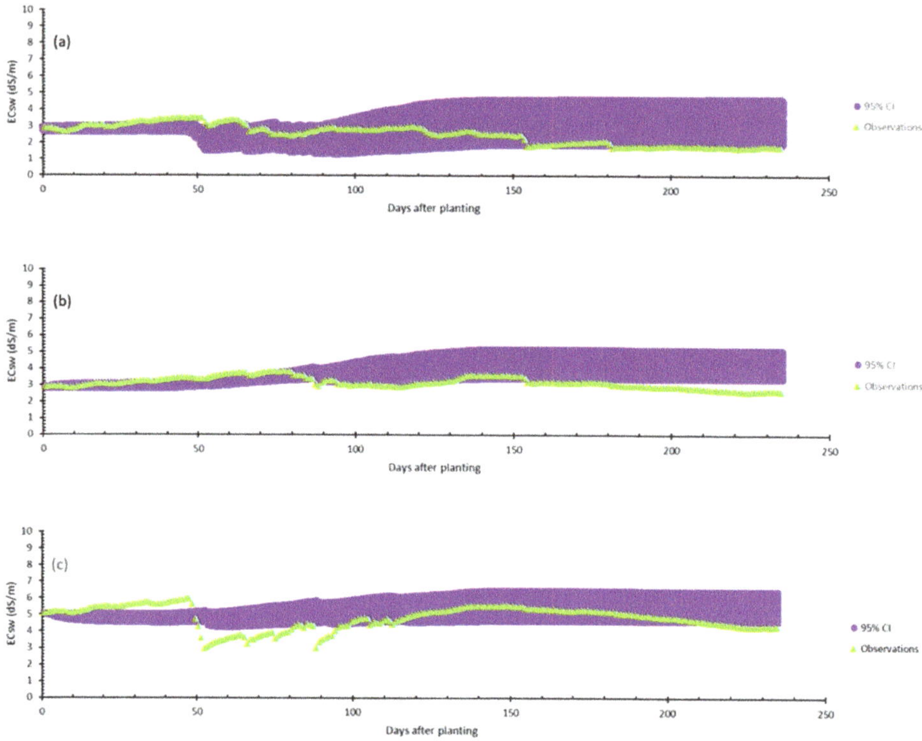

Figure 6. Predictive uncertainty of the HYDRUS-1D mode for simulating soil salinity of corn root zone at (**a**) 16, (**b**) 46 and (**c**) 76 cm depths.

Additionally, the increase in predictive uncertainty can be inferred for the drier period of the field due to the magnification of the unsaturated condition, which increases the complexity of simulating water flow and solute transport process. The model outputs' uncertainty band was substantially lower for the first 50 days of the growing season (the corn reached the V8 growth stage) compared to the other days. This could be because the corn was at its initial growth stages, and root water uptake did not considerably affect the soil water content and, consequently, the ECsw at the measurement depths. Increasing the output uncertainty after corn's initial growth stages could be explained by remaining uncertainties in the root water uptake reduction parameters for water stress, as they were relatively higher than the other parameters, and the 95% CI band covered the observational points. Moreover, the predictive uncertainty band only covers a portion of the observational points for ECsw simulations at 76 cm soil depth, which could be related to existing errors in boundary conditions. Due to the scale of the study, the irrigation application efficiency of linear move irrigation systems could sometimes vary due to the prevailing windy conditions of western Kansas. It is probable that during some of the irrigation events, the unexpected wind changed the amount of water received by the soil, which consequently caused errors in the boundary condition of the study.

3.3. The HYDRUS-1D Model Performance

The performance of the calibrated HYDRUS-1D model based on 50 and 97.5% quantiles of parameters posterior distributions (Table 5) and optimum parameters set based on likelihood scatterplots (Figure 5) are presented in Table 6. The model's excellent accuracy has been observed in simulating soil water content and ECsw during and out of the corn growing season for all three series of parameters.

Table 6. Performance of the calibrated HYDRUS-1D model using the GLUE algorithm.

	SWC			ECsw		
	RMSE ($cm^3.cm^{-3}$)	NRMSE	R^2	RMSE (dS/m)	NRMSE	R^2
			Q50%			
16 cm	0.003	0.01	0.84	0.30	0.11	0.41
46 cm	0.001	0.005	0.86	0.29	0.09	0.22
76 cm	0.0006	0.003	0.72	0.42	0.09	0.50
			Q97.5%			
16 cm	0.008	0.03	0.87	0.30	0.12	0.31
46 cm	0.006	0.02	0.78	0.16	0.05	0.13
76 cm	0.003	0.01	0.47	0.46	0.1	0.58
			OptL			
16 cm	0.004	0.02	0.82	0.23	0.09	0.6
46 cm	0.002	0.01	0.90	0.35	0.11	0.3
76 cm	0.0006	0.003	0.86	0.53	0.11	0.32

SWC = soil water content, ECsw = electrical conductivity of soil water, Q50% = the calibrated model based 50% quantiles of the parameters, Q97.5% = the calibrated model based 97.5% quantiles of the parameters, OptL = Calibrated model based on optimum parameter set obtained from scatterplots of the likelihood values vs. parameter values.

The obtained RMSE values for simulating SWC were from 0.0006 and 0.008 cm^3cm^{-3} for parameters' sets based on the 50% and 97.5% quantiles of parameters posteriors and likelihood scatterplots used as calibrated values. The NRMSEs for simulating SWC were from 0.003 to 0.01. The highest accuracy of the model was detected for simulating SWC at 76 cm depth during and out of the growing season. In addition, good adequacy of the model was noticed based on R2 values between simulated and measured SWCs. The calibrated HYDRUS-1D model based on 50% quantiles posteriors resulted in the highest accuracy of the model for simulating SWC compared with the other two series of calibration values.

Furthermore, the highest adequacy of the model was obtained for optimum parameters' values based on likelihood scatterplots to simulate SWC. The overall performance of the calibrated model, based on all three calibration scenarios, was excellent based on statistical indices (Table 6). It has been detected that RMSE values for predicted ECsw were from 0.29 to 0.42 dS/m and 0.16 to 0.46 dS/m for the calibrated model based on 50% and 97.5% quantiles of the parameters' posteriors. Moreover, RMSE values were from 0.23 to 0.53 dS/m for simulating ECsw using calibrated model based on optimum values obtained from scatterplots of the likelihood values. The performance results show that the adequacy of the model for simulating ECsw was not as satisfying as SWC simulations during and out of the corn growing season in our study region. The R^2 values were from 0.13 to 0.60 for both parameters' sets. The highest adequacy of the model for ECs predictions was observed at 16 cm soil depth for calibrated model based on likelihood values, and the lowest one was noticed for ECsw simulations at 46 cm soil depth for calibrated model based on 97.5% of parameters posterior distributions.

The results of this study indicated that using 50% quantiles of parameters posterior distributions could be introduced as acceptable calibration results for implementation of the GLUE algorithm for the HYDRUS-1D model to simulate salinity of corn root zone at field scale under linear move irrigation system. The time series of simulated soil water content and ECsw, along with observational for different soil depths, are presented in Figures 7 and 8 to obtain further details on the performance of the calibrated model based on 50% quantiles of parameters posteriors. The good performance of the model during the corn growing season and the excellent performance of the model for simulating SWC was observed for all three soil depths. The model outputs follow the observational SWC data trend.

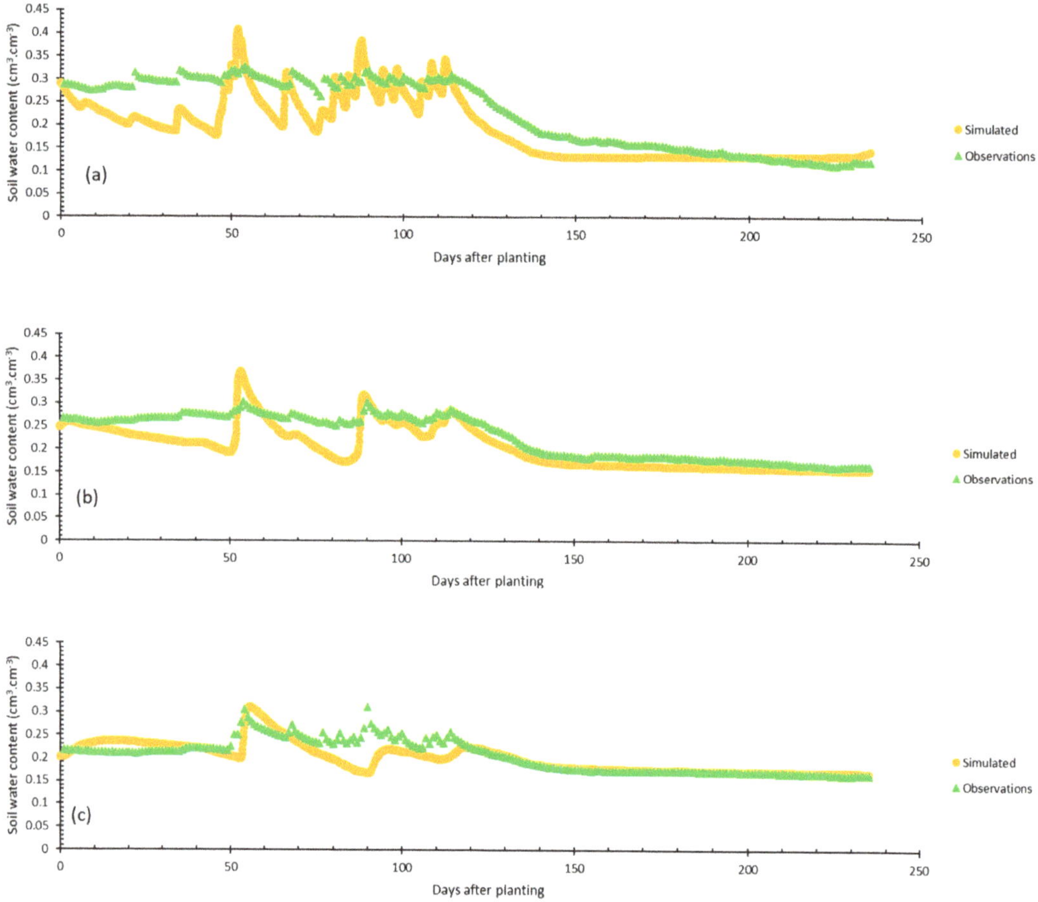

Figure 7. The simulated and observed soil water content at (**a**) 16, (**b**) 46, and (**c**) 76 cm soil depths.

Similar performance of the model was observed for ECsw simulation. The discrepancies between observations and simulated were not noticeable for salinity simulations. However, some noticeable deviations were detected between simulated and observed ECsw values at 76 cm soil depth, which was expected as another source of uncertainty detected for this soil depth in the analysis of the model predictive uncertainties.

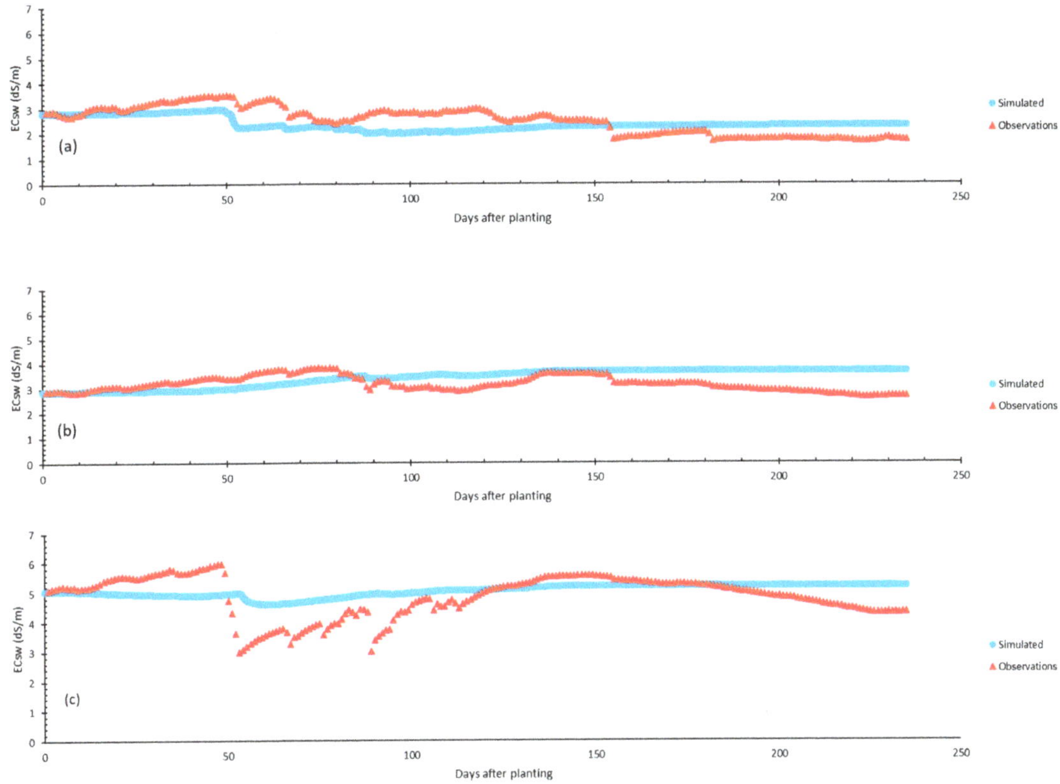

Figure 8. The simulated and observed electrical conductivity of soil water at (**a**) 16, (**b**) 46, and (**c**) 76 cm soil depths.

4. Conclusions

A study was conducted to investigate calibration and the uncertainty in parameters and outputs of the HYDRUS-1D model to simulate soil salinity of corn root zone under irrigation with saline water using a linear move irrigation system. In this research, the generalized likelihood uncertainty estimation (GLUE) algorithm was implemented for the HYDRUS-1D model in the R environment to achieve the goals of the study. The results have found a lower level of uncertainty in θs, n, and α among the soil water flow simulations parameters, adsorption isotherm coefficient (Kd) and dispersivity (λ) among the solute transport parameters, and threshold (a) and slope (b) parameters of root water uptake reduction function for salinity stress compare to uncertainty in the other parameters. This study detected a minor contribution of the coefficient of diffusion in the solute transport process at the field scale compared to advection and hydrodynamic dispersion because its posterior distribution was not noticeably different from its corresponding priors. The mean value of the posterior distribution of the root water uptake threshold to salinity was 1.662 dS/m, which was close to the value reported by Mass and Hoffman, 1977 [41] for corn (ECe = 1.7 dS/m). The relative sensitivity analysis of parameters has revealed α and P1 as the most sensitive and least sensitive parameters, respectively. Predictive uncertainty analysis showed that the uncertainty in the HYDRUS-1D parameters is the main source of uncertainty in the model outputs. In addition, it was illustrated that uncertainty in model outputs for simulating ECsw was relatively lower in the initial growth stages (emergence to V8) of corn compared to the other growth stages and out of the growing season. The model was able to successfully simulate the soil water content and electrical conductivity of soil water (ECsw) with calibrated parameters using 50% quantiles of the parameters' posterior distributions. The ranges of NRMSE values were 0.003 to 0.01 for simulating soil water content and 0.09 to 0.11 for simulating ECsw, which specify the excellent performance of the calibrated model based on posterior distributions of parameters. The results of this study have proved the reliability of the GLUE algorithm to explore uncertainty analysis of the HYDRUS-1D model and its calibration for reproducing soil salinity of corn root zone under saline water irrigation at field scale under linear move sprinkle irrigation system.

Author Contributions: Conceptualization, F.M. and B.G.; methodology, F.M., A.M., M.C.G. and B.G.; supervision, A.M., J.A. and B.G.; software, F.M. and J.A.; investigation, F.M., data curation, J.A.; resources, J.A. and H.A., writing-the original draft preparation, F.M., writing-review and editing, F.M., A.M. and J.A. All authors have read and agreed to the published version of the manuscript.

Funding: The APC was funded by Kansas State University and USDA-ARS, Ogallala Aquifer Program.

Institutional Review Board Statement: Not applicable.

Informed Consent Statement: Not applicable.

Data Availability Statement: The data could be available by requesting Dr. J.A.

Acknowledgments: The group of authors would like to thank Kansas State University for financially supporting the APC for this study. Special thanks to Mojtaba Shafiei, the astonishing alumni of Ferdowsi University of Mashhad for being pioneer of implementing Bayesian statistics concepts in the department of water science and engineering.

Conflicts of Interest: The authors declare no conflict of interest.

References

1. Skaggs, T.H.; Anderson, R.G.; Corwin, D.L.; Suarez, D.L. Analytical steady-state solutions for water-limited cropping systems using saline irrigation water. *Water Resour. Res.* **2014**, *50*, 9656–9674. [CrossRef]
2. Bradford, S.; Letey, J. Cyclic and blending strategies for using nonsaline and saline waters for irrigation. *Irrig. Sci.* **1992**, *13*, 123–128. [CrossRef]
3. Rhoades, J.D. Use of saline drainage water for irrigation. *Agric. Drain.* **1999**, *38*, 615–657.
4. Sharma, D.P.; Rao, K. Strategy for long term use of saline drainage water for irrigation in semi-arid regions. *Soil Tillage Res.* **1998**, *48*, 287–295. [CrossRef]
5. Mendoza-Espinosa, L.G.; Burgess, J.E.; Daesslé, L.; Villada-Canela, M. Reclaimed water for the irrigation of vineyards: Mexico and South Africa as case studies. *Sustain. Cities Soc.* **2019**, *51*, 101769. [CrossRef]
6. Kramer, I.; Bayer, Y.; Mau, Y. The sustainability of treated wastewater irrigation: The impact of hysteresis on saturated soil hydraulic conductivity. *Water Resour. Res.* **2022**, *58*, e2021WR031307. [CrossRef]
7. Sheng, F.; Xiuling, C. Using shallow saline groundwater for irrigation and regulating for soil salt-water regime. *Irrig. Drain. Syst.* **1997**, *11*, 1–14. [CrossRef]
8. Zhang, J.; Li, K.; Gao, Y.; Feng, D.; Zheng, C.; Cao, C.; Sun, J.; Dang, H.; Hamani, A.K.M. Evaluation of saline water irrigation on cotton growth and yield using the AquaCrop crop simulation model. *Agric. Water Manag.* **2022**, *261*, 107355. [CrossRef]
9. Singh, A. Soil salinization management for sustainable development: A review. *J. Environ. Manag.* **2021**, *277*, 111383. [CrossRef]
10. Fathizad, H.; Ali Hakimzadeh Ardakani, M.; Sodaiezadeh, H.; Kerry, R.; Taghizadeh-Mehrjardi, R. Investigation of the spatial and temporal variation of soil salinity using random forests in the central desert of Iran. *Geoderma* **2020**, *365*, 114233. [CrossRef]
11. Mostafazadeh-Fard, B.; Mansouri, H.; Mousavi, S.-F.; Feizi, M. Effects of Different Levels of Irrigation Water Salinity and Leaching on Yield and Yield Components of Wheat in an Arid Region. *J. Irrig. Drain. Eng.* **2009**, *135*, 32–38. [CrossRef]
12. Kang, Y.; Chen, M.; Wan, S. Effects of drip irrigation with saline water on waxy maize (*Zea mays* L. var. ceratina Kulesh) in North China Plain. *Agric. Water Manag.* **2010**, *97*, 1303–1309. [CrossRef]
13. Li, D.; Wan, S.; Li, X.; Kang, Y.; Han, X. Effect of water-salt regulation drip irrigation with saline water on tomato quality in an arid region. *Agric. Water Manag.* **2022**, *261*, 107347. [CrossRef]
14. Corwin, D.L.; Waggoner, B.L. TETrans: A user-friendly, functional model of solute transport. *Water Sci. Technol.* **1991**, *24*, 57–65. [CrossRef]
15. Oosterbaan, R.J. *SALTMOD: Description of Principles, User Manual, and Examples of Application, Version 1.1*; ILRI: Wageningen, The Netherlands, 2001.
16. Shahrokhnia, H.; Wu, L. SALEACH: A new web-based soil salinity leaching model for improved irrigation management. *Agric. Water Manag.* **2021**, *252*, 106905. [CrossRef]

17. Liu, Y.; Zhu, Y.; Mao, W.; Sun, G.; Han, X.; Wu, J.; Yang, J. Development and Application of a Water and Salt Balance Model for Well-Canal Conjunctive Irrigation in Semiarid Areas with Shallow Water Tables. *Agriculture* **2022**, *12*, 399. [CrossRef]
18. Corwin, D.L.; Waggoner, B.L. *TETRANS: Solute Transport Modeling Software User's Guide*; Report 121. USDA-ARS; U.S. Salinity Laboratory: Riverside, CA, USA, 1990.
19. Chang, X.; Gao, Z.; Wang, S.; Chen, H. Modelling long-term soil salinity dynamics using SaltMod in Hetao Irrigation District, China. *Comput. Electron. Agric.* **2019**, *156*, 447–458. [CrossRef]
20. Letey, J.; Hoffman, G.J.; Hopmans, J.W.; Grattan, S.R.; Suarez, D.; Corwin, D.L.; Oster, J.D.; Wu, L.; Amrhein, C. Evaluation of soil salinity leaching requirement guidelines. *Agric. Water Manag.* **2011**, *98*, 502–506. [CrossRef]
21. Simunek, J.; Van Genuchten, M.T.; Sejna, M. The HYDRUS-1D software package for simulating the one-dimensional movement of water, heat, and multiple solutes in variably-saturated media. *Univ. Calif.-Riverside Res. Rep.* **2005**, *3*, 240.
22. Liu, B.; Wang, S.; Liu, X.; Sun, H. Evaluating soil water and salt transport in response to varied rainfall events and hydrological years under brackish water irrigation in the North China Plain. *Geoderma* **2022**, *422*, 115954. [CrossRef]
23. Noshadi, M.; Fahandej-Saadi, S.; Sepaskhah, A.R. Application of SALTMED and HYDRUS-1D models for simulations of soil water content and soil salinity in controlled groundwater depth. *J. Arid Land* **2020**, *12*, 447–461. [CrossRef]
24. Kanzari, S.; Ben Nouna, B.; Ben Mariem, S.; Rezig, M. Hydrus-1D model calibration and validation in various field conditions for simulating water flow and salts transport in a semi-arid region of Tunisia. *Sustain. Environ. Res.* **2018**, *28*, 350–356. [CrossRef]
25. Shaygan, M.; Baumgartl, T.; Arnold, S.; Reading, L.P. The effect of soil physical amendments on reclamation of a saline-sodic soil: Simulation of salt leaching using HYDRUS-1D. *Soil Res.* **2018**, *56*, 829–845. [CrossRef]
26. Ramos, T.B.; Šimůnek, J.; Gonçalves, M.C.; Martins, J.C.; Prazeres, A.; Castanheira, N.L.; Pereira, L.S. Field evaluation of a multicomponent solute transport model in soils irrigated with saline waters. *J. Hydrol.* **2011**, *407*, 129–144. [CrossRef]
27. Askri, B.; Ahmed, A.T.; Abichou, T.; Bouhlila, R. Effects of shallow water table, salinity and frequency of irrigation water on the date palm water use. *J. Hydrol.* **2014**, *513*, 81–90. [CrossRef]
28. Skaggs, T.H.; Shouse, P.J.; Poss, J.A. Irrigating Forage Crops with Saline Waters: 2. Modeling Root Uptake and Drainage. *Vadose Zone J.* **2006**, *5*, 824–837. [CrossRef]
29. Ali, A.; Bennett, J.M.; Biggs, A.A.J.; Marchuk, A.; Ghahramani, A. Assessing the hydraulic reduction performance of HYDRUS-1D for application of alkaline irrigation in variably-saturated soils: Validation of pH driven hydraulic reduction scaling factors. *Agric. Water Manag.* **2021**, *256*, 107101. [CrossRef]
30. Phogat, V.; Yadav, A.K.; Malik, R.S.; Kumar, S.; Cox, J. Simulation of salt and water movement and estimation of water productivity of rice crop irrigated with saline water. *Paddy Water Environ.* **2010**, *8*, 333–346. [CrossRef]
31. He, J.; Jones, J.W.; Graham, W.D.; Dukes, M.D. Influence of likelihood function choice for estimating crop model parameters using the generalized likelihood uncertainty estimation method. *Agric. Syst.* **2010**, *103*, 256–264. [CrossRef]
32. Vrugt, J.A.; Gupta, H.V.; Bouten, W.; Sorooshian, S. A Shuffled Complex Evolution Metropolis algorithm for optimization and uncertainty assessment of hydrologic model parameters. *Water Resour. Res.* **2003**, *39*. [CrossRef]
33. Braak, C.J.F. Ter A Markov Chain Monte Carlo version of the genetic algorithm Differential Evolution: Easy Bayesian computing for real parameter spaces. *Stat. Comput.* **2006**, *16*, 239–249. [CrossRef]
34. Vrugt, J.A.; Ter Braak, C.J.F.; Diks, C.G.H.; Robinson, B.A.; Hyman, J.M.; Higdon, D. Accelerating Markov chain Monte Carlo simulation by differential evolution with self-adaptive randomized subspace sampling. *Int. J. Nonlinear Sci. Numer. Simul.* **2009**, *10*, 273–290. [CrossRef]
35. Zhang, J.; Vrugt, J.A.; Shi, X.; Lin, G.; Wu, L.; Zeng, L. Improving Simulation Efficiency of MCMC for Inverse Modeling of Hydrologic Systems With a Kalman-Inspired Proposal Distribution. *Water Resour. Res.* **2020**, *56*, e2019WR025474. [CrossRef]
36. Laloy, E.; Vrugt, J.A. High-dimensional posterior exploration of hydrologic models using multiple-try DREAM (ZS) and high-performance computing. *Water Resour. Res.* **2012**, *48*. [CrossRef]
37. Vrugt, J.A.; Ter Braak, C.J.F. DREAM (D): An adaptive Markov Chain Monte Carlo simulation algorithm to solve discrete, noncontinuous, and combinatorial posterior parameter estimation problems. *Hydrol. Earth Syst. Sci.* **2011**, *15*, 3701–3713. [CrossRef]
38. Beven, K.; Binley, A. The future of distributed models: Model calibration and uncertainty prediction. *Hydrol. Process.* **1992**, *6*, 279–298. [CrossRef]
39. Makowski, D.; Wallach, D.; Tremblay, M. Using a Bayesian approach to parameter estimation; comparison of the GLUE and MCMC methods. *Agronomie* **2002**, *22*, 191–203. [CrossRef]
40. Shafiei, M.; Ghahraman, B.; Saghafian, B.; Davary, K.; Pande, S.; Vazifedoust, M. Uncertainty assessment of the agro-hydrological SWAP model application at field scale: A case study in a dry region. *Agric. Water Manag.* **2014**, *146*, 324–334. [CrossRef]
41. Li, Z.; He, J.; Xu, X.; Jin, X.; Huang, W.; Clark, B.; Yang, G.; Li, Z. Estimating genetic parameters of DSSAT-CERES model with the GLUE method for winter wheat (Triticum aestivum L.) production. *Comput. Electron. Agric.* **2018**, *154*, 213–221. [CrossRef]
42. Sun, M.; Zhang, X.; Huo, Z.; Feng, S.; Huang, G.; Mao, X. Uncertainty and sensitivity assessments of an agricultural–hydrological model (RZWQM2) using the GLUE method. *J. Hydrol.* **2016**, *534*, 19–30. [CrossRef]
43. Sheng, M.; Liu, J.; Zhu, A.-X.; Rossiter, D.G.; Liu, H.; Liu, Z.; Zhu, L. Comparison of GLUE and DREAM for the estimation of cultivar parameters in the APSIM-maize model. *Agric. For. Meteorol.* **2019**, *278*, 107659. [CrossRef]
44. Klocke, N.L.; Currie, R.S.; Tomsicek, D.J.; Koehn, J. Corn yield response to deficit irrigation. *Trans. ASABE* **2011**, *54*, 931–940. [CrossRef]

45. Araya, A.; Kisekka, I.; Gowda, P.H.; Prasad, P.V.V. Grain sorghum production functions under different irrigation capacities. *Agric. Water Manag.* **2018**, *203*, 261–271. [CrossRef]
46. Helalia, S.A.; Anderson, R.G.; Skaggs, T.H.; Jenerette, G.D.; Wang, D.; Šimůnek, J. Impact of Drought and Changing Water Sources on Water Use and Soil Salinity of Almond and Pistachio Orchards: 1. Observations. *Soil Syst.* **2021**, *5*, 50. [CrossRef]
47. Helalia, S.A.; Anderson, R.G.; Skaggs, T.H.; Šimůnek, J. Impact of Drought and Changing Water Sources on Water Use and Soil Salinity of Almond and Pistachio Orchards: 2. Modeling. *Soil Syst.* **2021**, *5*, 58. [CrossRef]
48. Allen, R.G.; Pereira, L.S.; Raes, D.; Smith, M. Crop evapotranspiration-Guidelines for computing crop water requirements-FAO Irrigation and drainage paper 56. *Fao Rome* **1998**, *300*, D05109.
49. Van Genuchten, M.T. A closed-form equation for predicting the hydraulic conductivity of unsaturated soils. *Soil Sci. Soc. Am. J.* **1980**, *44*, 892–898. [CrossRef]
50. Van Genuchten, M.T. A Numerical Model for Water and Solute Movernent 1987. Available online: https://www.pc-progress.com/Documents/RVGenugten/M1987,vG,%20Worm%20USSL%20report.pdf (accessed on 9 November 2022).
51. Ritchie, J.T. Model for predicting evaporation from a row crop with incomplete cover. *Water Resour. Res.* **1972**, *8*, 1204–1213. [CrossRef]
52. Radcliffe, D.E.; Simunek, J. *Soil Physics with HYDRUS: Modeling and Applications*; CRC Press: Boca Raton, FL, USA, 2018; ISBN 1420073818.
53. Klauer, B.; Brown, J.D. Conceptualising imperfect knowledge in public decision-making: Ignorance, uncertainty, error and risk situations. *Environ. Res. Eng. Manag.* **2004**, *1*, 124–128.
54. Maas, E.V.; Hoffman, G.J. Crop salt tolerance—Current assessment. *J. Irrig. Drain. Div.* **1977**, *103*, 115–134. [CrossRef]
55. Grieve, C.M.; Grattan, S.R.; Maas, E.V. Plant salt tolerance. *ASCE Man. Rep. Eng. Pract.* **2012**, *71*, 405–459.

water

MDPI

Article

Noninvasive Monitoring of Subsurface Soil Conditions to Evaluate the Efficacy of Mole Drain in Heavy Clay Soils

Akram Aziz [1], Ronny Berndtsson [2,*], Tamer Attia [1], Yasser Hamed [3] and Tarek Selim [3,*]

[1] Geology Department, Faculty of Science, Port Said University, Port Said 42522, Egypt
[2] Centre for Advanced Middle Eastern Studies & Division of Water Resources Engineering, Lund University, SE-22100 Lund, Sweden
[3] Civil Engineering Department, Faculty of Engineering, Port Said University, Port Said 42523, Egypt
* Correspondence: ronny.berndtsson@tvrl.lth.se (R.B.); eng_tarek_selim@yahoo.com (T.S.)

Abstract: Soil degradation and low productivity are among the major agricultural problems facing farmers of the newly reclaimed agricultural area in the Nile Delta region, Egypt. High content of clay and silt characterizes the soil texture of all farms in the area, while farmers still rely on the traditional mole drainage (MD) system to reduce the salinity of the farm soil. We present a comparison of innovative geo-resistivity methods to evaluate mole drains and the salinity affected clay soils. Geoelectrical surveys were conducted on three newly reclaimed farms to image the subsurface soil drainage conditions and to evaluate the efficiency of using the traditional MD systems in these heavy clay environments. The surveys included measuring the natural spontaneous potential (SP), apparent resistivity gradient (RG), and electrical resistivity tomography (ERT). Integrating the results of the three methods reduced the ambiguous interpretation of the inverted ERT models and allowed us to determine the subsurface soil structure. The inverted ERT models were suitable for locating the buried MDs and delineating the upper surface of the undisturbed clay beds. The proximity of these layers to the topsoil reduces the role played by MDs in draining the soil in the first farm and prevents the growth of deep-rooted plants in the second farm. Time-lapse ERT measurements on the third farm revealed a defect in its drainage network where the slope of the clay beds opposes the main direction of the MDs. That has completely obstructed the drainage system of the farm and caused waterlogging. The presented geo-resistivity methods show that integrated models can be used to improve the assessment of in situ sub-surface drainage in clay-rich soils.

Keywords: water logging; resistivity gradient; ERT; electrical conductivity; soil drainage

1. Introduction

Most countries in the Middle East include large arid and semi-arid climate regions [1]. They all face the challenge of securing enough food supplies in unfavorable dry climate conditions with water scarcity. They need not only to conserve their existing resources (e.g., water, cultivated areas, etc.) but also to maximize the utilization of their resources. In Egypt, the government has launched a national project for land reclamation to meet the increasing food demands of its 110 million population. The area south of Port Said city (Figure 1a,b) is one of the main sectors in that project, as its production serves the districts of the Suez Canal and northern Sinai regions [2,3]. This area was once a fertile region in ancient Egypt [4–6]. Currently, the area is characterized by low precipitation—around 150 mm annually—and high evapotranspiration rates. The temperature ranges from 31° to 36 °C during July/August, and from 8° to 19 °C during December/January [7]. Local farmers mainly rely on the use of the traditional mole drain (MD) method to increase the drainage of the clay-rich soils. Mole drains are sets of unlined soil channels dug by a mole plow. The plow consists of a cylindrical foot attached to a narrow shank followed by a cylindrical expander. The shank creates a narrow slot extending from the soil surface down to the mole channel. Each channel is 10 cm in diameter and 50 to 70 cm deep below the soil surface. Heavy soils with low hydraulic conductivity need a closely spaced drainage system (2–4 m). The MD technique is used in combination with open surface drains at the boundaries of the reclaimed farms to facilitate the soil internal drainage, minimize deep percolation, and control groundwater table level, thereby increasing crop production.

Recently, many of the newly reclaimed farms within this area are facing serious agricultural problems that have led to soil degradation [8–11]. Previous studies [12,13] have shown that about 35% of the Egyptian irrigated lands suffer from salinization problems due to improper agricultural practices and inefficient drainage systems. In view of this, there is an urgent need to develop simple and non-invasive techniques for the evaluation of existing drainage systems and the state of salinity in the heavy clay soils. Thus, the objective of this study was to compare three innovative geo-resistivity methods for the clay-rich and salinity-affected farm soils. Since salinity may affect the readings from the different resistivity methods, it important to establish which methods give the most reliable results for varying salt content of the soil and how readings from different methods can be combined to improve results.

2. Methods & Materials

2.1. Site Description and Location

The geological map in Figure 1a shows that Holocene silty clay sediments cover the entire region of the Nile Delta. Moreover, as shown in Figure 1b, the location of the area under investigation in this research is bounded from the east by the Suez Canal and in the west by El-Manzala Lake, where sabkha deposits and salt crusts dominate the region. The experimental area is located between latitude 31°06′30″ and 31°06′50″ N, and longitude 32°17′15″ E and 32°17′50″ E (Figure 1c). The high content of clay in soils located within the Nile Delta region complicates water seepage and the provision of efficient drainage and sustainable soil productivity [14].

Three farms A–C located within the reclamation area were chosen to conduct the ground geophysical surveys aiming to decipher the subsurface conditions that may affect their productivity and to monitor the lateral and vertical variations occurring during different seasons. The vegetation cover on the first farm (Farm A) is heterogeneous, with plants dying after a short period of growth in certain areas (Figure 2A). No agronomic problems can be observed on the second farm (Farm B) except for the inability of deep-rooted vegetables to grow (Figure 2B). On the third farm (Farm C), all attempts to cultivate have failed because the soil constantly suffers from waterlogging problems (Figure 2C).

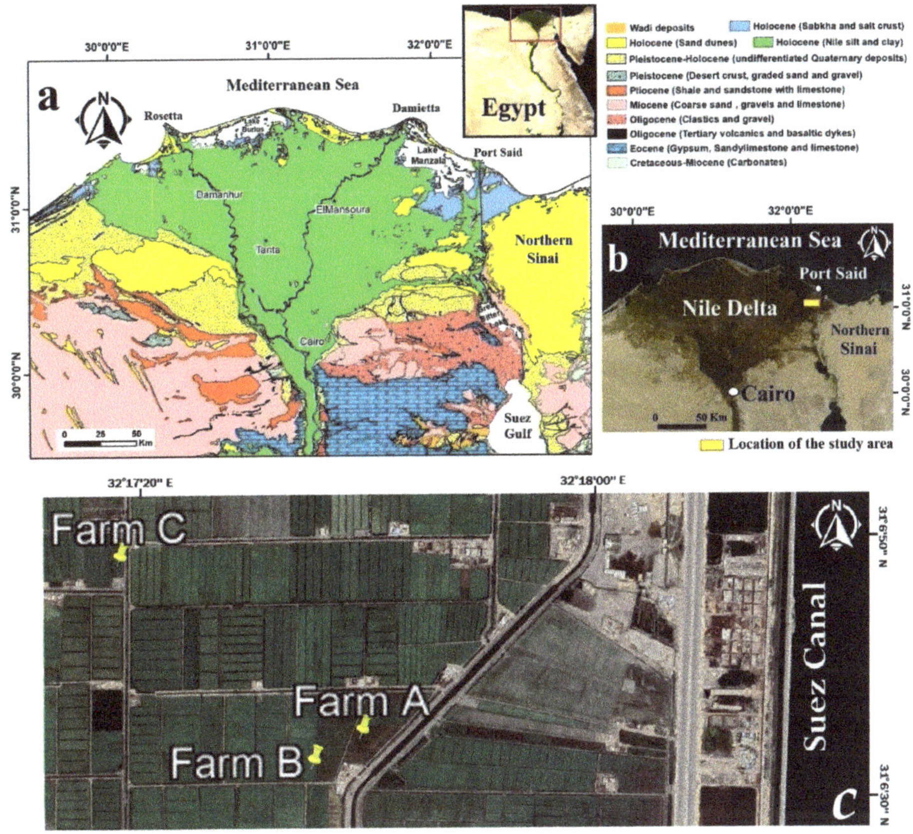

Figure 1. (**a**) Geological map of the Nile Delta (modified after Conoco [15] and Hassan et al. [16]), (**b**) location map of the recently reclaimed area denoted by the white rectangle, and (**c**) location of the three farms.

Figure 2. Agricultural conditions of the three investigated farms: (**A**) Heterogeneous and partially dead vegetation cover, (**B**) no deep-rooted plants can successfully grow, and (**C**) barren saline soil and frequent waterlogging.

2.2. Methodology

Geoelectrical methods have a wide range of applications in agriculture [17–20] because of their ability to monitor lateral and vertical variations in soil conductivity as a function of the electrical resistivity [21–23]. They have been efficiently used for imaging clay-pan [24,25], peatland stratigraphy [26], root zones [27–31], and for monitoring contamination [32,33]. They have also been applied in various post-reclamation problems [34], soil characterization [35,36], the monitoring of clay behavior during seasonal water content variations [37,38], and various other near-surface investigations [39]. In this study, three geoelectrical methods, namely electrical resistivity tomography (ERT), spontaneous potential (SP), and apparent resistivity gradient (RG), were employed.

2.2.1. Electrical Resistivity Tomography (ERT)

This method relies on inverting the apparent resistivity data, measured along a profile, to create a 2D model that shows vertical and horizontal resistivity changes underneath the surveyed soil surface [40]. Because of the high background noise levels in cultivated soils, the Wenner-alpha array was chosen to implement the survey. The array is less sensitive to noise contamination because it has the highest signal strength compared to other arrays [41]. It is very sensitive to vertical changes in apparent resistivity. Therefore, it can resolve subsurface horizontal structures even at noisy sites [40]. Figure 3 shows the design of the ERT survey used in the current study. It consisted of 26 electrodes arranged along a profile and spaced by a unit distance (a). The apparent resistivity (ρ_a) of each station was calculated according to:

$$\rho_{a_{(\Omega.m)}} = 2\pi a \frac{\Delta V}{I} \tag{1}$$

where I is the intensity of direct current introduced into the ground via electrodes A and B and ΔV is the potential difference measured between the inner electrodes M and N (Figure 3). The survey starts by measuring apparent resistivity along the profile, using the Wenner-alpha array with electrode spacing of '1a'. The profile is re-surveyed several times. Each time the spacing between electrodes is increased by n factor to map different soil depths [42]. As the spacing between electrodes increases, the number of measurements will decrease and consequently the width of the apparent resistivity pseudo-section (Figure 3). The whole set of apparent resistivity measurements is checked to remove abnormal (noise) data. Noise is removed in the processing steps before final model calculations. Then the data can be inverted using RES2DINV software [43,44] to construct ERT models.

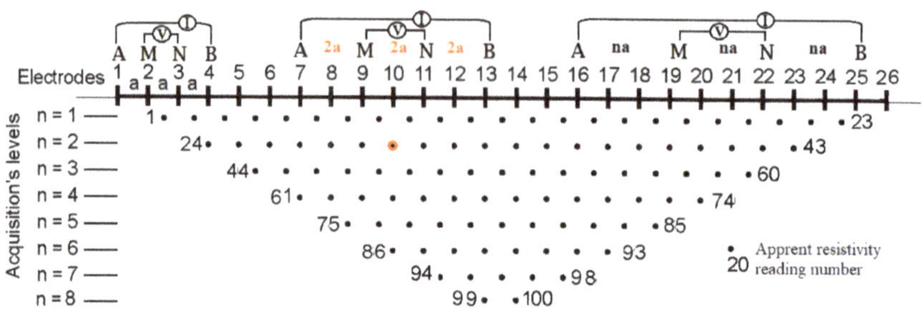

Figure 3. Schematic of electrode arrangement and the sequence of measurements used to build apparent resistivity pseudo-sections.

The inversion process aims to determine the true resistivity of the subsurface layers from the apparent resistivity pseudo-section. Thus, the process begins with constructing apparent resistivity pseudo-sections from a series of readings measured at the ground surface by using finite-difference or finite-element methods to divide the section into many rectangular blocks with different resistivity values [40,45]. Then the Jacobian matrix of partial derivatives is calculated using Gauss–Newton or quasi-Newton methods. Finally, the process ends with solving the least-square equation of the Gauss–Newton and quasi-

Newton methods using a regularized least-square optimization method [43,44]. Two algorithms are used to constrain the regularized least-square optimization. The first is the blocky constrained least-square method (L1-norm); the second is the smooth constrained least-square method (L2-norm) [46].

Some ERT profiles were repeated at different times during the field investigations to monitor changes that occurred between different seasons. Before inverting these time-lapse measurements, it was necessary to remove the effect of temperature changes and express all apparent resistivity values at a standardized temperature of 25 °C as soil conductivity increases by 1.9% [47] to 2.02% [48] per one-degree Celsius increase in temperature [49,50]. The following equation [47,51] was used to remove the effect of temperature changes from apparent resistivity data:

$$\rho_{25\,^{\circ}\mathrm{C}} = \rho_{a_T} / \left[04470 + 1.4034 e^{(T/26.815)}\right] \tag{2}$$

where ρ_{a_T} is the apparent resistivity measured at a temperature T in °C, and $\rho_{25\,^{\circ}\mathrm{C}}$ is the standardized apparent resistivity referenced to 25 °C.

Inverting these time-lapse datasets should be carried out using a joint inversion technique where the model obtained from the initial dataset is used to constrain the inversion of the later time datasets [45].

2.2.2. Spontaneous Potential (SP)

This is a passive geophysical method used for mapping galvanic currents that are naturally developed in soils. These currents originate either due to the difference in concentrations between soil fluids (diffusion or electrochemical potential), movement of fluids in porous soils (streaming or electrokinetic potential), electronic conduction occurring in sulfide ores (mineralization potential), or due to the presence of clay that adsorbs ions on its surface (shale potential) [32,52–55]. The low cost of equipment required to conduct SP surveys in the field is a major advantage. Only a portable high impedance millivolt meter and an insulated wire reel are needed to measure potential differences directly between two non-polarizable electrodes [39,51,56]. Non-polarizable electrodes are copper rods immersed in porous pots filled with a saturated copper sulfate solution (Figure 4).

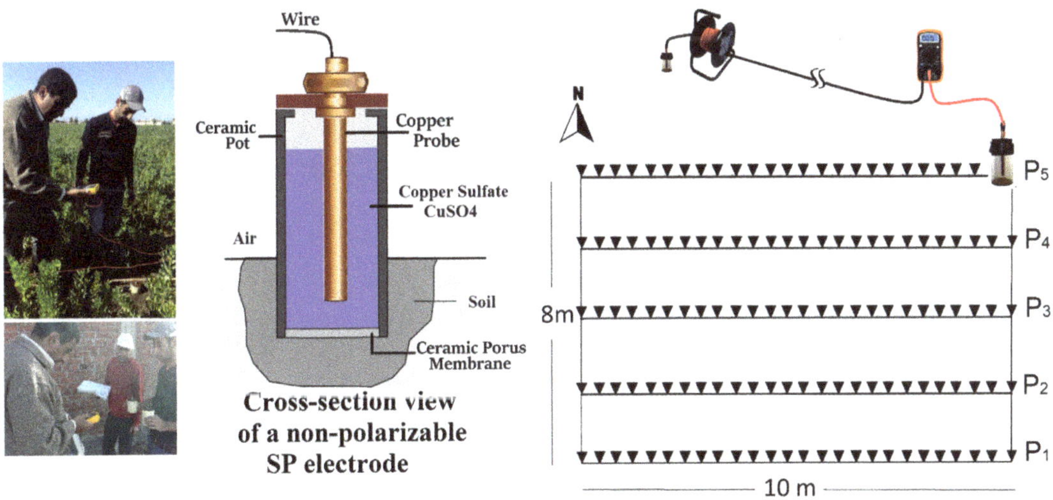

Figure 4. Placing the roving non-polarizable electrode in small well-watered holes to ensure efficient contact with the soil and layout of the grid used to conduct the SP survey using a fixed-based configuration.

One of these electrodes will be set as the remote reference station for all readings, while the other electrode will move along parallel profiles. For better results, the SP profiles should lay perpendicular to the direction of the subsurface flow [57]. After the roving electrode is implanted at each station, several minutes are allowed to pass during which the electrode stabilizes. Then the potential difference is measured relative to the reference electrode. During the work, contact resistance between the two electrodes should be tested regularly; a base station must be reoccupied with every constant interval of time for noise reduction. Linear corrections are applied to compensate for the effects of leakage or increased solution temperature inside the pots [58].

2.2.3. Apparent Resistivity Gradient

The apparent resistivity gradient (RG) method uses two metal electrodes, A and B, to introduce a direct current (DC) of intensity (I) into the ground. The current electrodes are kept at fixed positions while the voltage difference (v) is measured between another two rover electrodes M and N (Figure 5). After each measurement, only the voltage electrodes will leap by a distance equal to MN to measure voltage difference at the next stations (Figure 5). The acquisition procedures will be repeated along a set of profiles laying parallel to AB [59]. The apparent resistivity (ρ_a) in ohm meter (Ωm) at each station is calculated according to [60]:

$$\rho_{a_{(\Omega.m)}} = \frac{2\pi V}{I}\left[\frac{1}{AM} - \frac{1}{BM} - \frac{1}{AN} + \frac{1}{BN}\right]^{-1} \quad (3)$$

where AM, BM, AN, and BN are the distances between electrodes at each station. All measurements are conducted within a rectangular area centered on the midpoint between A and B (Figure 5). The width and the length of the surveyed rectangular area should not exceed 0.3 and 0.5 of the distance between A and B, respectively, where the current electrodes produce a quasi-homogeneous field [61].

It is worth mentioning here that all of the ERT and RG surveys were performed using the IRIS Syscal-R2 resistivity instrument. The injection current duration was set to the maximum of 2000 ms and the stack (min/max) numbers were set to (3/10) to ensure the best quality of measurements in the highly conductive soil.

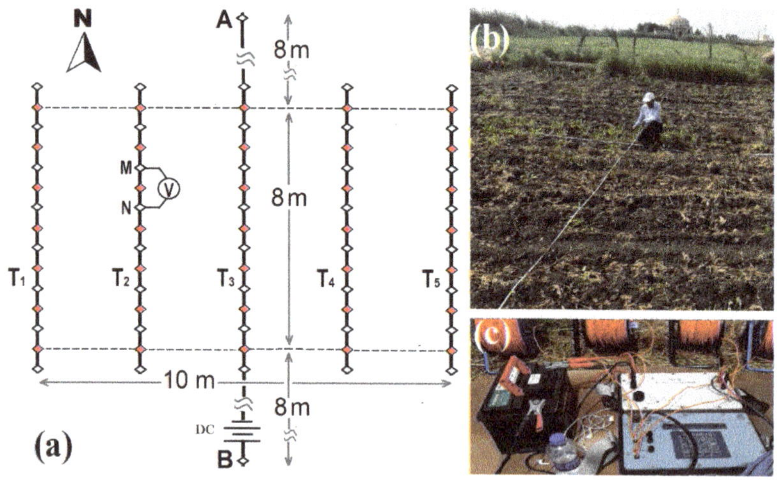

Figure 5. (**a**) Grid pattern used to conduct the apparent resistivity gradient (RG) survey, measurements were recorded at the midpoint, denoted by red between the two potential electrodes, (**b**) laying out the grid in the field, and (**c**) instrument used for measuring resistivity (SYSCAL-R2).

2.3. Field Work

2.3.1. Pilot Electrical Conductivity Survey

Prior to conducting the geoelectrical investigations according to the above, a pilot electrical conductivity (EC) survey was conducted. The WET Sensor (Delta-T Devices Ltd., Cambridge, UK) was used to measure the EC of the topsoil at the three farms. Soil samples at different depths (10–60 cm) were collected from the studied area within each farm. The first two farms (A and B) showed average EC levels ranging between 4 and 8 dS/m, while the EC of the barren farm (C) ranged between 22 and 30 dS/m. In addition to that, Table 1 shows that these samples had approximately the same composition and were characterized by their high clay content. The percentage of the soil fractions was determined using standard methods (i.e., sieve analysis and hydrometer).

Table 1. Summary of soil texture and density at the three investigated farms.

Location	Depth (cm)	Sand (%)	Silt (%)	Clay (%)	Bulk Density ($g{\cdot}cm^{-3}$)
Farm A	0–10	3.7	25.8	70.5	1.8
	10–20	3.5	30.0	66.5	1.9
	20–40	2.1	32.1	65.8	1.8
	40–60	2.9	29.0	68.1	1.9
Farm B	0–10	4.0	30.8	65.2	1.8
	10–20	4.0	28.5	67.6	1.9
	20–40	2.4	31.1	66.5	1.7
	40–60	3.5	23.6	72.9	1.7
Farm C	0–10	3.2	32.7	64.1	1.9
	10–20	2.9	33.0	64.1	1.7
	20–40	5.6	25.8	66.6	1.9
	40–60	4.8	29.2	66.9	1.8

2.3.2. Geoelectrical Field Work

All the aforementioned geoelectrical methods were tested on Farm A. The investigation began with an ERT survey across an area characterized by poor and heterogeneous vegetation cover at the center of the farm plot. The length of the ERT profile was 10 m, and its direction was chosen to be orthogonal to the main path of the subsurface mole drains of the farm. The initial spacing between electrodes implanted along the profile was 0.4 m. To create a continuous cross-section of the subsoil, the profile was re-surveyed eight times. Each time, the spacing between electrodes increased by 0.4 m (Figure 3). This configuration sufficed to image 1.6 m depth of the soil, exceeding the depth of the mole drains. The apparent resistivity was measured at each station individually (manually), along the designed profile shown in Figure 3. This provided a good opportunity to check noise levels directly during the field acquisition and to repeat measurements. The whole set of apparent resistivity measurements was rechecked again for abnormal data points before inverting it, using both robust inversion (L1-norm) and smooth constrained least square (L2-norm) algorithms [43,44].

A set of five profiles were surveyed using the self-potential (SP) method. The survey covered a rectangular area of 10 m by 8 m. The SP profiles were spaced by 2 m, and the middle profile (P3) superimposed the exact location of the ERT profile. A reading was recorded every 40 cm along each profile. Figure 4 shows the layout of the grid used to conduct the SP survey. After correcting the SP measurements as discussed in Section 2.2, the measurements were plotted in a map showing the surface variations of the natural potential in soil.

The same rectangular area was re-surveyed using the RG method. The directions of RG surveyed profiles were chosen to be running perpendicular to the main azimuth of the detected high SP anomaly. Therefore, the length of the surveyed profiles was only 8 m long (Figure 5a). The distance between the fixed current electrodes A and B was 24 m, i.e., three times the length of the surveyed profile. Potential differences were recorded every 1 m along five profiles that lay parallel to AB and separated by 2.5 m. After calculating the ρ_a for each station measured within the RG grid according to Equation (3), the results were plotted in a map that shows the lateral variation of surface apparent resistivity within the same surveyed plot.

3. Results and Discussion

Figure 6 compares the two ERT models produced by the different inversion algorithms. Commonly, the L1-norm method is preferred when there are high contrasts in apparent resistivity readings [62] because it is less sensitive to noisy data points compared to the L2-norm method [41]. Since the smooth constrained L2-norm method relies on minimizing the square of the differences between the measured and calculated apparent resistivity, it shows a smooth distribution of the inverted resistivity of the subsoil (Figure 6b). The robust inverse L1-norm produced a model with sharper edges than that produced by the L2-norm method (Figure 6a).

It can be noted that the L2-norm inversion gives better results than the L1-norm as it is more suitable for the subsurface geology conditions that exhibit a smooth variation [41,46]. Therefore, the L2-norm was chosen to invert the measured apparent resistivity data.

Initial examination of the SP and RG maps shown in Figure 7 reveals a significant discrepancy between the results of the methods. For instance, the elongated high anomaly in the central part of the SP map has disappeared completely from the RG map. Its location is occupied by a gradual decrease in the apparent resistivity towards the southern parts of the surveyed plot. Similarly, it was clearly observed during the fieldwork that the apparent resistivity values recorded in areas characterized by dense vegetation were always high in contrast to SP.

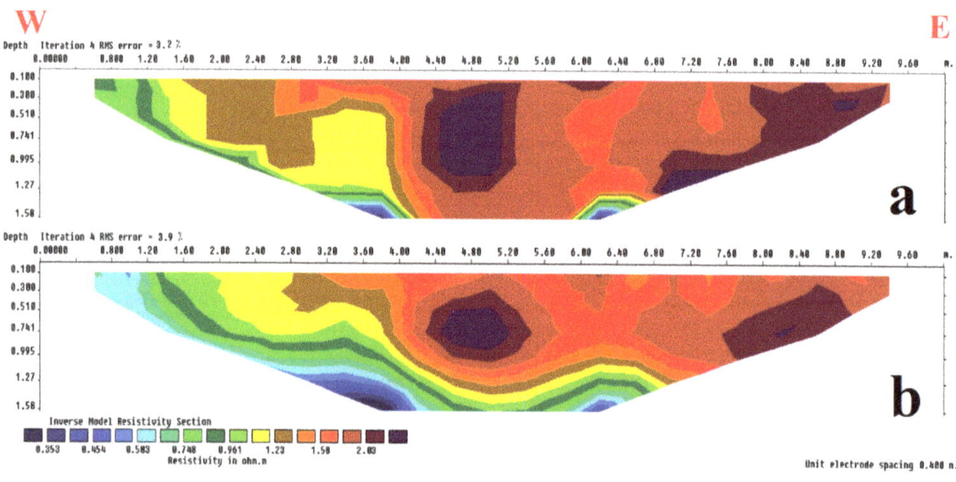

Figure 6. ERT models of the same section inverted using: (**a**) the robust inversion method, and (**b**) the least-squares smooth constrained method. Data were surveyed at Farm A in October 2021.

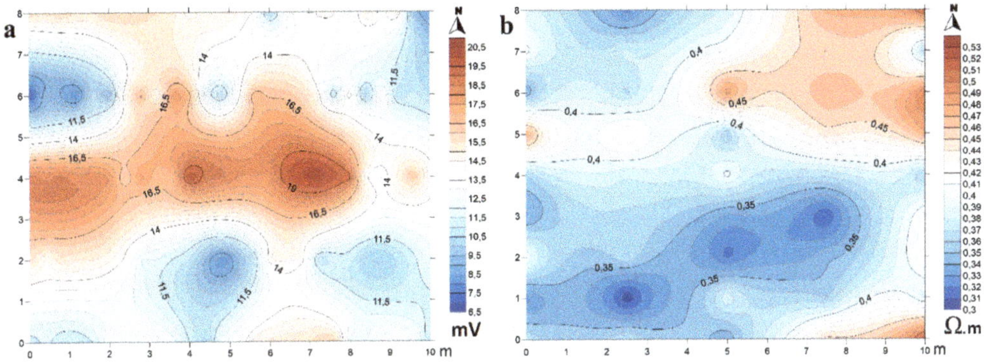

Figure 7. (**a**) Spontaneous potential map and (**b**) calculated apparent resistivity gradient map of the same plot surveyed at Farm A.

For better understanding, combining results of more than one geophysical method is a common practice to reduce the ambiguities in data interpretation. Thus, the results obtained from the three surveys on Farm A were compiled into one figure (Figure 8). The green and the light blue curves shown in the upper part of the figure represent lateral variations of SP and RG values, respectively. Both curves were plotted at their exact location along the ERT cross-section. This compilation revealed several important points regarding the response of each method to the changes in soil texture and structure.

First, the two curves (Figure 8) conspicuously confirm the discrepancy between the SP and RG readings, i.e., when SP shows high values at a particular location along the profile, GR correspondingly shows low values for the same location, and vice versa. Usually, the SP and GR curves diverge opposite permeable freshwater sand and converge in front of shale [63,64]. This relationship is a well-established principle used in interpreting SP and resistivity logs [65,66]. In this case, the high apparent resistivity (RG) values at 2 m, 4.8 m, and 8.2 m are accompanied by dips on the SP curve. These sites are located within regions characterized by permeable sandy soils. Conversely, the SP curve shows continuous high values along the first 1.5 m of the profile and a high peak centered at 6.4 m. At these locations, the RG readings are lower than normal, and the soil becomes more clayey.

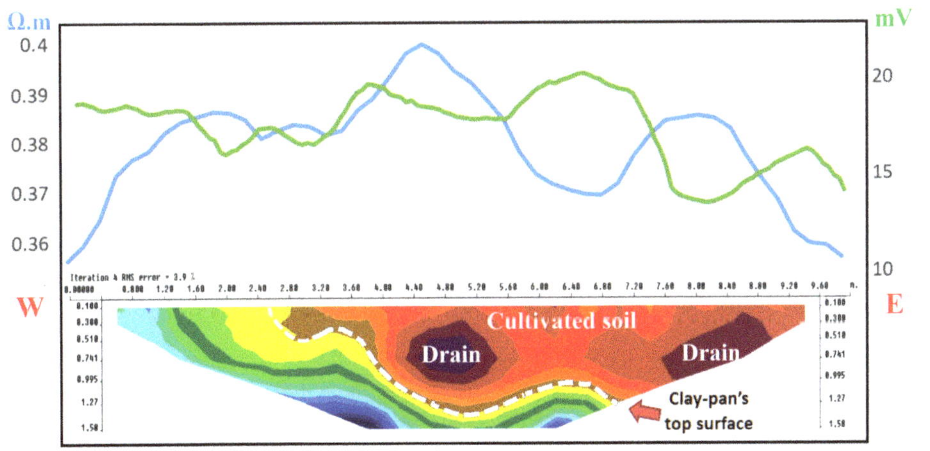

Figure 8. Spontaneous potential curve (green), apparent resistivity gradient curve (light blue), and the ERT model of the same profile surveyed at Farm A.

Second, there is a strong resemblance between the shape of the SP curve and the surface relief of the layer delineated by the white dashed line in the ERI model (Figure 8). As the surface of that layer approaches the ground surface of the soil, the readings on the SP curve increase. It is worth mentioning here that the abrupt changes in SP curves are commonly associated with changes in soil composition, especially with increasing clay content in the soil. Because clay particles have negative electrical charges on their surfaces, clay layers can attract and retain positively charged ions (cations) and prevent them from escaping from the soil [67]. Consequently, an electrochemical interaction occurs between clay surfaces and soil-water, especially when the latter has a different ionic activity. The electromotive force originated from this interaction causes currents to flow around the clay boundaries. Therefore, SP measurements record high positive voltages adjacent to clay layers [52,60,68], which in turn implies that this layer represents the top surface of the underlying clay-pan.

Finally, the divergence between the SP and RG curves that occurred at 4.8 m and 8.2 m is attributed to the lower clay content in the topsoil or to the change of soil texture that becomes sandier at these locations. The latter assumption fits well with the results obtained from the ERT section (Figure 8) which shows that the depth to the surface of the clay bed increases in the central and eastern parts of the farm. This allows the drainage system to function without hindrance in these parts of the farm and results in lowering the SP values due to infiltration from the drainage ditches [55]. The ERT section is also able to locate two mole drains (Figure 8). The active soil filtration nearby the locations of the drains, as the soil becomes more permeable, reduces the salinity and consequently increases its apparent resistivity. Therefore, the presence of these drains is responsible for the two high RG anomalies.

The rule of thumb in interpreting any near-surface geophysical results is that: without drilling, all interpretations are refutable. Accordingly, these results were substantiated by auguring holes along the surveyed profile. Based on the aforementioned notes, we can conclude that as depth to the clay-pan increases eastwardly, deep-rooted plants can grow in the upper permeable reworked soil. Poor vegetation on the western side of the farm can be attributed to the increase of silt and clay contents in the topsoil where the surface of the clay-pan approaches the ground surface. This has resulted in blocking the MD, thereby diminishing its efficiency in accelerating salt leaching. Moreover, soil compaction on the western side of the farm adversely affects crop production by hindering plant root growth. Compaction reduces soil aeration, lessens gaseous exchange in the soil, and limits water flow and hence drainage [69]. Thus, we can advise the farm owner to plow the western strip of the farm, whenever possible, to loosen the soil.

We plan to repeat the ERT survey along the same profile every three months. The main objective of repeating the survey is to establish the reliability of the technique in observing subsurface resistivity changes that occur on the farm after plowing. Thus, we left permanent wedges on both ends of the surveyed profile. The wedges enabled us to precisely locate the profile and repeat the ERT measurements along the same path.

It should be mentioned here that the first ERT survey was conducted in October 2021. The same profile was surveyed again in December 2021 and in April 2022. The effect of temperature changes was removed from all measurements before inversion [47,49]. The inversion of the datasets was performed using a joint inversion technique where the resistivity model of the first survey (Figure 6b) was used to constrain the inversion of the subsequent datasets. The inverted resistivity models of the time-lapse surveys are shown in Figure 9.

A qualitative comparison of the two inverted resistivity models, illustrated in Figures 6b and 9a, did not reveal any significant changes in the soil structure during the period between October and December 2021. However, it was obvious that surface plowing had slightly loosened the compacted soil on the western side of the farm, but it did not disturb the deep clay layer. As the soil became permeable for freshwater flows, its resistivity increased especially along the first 2.80 m of the profile (Figure 9a). This has improved the functionality of the central drain, which was able to reduce the salinity of larger cross-sectional areas around its course compared to that shown in the first model (Figure 6b). This result prompted the farm owner to continue plowing the soil whenever possible. After another three months, in April 2022, the situation improved further, and the cultivated soil of the farm extended westward (Figure 9b).

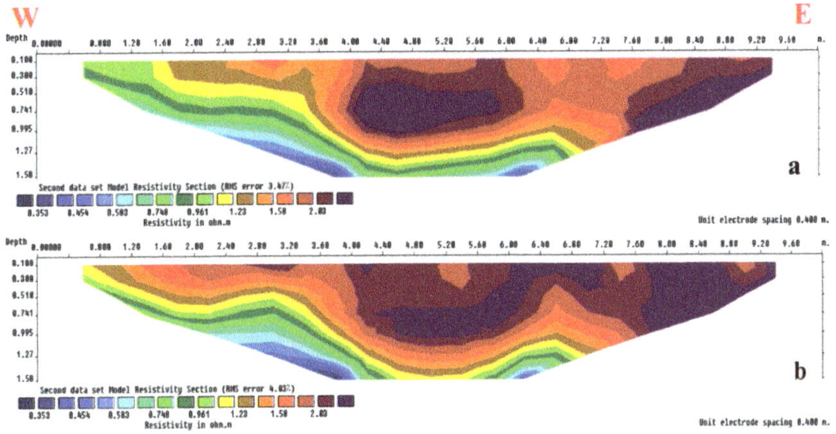

Figure 9. ERT models of the inverted apparent resistivity data measured along the same profile that was resurveyed in (**a**) December 2021 and (**b**) in April 2022.

The time-lapse sections shown in (Figure 10a,b) are helpful in illustrating quantitatively the changes occurring during the whole period of the survey, where the inverted resistivity model of the first survey (Figure 6b) was used as a reference to estimate the percentage changes in resistivity between the reference model and each one of the subsequent models. High percentages in these sections represent areas where maximum changes in soil resistivity occur, while low values indicate no significant changes. The figure also shows that the surface salinity of the topsoil decreased as its resistivity increased by 20% in December (Figure 10a), while it reached 80% in April (Figure 10b).

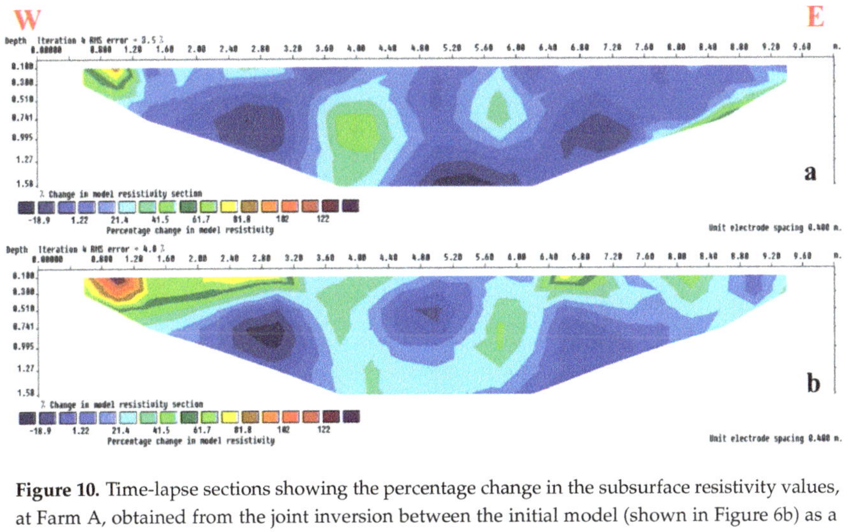

Figure 10. Time-lapse sections showing the percentage change in the subsurface resistivity values, at Farm A, obtained from the joint inversion between the initial model (shown in Figure 6b) as a reference with the lateral datasets collected in (**a**) December 2021 and (**b**) in April 2022.

The above experiments demonstrate the ability of the ERT method to visualize subterranean soil structure and resolve complex sources of resistivity anomalies that cannot be distinguished using RG or SP methods alone. The method was able to determine the location and depth of mole drains, trace the surface of the clay-pan layer, and estimate the thickness of the cultivable soil. This encouraged us to conduct ERT surveys directly on the other two farms (B and C).

Figure 11 shows the inverted resistivity model of the ERT survey performed at Farm B. The azimuth of the surveyed profile was chosen to run perpendicular to the direction of the buried mole drain system of the farm. High resistivity closures identify the location of the detected drains within the cultivable layer (Figure 11). Yellow arrows in the figure denote the location of these buried drains which are spaced at equal distances of 1.5 m. The average depth to the top of the clay layer is about 55 cm. This shallow depth limits the thickness of the active root zone. As a result, the thickness of the cultivable soil is not sufficient to grow vegetables with medium or deep roots such as beans, carrots, tomatoes, peas, or eggplants.

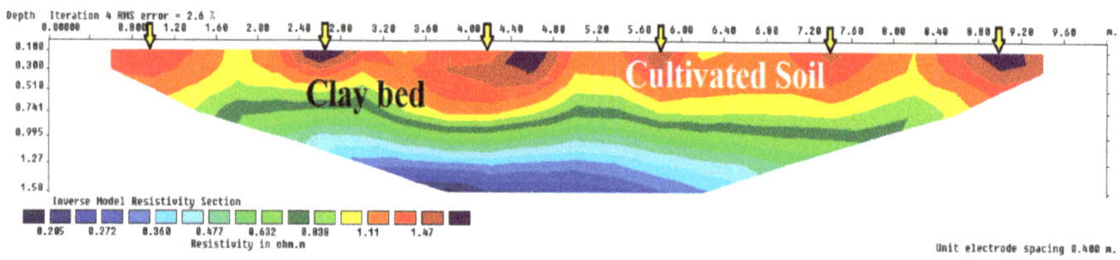

Figure 11. Inverted resistivity models of the ERT data conducted at plot B showing the thickness of its root zone. Yellow arrows are pointing at the locations of the drains.

As only traditional reclamation methods were practiced in Farm C, all cultivation efforts in this farm failed due to the frequent rise of water in the center of the farm (Figure 12c). EC measurements showed that the electrical conductivity of the topsoil reaches 30 dS/m. When the waterlogging stopped, we were able to conduct two ERT surveys along the same profile, parallel to the direction of the mole drains. The first survey took place in November 2021 and was then repeated in January 2022. The inverted resistivity models of the two surveys are shown in (Figure 12a,b). The first model (Figure 12a) shows the infiltration of irrigation water through the topsoil during November 2021. In January 2022, the poorly drained soil retains the infiltrated water in the deepest parts of the clay layer (Figure 12b) because the clay layer is gently inclined towards the middle part of the farm.

This inclination opposes the direction of the drainage system that was initially designed according to the surface levels of surface soil. This conflict led to the malfunctioning of the farm drainage system, which plays almost no role in reducing soil salts [70,71]. Therefore, the salinity levels of the soil and the water retained in it increased. This can be inferred from the time-lapse model (Figure 12c), as the resistivity of the saturated subsoil in Farm C has decreased by about 10%. We tried to conduct an additional ERT survey in February but were prevented by water logging (Figure 12d). Yellow arrows in Figure 12b,d indicate the location of the surface clayey soil that formed a sub-basin in the center of the farm.

It is noteworthy that designing the agricultural drainage system requires more information about the soil subsurface structure and should not be based only on the surface levels of the soil, especially for land with mild to no slope. It is recommended to image the subsurface soil structure along several arbitrary directions before designing the drainage system of any farm.

The role played by the drainage system in reducing soil salinity can be inferred through comparing the inverted ERT models of the two farms A and C. The active drainage system on the first farm (A) was able to reduce the salinity of the soil, as can be noticed from the increment of resistivity nearby the drain location (Figures 9 and 10). On the other hand, the malfunction of the drainage system in Farm C resulted in decreasing resistivity of the soil due to increased salt content. These results confirmed the reliability of the EC measurements during the reconnaissance surveys.

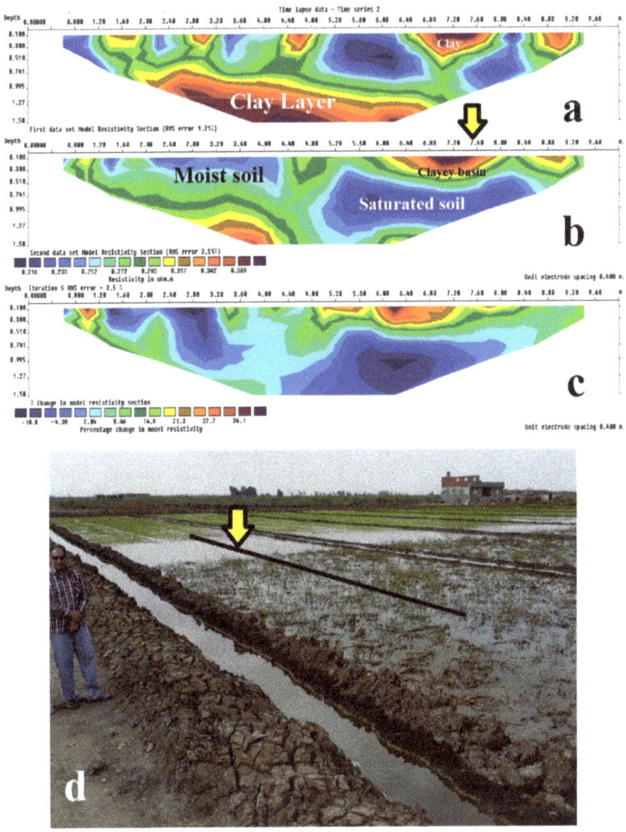

Figure 12. ERT models conducted at Farm C along the same profile in (**a**) November 2021 and (**b**) in January 2022. (**c**) The percentage change in the subsurface resistivity. (**d**) Waterlogs appeared on the farm in February 2022. The black line shows the location of the surveyed ERT profile. The yellow arrow points at the location of the compacted clay on the surface of the farm, which extends laterally to other plots and holds water after a rainfall.

4. Conclusions

Geoelectrical surveys were conducted on three newly reclaimed farms located in the northeastern corner of the Nile Delta, Egypt. The soils of all farms within this region are characterized by their high content of clay. Farmers usually rely on the traditional mole drainage (MD) on these farms. Most of the farms are currently suffering from waterlogging and salinization problems.

The present research has proven the applicability of geoelectrical methods in imaging subsurface soil structures, even for soil characterized by high clay contents and affected by salinity. Inverted ERT models can suffice to understand the subsurface structures but integrating results from SP and RG helped to understand the geology and to reduce the ambiguous interpretation that could result by using any of these methods individually. Results show that the SP measurements are mainly influenced by the contrasts in clay contents within the surface soil, while high apparent RG closures can define cultivable areas on the farm. The inverted ERT models were successful in determining the thickness of the root zone layers, locating subsurface drains, and tracing the upper surface of the clay bed (bedrock). The shallow depth of the undisturbed clay beds reduces the productivity of the farm as it limits the thickness of the active root zone and prevents the growth of deep-rooted plants. It is necessary to consider the slopes of the subsurface clay beds before planning the drainage network of any farm, even if the farm has a flat surface. We can conclude that the efficiency of MD depends on the local subsurface soil conditions. Thus, it is also important to image the subsurface conditions of the soil periodically to monitor the dynamics of agricultural soil processes. Time-lapse ERT surveys showed their ability to monitor the accumulation of groundwater over the course of different seasons, which is the main cause of waterlogging problems on the farms.

Author Contributions: Conceptualization, A.A., Y.H. and T.S.; Data curation, A.A., T.A., Y.H. and T.S.; Formal analysis, A.A., T.A. and T.S.; Funding acquisition, R.B., Y.H. and T.S.; Investigation, A.A., T.A., Y.H. and T.S.; Methodology, A.A., R.B., T.A., Y.H. and T.S.; Project administration, R.B., Y.H. and T.S.; Resources, R.B., Y.H. and T.S.; Software, A.A. and T.A.; Supervision, A.A., R.B., T.A., Y.H. and T.S.; Validation, A.A., T.A. and T.S.; Visualization, A.A., R.B., T.A., Y.H. and T.S.; Writing—original draft, A.A. and T.S.; Writing—review & editing, A.A., R.B., T.A., Y.H. and T.S. All authors have read and agreed to the published version of the manuscript.

Funding: This research received no specific grant from any funding agency in the public, commercial, or not-for-profit sectors.

Institutional Review Board Statement: Not applicable.

Informed Consent Statement: Not applicable.

Data Availability Statement: The data presented in this study is available on request from the corresponding author.

Acknowledgments: The authors acknowledge funding for this study from the Egyptian Academy of Scientific Research and Technology and Tunisian Institution of Agricultural Research and Higher Education through the SALTFREE project (ARIMNET2, Coordination of Agricultural Research in the Mediterranean Area). The research was also supported by the Strategic Research Area: The Middle East in the Contemporary World (MECW) at the Centre for Advanced Middle Eastern Studies, Lund University, Sweden. The authors wish to thank the Center of Environmental Studies and Consultancies at Port Said University for providing the necessary equipment and permissions to conduct the field surveys.

Conflicts of Interest: The authors declare that there is no conflict of interest.

References

1. Abahussain, A.A.; Abdu, A.S.; Al-Zubari, W.K.; El-Deen, N.A.; Abdul-Raheem, M. Desertification in the Arab Region: Analysis of current status and trends. *J. Arid. Environ.* **2002**, *51*, 521–545. [CrossRef]
2. Abo-El-Enein, S. *Towards Sustainable and Improved Water Productivity in the Old Lands of the Nile Delta*; International Center for Agricultural Research in the Dry Areas (ICARDA): Giza, Egypt, 2018.
3. Alary, V.; Aboul-Naga, A.; Osman, M.A.; Daoud, I.; Abdelraheem, S.; Salah, E.; Juanes, X.; Bonnet, P. Desert land reclamation programs and family land dynamics in the Western Desert of the Nile Delta (Egypt), 1960–2010. *World Dev.* **2018**, *104*, 140–153. [CrossRef]
4. Embabi, N.S. The Nile Delta. In *Landscapes and Landforms of Egypt*; Springer: Cham, Switzerland, 2018; pp. 57–68.
5. Shendi, E.-A.H.; Aziz, A.M. Discovery of an ancient pharaoh's temple on the Horus military road, Northern Sinai, Egypt. *Arab. J. Geosci.* **2010**, *3*, 249–255. [CrossRef]
6. Taha, M.M.; El-Asmar, H.M. Geo-Archeoheritage Sites Are at Risk, the Manzala Lagoon, NE Nile Delta Coast, Egypt. *Geoheritage* **2019**, *11*, 441–457. [CrossRef]
7. Rashed, A.; Khalifa, E.; Fahmy, H. Paddy rice cultivation in irrigated water managed saline sodic lands under reclamation, Egypt. In Proceedings of the 9th ICID International Drainage Workshop, Utrecht, Netherlands, 10–13 September 2003.
8. Mohamed, E.S.; Belal, A.; Saleh, A. Assessment of land degradation east of the Nile Delta, Egypt using remote sensing and GIS techniques. *Arab. J. Geosci.* **2013**, *6*, 2843–2853. [CrossRef]
9. Mohamed, E.S.; Belal, A.; Shalaby, A. Impacts of soil sealing on potential agriculture in Egypt using remote sensing and GIS techniques. *Eurasian Soil Sci.* **2015**, *48*, 1159–1169. [CrossRef]
10. El-Kady, R.Y.; El-Rayes, A.E.; Sultan, Y.M.; Aziz, A.M. Mapping of soil geochemistry in Port Said Governorate, Egypt utilizing GIS and remote sensing techniques. *Imp. J. Interdiscip. Res.* **2017**, *3*, 1261–1270.

11. El-Rayes, A.E.; Arnous, M.O.; Aziz, A.M. Morphotectonic controls of groundwater flow regime and relating environmental impacts in Northwest Sinai, Egypt. *Arab. J. Geosci.* **2017**, *10*, 401. [CrossRef]
12. Karajeh, F.; Oweis, T.; Swelam, A.; El-Gindy, A.; El-Quosy, D.; Khalifa, H.; El-Kholy, M.; El-Hafez, S.A. *Water and Agriculture in Egypt*; Technical Paper Based on the Egypt-Australia-ICARDA Workshop on On-farm Water-Use Efficiency; International Center for Agricultural Research in the Dry Areas (ICARDA): Giza, Egypt, 2011.
13. Satoh, M.; Aboulroos, S. *Irrigated Agriculture in Egypt: Past, Present and Future*; Springer: Cham, Switzerland, 2017.
14. Hoffman, G.J.; Durnford, D. Drainage design for salinity control. *Agric. Drain.* **1999**, *38*, 579–614.
15. Conoco, C. *Geological Map of Egypt, Scale 1: 500,000*; Egyptian General Petroleum Corporation-Conoco Coral, Geological Survey: Cairo, Egypt, 1987.
16. Hassan, S.; Sultan, M.; Sobh, M.; Elhebiry, M.S.; Zahran, K.; Abdeldayem, A.; Issawy, E.; Kamh, S. Crustal Structure of the Nile Delta: Interpretation of Seismic-Constrained Satellite-Based Gravity Data. *Remote Sens.* **2021**, *13*, 1934. [CrossRef]
17. Allred, B.; Daniels, J.J.; Ehsani, M.R. *Handbook of Agricultural Geophysics*; CRC Press: Boca Raton, FL, USA, 2008.
18. Cousin, I.; Besson, A.; Bourennane, H.; Pasquier, C.; Nicoullaud, B.; King, D.; Richard, G. From spatial-continuous electrical resistivity measurements to the soil hydraulic functioning at the field scale. *Comptes Rendus Geosci.* **2009**, *341*, 859–867. [CrossRef]
19. Ferronsky, V. *Nuclear Geophysics: Applications in Hydrology, Hydrogeology, Engineering Geology, Agriculture and Environmental Science*; Springer: Moscow, Russia, 2015.
20. Pasquier, C.; Bourennane, H.; Cousin, I.; Séger, M.; Dabas, M.; Thiesson, J.; Tabbagh, J. Comparison between thermal airborne remote sensing, multi-depth electrical resistivity profiling, and soil mapping: An example from Beauce (Loiret, France). *Near Surf. Geophys.* **2016**, *14*, 345–356. [CrossRef]
21. Coulouma, G.; Lagacherie, P.; Samyn, K.; Grandjean, G. Comparisons of dry ERT, diachronic ERT and the spectral analysis of surface waves for estimating bedrock depth in various Mediterranean landscapes. *Geoderma* **2013**, *199*, 128–134. [CrossRef]
22. Michot, D.; Benderitter, Y.; Dorigny, A.; Nicoullaud, B.; King, D.; Tabbagh, A. Spatial and temporal monitoring of soil water content with an irrigated corn crop cover using surface electrical resistivity tomography. *Water Resources Research* **2003**, *39*. [CrossRef]
23. Soupios, P.; Papadopoulos, N.; Papadopoulos, I.; Kouli, M.; Vallianatos, F.; Sarris, A.; Manios, T. Application of integrated methods in mapping waste disposal areas. *Environ. Geol.* **2007**, *53*, 661. [CrossRef]
24. Jeřábek, J.; Zumr, D.; Dostál, T. Identifying the plough pan position on cultivated soils by measurements of electrical resistivity and penetration resistance. *Soil Tillage Res.* **2017**, *174*, 231–240. [CrossRef]
25. Mathis, I.; Tucker-Kulesza, S.; Sassenrath, G. Electrical Resistivity Tomography of Claypan Soils in Southeastern Kansas. *Kans. Agric. Exp. Stn. Res. Rep.* **2018**, *4*, 13. [CrossRef]
26. Pezdir, V.; Čeru, T.; Horn, B.; Gosar, M. Investigating peatland stratigraphy and development of the Šijec bog (Slovenia) using near-surface geophysical methods. *CATENA* **2021**, *206*, 105484. [CrossRef]
27. Attia al Hagrey, S. Geophysical imaging of root-zone, trunk, and moisture heterogeneity. *J. Exp. Bot.* **2007**, *58*, 839–854. [CrossRef]
28. Carminati, A.; Vetterlein, D.; Weller, U.; Vogel, H.-J.; Oswald, S.E. When roots lose contact. *Vadose Zone J.* **2009**, *8*, 805–809. [CrossRef]
29. Furman, A.; Arnon-Zur, A.; Assouline, S. Electrical resistivity tomography of the root zone. *Soil–Water–Root Process. Adv. Tomogr. Imaging* **2013**, *61*, 223–245.
30. Xu, S.; Sirieix, C.; Riss, J.; Malaurent, P. A clustering approach applied to time-lapse ERT interpretation—Case study of Lascaux cave. *J. Appl. Geophys.* **2017**, *144*, 115–124. [CrossRef]
31. Zhao, P.-F.; Wang, Y.-Q.; Yan, S.-X.; Fan, L.-F.; Wang, Z.-Y.; Zhou, Q.; Yao, J.-P.; Cheng, Q.; Wang, Z.-Y.; Huang, L. Electrical imaging of plant root zone: A review. *Comput. Electron. Agric.* **2019**, *167*, 105058. [CrossRef]
32. Gallas, J.D.F.; Taioli, F.; Malagutti Filho, W. Induced polarization, resistivity, and self-potential: A case history of contamination evaluation due to landfill leakage. *Environ. Earth Sci.* **2011**, *63*, 251–261. [CrossRef]
33. Patra, H.; Adhikari, S.K.; Kunar, S. *Groundwater Prospecting and Management*; Springer: Singapore, 2016.
34. Olatinsu, O.; Oyedele, K.; Ige-Adeyeye, A. Electrical resistivity mapping as a tool for post-reclamation assessment of subsurface condition at a sand-filled site in Lagos, southwest Nigeria. *SN Appl. Sci.* **2019**, *1*, 24. [CrossRef]
35. Hossain, S.; Kibria, G.; Khan, S. *Site Investigation using Resistivity Imaging*; CRC Press: Boca Raton, FL, USA, 2018.
36. Soupios, P.; Georgakopoulos, P.; Papadopoulos, N.; Saltas, V.; Andreadakis, A.; Vallianatos, F.; Sarris, A.; Makris, J. Use of engineering geophysics to investigate a site for a building foundation. *J. Geophys. Eng.* **2007**, *4*, 94–103. [CrossRef]
37. Chrétien, M.; Lataste, J.; Fabre, R.; Denis, A. Electrical resistivity tomography to understand clay behavior during seasonal water content variations. *Eng. Geol.* **2014**, *169*, 112–123. [CrossRef]
38. Genelle, F.; Sirieix, C.; Riss, J.; Naudet, V. Monitoring landfill cover by electrical resistivity tomography on an experimental site. *Eng. Geol.* **2012**, *145–146*, 18–29. [CrossRef]
39. Butler, D.K. *Near-Surface Geophysics*; Society of Exploration Geophysicists: Houston, TX, USA, 2005.
40. Loke, M.H. Tutorial: 2-D and 3-D Electrical Imaging Surveys. Geotomo Software. 2004. Available online: https://sites.ualberta.ca/~{}\{\}unsworth/UA-classes/223/loke_course_notes.pdf (accessed on 8 August 2022).
41. Dahlin, T.; Zhou, B. A numerical comparison of 2D resistivity imaging with 10 electrode arrays. *Geophys. Prospect.* **2004**, *52*, 379–398. [CrossRef]
42. Edwards, L. A modified pseudosection for resistivity and IP. *Geophysics* **1977**, *42*, 1020–1036. [CrossRef]

43. Loke, M. *Rapid 2-D Resistivity & IP Inversion Using the Least-Squares Method*; RES2DINV ver. 3.55; Geotomo Software: Penang, Malaysia, 2006; Volume 139.
44. Loke, M.; Barker, R. Rapid least-squares inversion of apparent resistivity pseudosections by a quasi-Newton method 1. *Geophys. Prospect.* **1996**, *44*, 131–152. [CrossRef]
45. Loke, M. *Tutorial: 2-D and 3-D Electrical Imaging Surveys*; Geotomo Software SdnBhd: Penang, Malaysia, 2018.
46. Loke, M.H.; Acworth, I.; Dahlin, T. A comparison of smooth and blocky inversion methods in 2D electrical imaging surveys. *Explor. Geophys.* **2003**, *34*, 182–187. [CrossRef]
47. Sheets, K.R.; Hendrickx, J.M. Noninvasive soil water content measurement using electromagnetic induction. *Water Resour. Res.* **1995**, *31*, 2401–2409. [CrossRef]
48. Campbell, R.; Bower, C.; Richards, L. Change of electrical conductivity with temperature and the relation of osmotic pressure to electrical conductivity and ion concentration for soil extracts. *Soil Sci. Soc. Am. J.* **1949**, *13*, 66–69. [CrossRef]
49. Corwin, D.L.; Lesch, S.M. Apparent soil electrical conductivity measurements in agriculture. *Comput. Electron. Agric.* **2005**, *46*, 11–43. [CrossRef]
50. Samouëlian, A.; Cousin, I.; Tabbagh, A.; Bruand, A.; Richard, G. Electrical resistivity survey in soil science: A review. *Soil Tillage Res.* **2005**, *83*, 173–193. [CrossRef]
51. Corwin, R.F. The self-potential method for environmental and engineering applications. In *Geotechnical an Environmental Geophysics: Volume I: Review and Tutorial*; Society of Exploration Geophysicists: Houston, TX, USA, 1990; pp. 127–146.
52. Fensom, D. The bio-electric potentials of plants and their functional significance: I. An electrokinetic theory of transport. *Can. J. Bot.* **1957**, *35*, 573–582. [CrossRef]
53. Lénat, J.-F. Retrieving self-potential anomalies in a complex volcanic environment: An SP/elevation gradient approach. *Near Surf. Geophys.* **2007**, *5*, 161–170. [CrossRef]
54. Naudet, V.; Revil, A.; Bottero, J.Y.; Bégassat, P. Relationship between self-potential (SP) signals and redox conditions in contaminated groundwater. *Geophys. Res. Lett.* **2003**, *30*. [CrossRef]
55. Revil, A.; Fernandez, P.; Mao, D.; French, H.K.; Bloem, E.; Binley, A. Self-potential monitoring of the enhanced biodegradation of an organic contaminant using a bioelectrochemical cell. *Lead. Edge* **2015**, *34*, 198–202. [CrossRef]
56. Dentith, M.; Mudge, S.T. *Geophysics for the Mineral Exploration Geoscientist*; Cambridge University Press: Cambridge, UK, 2014.
57. Skianis, G.A. The Self-Potential Anomaly Produced by a Subsurface Flow at the Contact of Two Horizontal Layers and Its Quantitative Interpretation. *Int. J. Geophys.* **2012**, *2012*, 483590. [CrossRef]
58. Eppelbaum, L.V. Review of Processing and Interpretation of Self-Potential Anomalies: Transfer of Methodologies Developed in Magnetic Prospecting. *Geosciences* **2021**, *11*, 194. [CrossRef]
59. Bhattacharya, B.; Dutta, I. Depth of investigation studies for gradient arrays over homogeneous isotropic half-space. *Geophysics* **1982**, *47*, 1198–1203. [CrossRef]
60. Reynolds, J.M. *An Introduction to Applied and Environmental Geophysics*; John Wiley & Sons: Hoboken, NJ, USA, 2011.
61. Schulz, R. Interpretation and depth of investigation of gradient measurements in direct current geoelectrics. *Geophys. Prospect.* **1985**, *33*, 1240–1253. [CrossRef]
62. Ibraheem, I.M.; Tezkan, B.; Bergers, R. Integrated Interpretation of Magnetic and ERT Data to Characterize a Landfill in the North-West of Cologne, Germany. *Pure Appl. Geophys.* **2021**, *178*, 2127–2148. [CrossRef]
63. Fujii, K.; Ohmiya, T. Ion solubilization in surface soil layers of montane region with heavy snowfall. *CATENA* **2021**, *206*, 105490. [CrossRef]
64. Throner, R.H. *Engineering Geology Field Manual*; U.S. Department of the Interior, Bureau of Reclamation: Washington, DC, USA, 2001; Volume II.
65. Ellis, D.V.; Singer, J.M. *Well Logging for Earth Scientists*; Springer: Dordrecht, The Netherlands, 2007.
66. Liu, H. *Principles and Applications of Well Logging*; Springer: Berlin/Heidelberg, Germany, 2017.
67. Gallas, J.D.F. Self-potential (SP) generated by electrokinesis—Efficiency and low cost dam safety. *J. Appl. Geophys.* **2020**, *180*, 104122. [CrossRef]
68. Sharma, P.V. *Environmental and Engineering Geophysics*; Cambridge University Press: Cambridge, UK, 1997.
69. Nawaz, M.F.; Bourrié, G.; Trolard, F. Soil compaction impact and modelling. A review. *Agron. Sustain. Dev.* **2013**, *33*, 291–309. [CrossRef]
70. El-Ramady, H.; Alshaal, T.; Yousef, S.; Elmahdy, S.; Faizy, S.E.-D.; Amer, M.; Shams El-Din, H.; El-Ghamry, A.M.; Mousa, A.A.; Prokisch, J.; et al. Soil Fertility and Its Security. In *The Soils of Egypt*; El-Ramady, H., Alshaal, T., Bakr, N., Elbana, T., Mohamed, E., Belal, A.-A., Eds.; Springer International Publishing: Cham, Switzerland, 2019; pp. 137–157. [CrossRef]
71. Singh, A. Soil salinization and waterlogging: A threat to environment and agricultural sustainability. *Ecol. Indic.* **2015**, *57*, 128–130. [CrossRef]

water

MDPI

Article

The Use of Mixed Composed Amendments to Improve Soil Water Content and Peach Growth (*Prunus persica* (L.) Batsch) in a Mediterranean Environment

Onofrio Cappelluti [1,†], Maria Roberta Bruno [2,*,†], Anna Francesca Modugno [2], Rossana Monica Ferrara [2], Liliana Gaeta [2], Gabriele De Carolis [2] and Pasquale Campi [2,*]

1 Department of Soil, Plant and Food Sciences (Di.S.S.P.A.), University of Bari Aldo Moro, 70125 Bari, Italy; onofrio.cappelluti@uniba.it

2 Research Centre for Agriculture and Environment, CREA—Council for Agricultural Research and Economics, 70125 Bari, Italy; francesca.modugno@crea.gov.it (A.F.M.); rossana.ferrara@crea.gov.it (R.M.F.); liliana.gaeta@crea.gov.it (L.G.); gabriele.decarolis@crea.gov.it (G.D.C.)

* Correspondence: mariaroberta.bruno@crea.gov.it (M.R.B.); pasquale.campi@crea.gov.it (P.C.)

† These authors contributed equally to this work.

Abstract: Reduction of water availability imposes the agronomic issues of increasing the storage capacity of the soil and improving the use of rainwater or irrigation water. A field experiment in 2021 was conducted in a 5-year-old peach orchard in a Mediterranean environment to study the effect of mixed composed amendments (ACM), applied in different amounts, on the dynamics of soil water status. Water balance was monitored during the peach vegetative reproductive cycle on a daily scale. Three treatments of mixed composed amendments (ACM) were compared: A0, control; A1, with amendment (10 t ha^{-1}); and A2, with half dose of amendment (5 t ha^{-1}). On a seasonal scale, soil water content increased by 27% and 33% in A1 and A2 compared to A0, while relative extractable water varied between 0.41 (A0) and 0.65 (A1 and A2). Both soil water balance indicators show that storage capacity increases with the addition of amendment. Improved soil storage capacity was associated with higher values of stem water potential (throughout the growing season) and stomatal conductance (at the end of the season). Shoot and fruit growth observations were consistent with soil water content dynamics.

Keywords: relative extractable water; stem water potential; stomatal conductance; fruit volume; shoot growth

1. Introduction

The observed reduction of water availability [1] imposes water saving in every human activity and every production sector. Climate change has led to an increase in temperatures of +1.5 °C [2] and a decrease in total annual rainfall [3,4], with an increase, in general, in the intensity of rainfall events [5]. According to Trenberth [6], the increase in rainfall intensity is mainly caused by air humidity, which affects rain or snow rates, but not the total annual precipitation, at least locally. Very intense but short-lasting rainfall does not allow soils to store water [7], and 40–50% of rainfall [8] is lost through runoff [9]. In areas where rain is the primary source of water, a decrease in soil organic matter (SOM) can negatively impact the effective use of intense limited precipitation, due to a decrease in infiltration and hydraulic conductivity and an increase in runoff and erosion [10].

Extreme precipitation events and an increase in the drought period have occurred in the Mediterranean region in recent decades [11–13]. In the Mediterranean region, where rainfall is characterized by scarcity and extreme variability in space and time [14], daily storms of hundreds of millimeters are common [15,16]. Zittis et al. [17] reported that the frequency of extreme rainfall events in the region has increased since the 1980s.

Additionally, Tramblay et al. [18] found that drought duration in the Mediterranean has increased by up to 60% over the past century, with the most significant changes occurring in southern Europe and the Middle East. As reported by Rashid, A. and Ryan, J. [19], Mediterranean-type soils typically have a high pH and low organic matter content due to the presence of free $CaCO_3$ [20] and the effect of high air temperature; as a result, nutrients deficiencies are the primary limiting factor for crop production in these kinds of soils, followed by soil water stress. According to Umer et al. [21], calcareous soils dominate in arid and semi-arid regions due to low leaching; they contain excessive amounts of calcium carbonate ($CaCO_3$) that alter soil properties associated with plant growth, such as water-holding capacity and nutrient availability. Soil pH is an important factor that affects plant growth and development, particularly under drought conditions. It influences nutrient availability, soil structure, and microbial activity in the rhizosphere [22]. In alkaline soils, for example, nutrient uptake is hindered due to the formation of insoluble metal hydroxides [23]. Drought stress, on the other hand, can have significant impacts on the physical, chemical, and biological properties of soil. It can alter soil structure, reduce water infiltration, and increase soil compaction [24]. According to Fernàndez et al. [25], agricultural soils in many Mediterranean regions are often subject to severe degradation, which includes a decrease in soil organic matter and an increased risk of erosion and desertification. As SOM decreases, the fertility of the soil negatively impacts its physical, chemical, and biological properties. As a consequence, soil limits its capacity to store water in the soil profile, and agronomy prescribes measures to improve water storage in the soil [10,25]. Several studies [26–29] have shown how agronomic techniques such as tillage, the use of mulches, and crop residue management can improve soil water storage. Moreover, the addition of organic matter can improve soil structure, water infiltration, and soil porosity [30,31]. Increasing the amount of soil organic matter in the soil can be achieved using organic amendments; in particular, mixed amendment, made from diverse organic materials such as animal manure and plant residues, contributes to the enhancement of soil's physical, chemical, and biological properties [32], resulting in a crucial input for sustainable agricultural production, promoting long-term soil conservation and restoration [33]. Municipal waste has been used for many years as a soil conditioner for agricultural soils, is an economically attractive alternative to disposal by landfill and/or incineration, and it also constitutes an important organic mass for the formation of stable humus [34] and contributes to the improvement of soil fertility [35]. Several studies [36,37], however, have shown a possible negative impact from the use of organics, causing land and water pollution. As a function of this potential risk of pollution, in Italy, the ACM used in agriculture must comply with the legislation (Legislative Decree n. 75 of 29 April 2010) that regulates its sourcing, production, and application amounts (see Table 2). Indeed, Legislative Decree n. 75/2010 also considers the possible emissions of pollutants on a large scale, attending to the possible negative effects on biodiversity but allowing its use in biological agriculture. The employment of organic amendments could be a farm-scale solution to the problem of soil water storage capacity. This hydrological parameter indicates the capacity of the soil to accumulate water (rain or irrigation water) and then make water available for crops. On a laboratory scale, studies have shown that the capacity of soil to accumulate water is proportional to its organic matter [38,39] content. Moving from the laboratory to the field scale, rather than the intrinsic hydrophilic capacity [40], the benefit of applying organic matter to the soil depends on the ability of organic matter to structure the soil [41], form aggregates [42,43], and, consequently, increase soil porosity [44]. In addition to these effects, which occur as a result of repeated treatments over years, under actual growing conditions, there is a further immediate benefit resulting from the application of amendment to the soil, which reduces water loss through evaporation [45].

To monitor the effect of amendment on SOM and water retention, total organic carbon (TOC) can be taken into account [46–48]. Moreover, the effect of amendment supply on soil water status can be evaluated via a simple index of drought stress, the relative extractable soil water (REW), which describes the soil water reserve in terms of relative value, together with plant-based indices, such as stem water potential and stomatal conductance [49–51].

Then, the improvement in the nutrient conditions of the soil and the amount of water available due to the employment of organic amendment can enhance the growth of the plants. Even if several studies analyzed the use of soil amendment combined with chemical/organic fertilizer in peach orchards, focusing on its effects on (i) physical, chemical, and biological soil properties [52–54]; (ii) carbon dynamics [55]; and (iii) nitrous oxide emission [56], few studies had analyzed the use of soil amendment alone to promote peach tree growth [57].

The hypothesis posed in this study involves adding organic amendments to the soil. Two different quantities of mixed amendment—without adding chemical/organic fertilizers—were employed in a peach orchard located in a Mediterranean area to test (i) the increase of water storage and TOC along the soil profile; and (ii) the improvement of the peach orchard performance.

2. Materials and Methods

2.1. Experimental Site and Crop Management

The study was carried out during the 2021 growing season (May to September) in southern Italy (Rutigliano, lat: 40°59′, long: 17°02′) in an experimental farm of the Council for Agricultural Research and Economics (CREA). The experimental site is under the Mediterranean climate, characterized by warm and dry summers, with minimum and maximum annual air temperatures ranging from 0 to 5 °C and 32 to 43 °C [26], respectively. The annual rainfall is 560 mm [58]. Rains are distributed mainly in autumn and late winter, and they are negligible in the spring–summer period [59]. No significant difference was identified between experimental fields, and the average physicochemical characteristics of soil were reported in Table 1. Soil texture was classified as clay–loam [60].

Table 1. Physical–chemical properties of the soil collected at the experimental site.

Parameter	Average	±sd
Sand (g 100 g^{-1})	21	0.6
Silt (g 100 g^{-1})	37	2.9
Clay (g 100 g^{-1})	42	3.6
E.C. (dS m^{-1})	0.6	0.05
Field Capacity (m^3 m^{-3})	0.36	0.03
Wilting Point (m^3 m^{-3})	0.22	0.02
SOC (g kg^{-1})	14	1.1
Total N (g kg^{-1})	1.5	0.2
Available P (mg kg^{-1})	71	3.1
Exchangeable K (mg kg^{-1})	540	61

Observations were carried out on a 5-year-old peach orchard of late ripening cv. Redcall, grown in a traditional pot, and grafted onto rootstock GF677, spaced 5.0 m × 5.0 m, and managed according to standard agricultural practices. Soil water content in volume at the field capacity and the wilting point are 0.36 and 0.22 m^3 m^{-3}, respectively (measured in Richards chambers). The soil water reserve was low (70 mm) because the root system did not develop below 0.5 m in this site. At 0.5 m of depth, there is a parent rock that reduces the capacity of the root systems to expand beyond this layer. Water was provided by a drip irrigation system with two drippers per tree and a flow rate of 16 L h^{-1} per dripper. The scheduling irrigation was performed using the FAO56 approach reported by Allen et al. [61] The required meteorological data were measured by a standard meteorological station near the experimental field. The seasonal irrigation volume of 116 mm was supplied to restore 100% of the crop evapotranspiration. A total of around 20 irrigations were carried out—2 irrigations per week, with an average duration of 4 h per irrigation session. Other agricultural practices, such as weed and pest control, were executed according to the local farmers' best practices for production.

2.2. Experimental Design

The choice of the soil amendment to be used and the quantities to be administered was made according to the prescriptions of the Legislative Decree n. 75 of 29 April 2010 (Table 2).

Table 2. D.Lgs n. 75/2010 "Reorganization and revision of the regulations of fertilizers".

Type of Amendment	Component Preparation Method	Requirements and Minimum Titer in Useful Elements and/or Substances	Other Requirements and Useful Substances to Clarify	Notes
ACM	Product obtained through a controlled process of transformation and stabilization of organic waste which may consist of the organic fraction of municipal solid waste from separate collection from animal waste including livestock slurry, waste from agro-industrial activities and untreated wood and natural textile processing, sewage and sludge, as well as the matrices provided for green composted soil amendment.	Maximum moisture: 50% pH: 6.5 to 8.5 Organic C on dry matter minimum: 20% C humic and fulvic on dry matter minimum: 7%. Organic nitrogen on dry matter: >80% of total nitrogen C/N maximum: 25	Moisture pH Organic C on dry C humic and fulvic on dry Organic nitrogen on dry C/N Salt content	The following parameters of biological nature are also set: − Salmonella: absence in 25g is sample as is; − Escherichia coli in 1g of sample as is; Germination index (30% dilution) must be $\geq$ 60%; − Thallium: less than 2 mg kg^{-1} on dry weight (only for soil conditioners with algae). Maximum heavy metal contents (expressed as mg/kg dry matter): Cd 0.7; Cu 70; Ni 25; Pb 45; Zn 200; Hg 0.4; Cr (total) 70; Cr (VI) 0. Recommendations for the use of ACM in arboriculture is on average from maximum 25 t ha^{-1} to minimum 1.5 t ha^{-1}

Three different amounts of soil amendment (Table 3) (treatments) were applied (ACM, Fertileva srl, Evainfruit: Amendmented Mixed) at the beginning of the vegetative season (12 April 2021) along the rows: no ACM—control (A0); 10 t ha^{-1} of ACM (A1); and 5 t ha^{-1} of ACM (A2). Treatments were arranged under a randomized complete block design (RCBD) with three replicates. The ACM was spread and buried manually in the top 10 cm of soil.

Table 3. Determined values of ACM, Fertileva srl, Evainfruit: Amendmented Mixed.

Source-Determined Values: Product Complying with the D.Lgs n. 75/2010	
Moisture (%)	31.80
pH (unit)	7.66
Organic carbon [C] (% DM)	35.90
Humic and fulvic carbon (% DM)	12.40
Organic nitrogen [N] (% DM)	2.60
Carbon/Nitrogen ratio [C/N]	13
Copper [Cu] (mg/kg DM)	57.8
Zinc [Zn] (mg/kg DM)	142
Salt content (meq/100 g)	22.40

To evaluate the effect of the different amounts of ACM on the soil, as well as on the tree performances, within each treatment, three plants, similar in terms of dimensional vigor and health status, were chosen in correspondence with the soil moisture probes.

2.3. Soil Water Monitoring

2.3.1. Soil Water Content

Soil water content (SWC) was measured using capacitive probes (TEROS11, Decagon Devices Inc., Pullman, WA, USA), starting from 1 May 2021. The daily soil water content is determined by measuring the water content by volume using probes connected to a data-logger (TE-CR1000, Campbell, Kenton, NJ, USA), and data were transmitted to a web server via LAN or GSM mode. Data download is available through an online platform to which the data-logger is connected. Soil-specific calibration functions were used to calculate volumetric SWC according to Mastrorilli et al. [62]. Three plants were monitored for each treatment. For each plant, three capacitive probes were installed horizontally in the soil profile and transversely to the row at −0.1, −0.3, and −0.45 m from the soil surface to intercept the dynamics of the SWC below the drip lines. Soil water content was determined daily for the soil profile (0.5 m) by integrating the values measured at each depth [26].

2.3.2. Relative Extractable Soil Water (REW)

Relative extractable soil water (REW) describes the soil water reserve in terms of relative value [63]. REW describes the availability of soil water in the root zone thoroughly, as it is derived from data that are estimated through probes set up in the soil in the root zone. It is most often used as a simple index of drought stress, as the REW can be calculated from the soil water content in the root zone at a given time, as follows:

$$REW = SWC_{day} - SWC_{min}/SWC_{max} - SWC_{min}.$$

SWC_{day} is the daily soil water content ($m^3\ m^{-3}$), SWC_{min} ($m^3\ m^{-3}$) is the minimum water content detected, while SWC_{max} ($m^3\ m^{-3}$) is the maximum water content consumed by plants throughout the root zone during the irrigation season, or the water field capacity. The REW ranges from 1.0 (maximum soil water content) to 0 (minimum soil water content). Daily REW values for the experiments were calculated from daily SWC measurements.

Because a critical, site-specific value of matrix potential was not available for assessing soil water deficits, it was assumed that water supply stress occurs when REW falls below the threshold of 0.4 (REW_c), triggering stomatal regulation [63–65]. The REW threshold < 0.4 is commonly used in various ecosystems [66]. In addition, the duration of water supply stress was calculated as the percentage of days in the growing season with a REW less than 0.4.

2.4. Soil Laboratory Measurements

Undisturbed soil cores were collected in 2021 during the entire vegetative period and the plant's vegetative rest period—4 April, 31 May, 5 July, 13 September, and 8 November—within each experimental treatment in triplicate, and the average value was reported. Soil samples were collected at 0–0.10 m depth. Total organic carbon (TOC) and physical indicators were measured on the soil samples.

TOC was quantified on dried and 2-mm sieved samples, following protocols reported in Ferrara et al. [67,68]. In detail, for TOC quantification, soil samples were ground to a fine powder (0.5 mm) using an agate ball mill. TOC was determined by the TOC Vario Select analyzer (Elementar, Hanau, Germany) [69], which performs catalytic oxidation of the specimen at high temperatures in the presence of air.

Soil physical indicators resulting from bulk density were not statistically different among the treatments.

2.5. Plant Water Monitoring

2.5.1. Stem Water Potentials and Stomatal Conductance

The stem water potential is the ecophysiological parameter that is directly related to the soil water status. Plant water status was characterized for each treatment by stem water potential (Ψ_{st}, MPa) and stomal conductance (gs, mmol $m^{-2}\ s^{-1}$), measured at midday since stem water potential and stomatal conductance are more closely correlated with leaf water status at midday [49,50].

Stem water potential was measured using Scholander-type pressure chamber (Soil Moisture Equipment Corp., Santa Barbara, CA, USA) at 12:00 p.m., once or twice a month on six plants per treatment, selecting two leaves per plant [70].

According to Gaeta et al. [51], in a late ripening peach cultivar such as Redcal, only stem water potential is not completely informative of plant water status because of its conservative or iso-hydric behaviour. The stomatal conductance was considered as another plant-based index and was measured using an open-circuit infrared gas analyzer with an LED light source (Li-COR6400XT, LI-COR, Lincoln, NE, USA). For each treatment, well-exposed leaves were selected in three replicates to the east and west sides of the canopy. Light intensity was held constant throughout the three treatments by adjusting the light source LED to the natural irradiance experienced by the leaf immediately before measurement. The values observed on the west and east sides of the canopies were averaged for each plant.

2.5.2. Fruit and Shoots Growth

The fruit growth trend was monitored during the season through a digital gauge implemented with a datalogger capable of memorizing and conserving data (HK-Horticultural Knowledge s.r.l. Bologna, Italy) for twelve fruits per treatment in the triple replication. Fruit volume (V, cm^3) and absolute growth rate (AGR, cm^3 day^{-1}) were calculated by considering the form of the peach as a spheroid and by measuring the three axes of each peach [26]. Absolute growth rate (AGR, cm^3 day^{-1}) was calculated using the following formula:

$$AGR = \frac{V1 - V0}{t1 - t0}$$

where $V1$ and $V0$ are volumes measured at time $t1$ and $t0$, respectively.

The mean shoot length was assessed using a meter. For each treatment, two trees were considered on which four shoots were measured along the four cardinal points (N, S, W, E).

2.6. Statistical Analyses

The data were analyzed via a one-way ANOVA (A0, A1, and A2) per season (2021). The differences in each treatment were assessed using Tukey's honestly significant difference (HSD) test.

To verify the correlations among stem water potential and relative extractable water, stem water potential and stomal conductance, and relative extractable water and stomal conductance, the Pearson correlation coefficient was determined. The confidence limits used in this study were based on 95% ($p < 0.05$).

The statistical analyses were computed using the statistical software R (R Development Core Team, http://www.r-project.org. accessed 10 March 2023).

3. Results and Discussion

This study, carried out here in a peach cropping system, is a rare example [52] of a field evaluation of the benefits of amendment application to soil water status. The study methodology included three agronomic criteria for evaluating the effects of amendment supply to the soil: soil water status, total organic carbon, vegetation water status and growth analysis of vegetation.

3.1. Weather Conditions

Figure 1a shows the evolution of daily air temperature values (°C) (average, minimum, and maximum) and daily precipitation (mm d^{-1}) during the observation period from bud opening (1 May 2021) to the end of the productive season (30 September 2021).

The average air temperature during the observation period was 23 °C. The minimum temperature fell below 10 °C only five times: at the beginning of the growing season (9 May 2021), during the first stage of the second phase of fruit growing (3 June 2021), and at the end of the vegetative cycle (24 September 2021), which did not affect peach tree productivity. The maximum temperature reached values between 35 and 40 °C several times during the growing season between 20th of June and 20th of August.

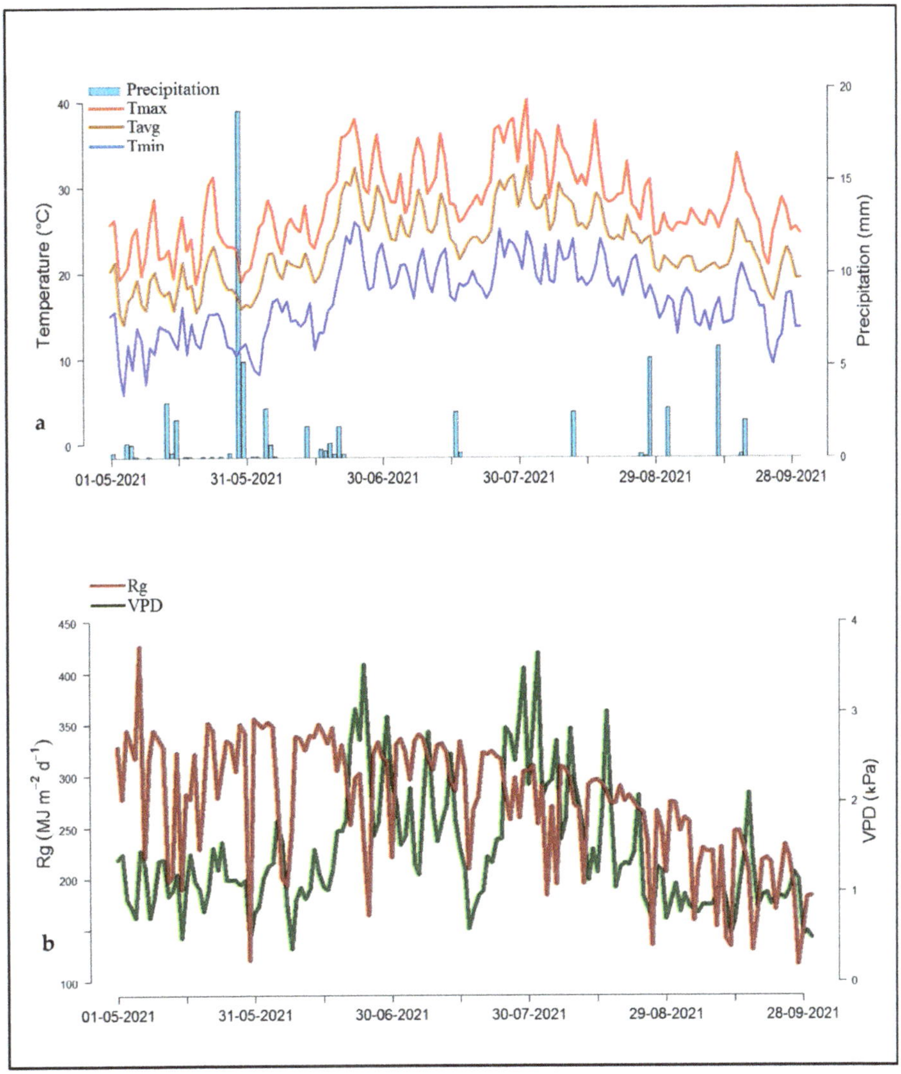

Figure 1. (**a**) Daily air temperature values (average, minimum, and maximum) and total precipitation; (**b**) daily values of global radiation (Rg) and vapor pressure deficit (VPD).

The five drops below 10 °C, as mentioned above, did not affect peach tree productivity as they did not result in any significant stress or frost damage [71].

Rainfall recorded during the observation period amounted to 124 mm. The daily values of global radiation (Rg) and vapor pressure deficit (VPD) are shown in Figure 1b. Radiation follows the circadian pattern and decreases in daily values from June to September, with only a few days of cloud cover during the first 20 days of June. The daily mean values of vapor pressure deficit (VPD) during the peach crop cycle ranged from 1 to 1.5 kPa, with higher values occurring regularly between mid-June and the end of August. VPD is a measure of the evaporative demand of the atmosphere and is related to the plant's ability to transpire water. As VPD increases, it is likely that the plant is more sensitive to water stress, especially during the most important stages of fruiting. Previous studies have shown that high VPD values can negatively impact peach tree growth and fruit quality [72].

Therefore, the observed higher VPD values in this study could have affected the plant's water use and productivity. Studies conducted on peach trees in similar Mediterranean environments have shown [73–75] that peach trees have good resistance to irrigation deficit conditions. This characteristic can be advantageous for saving high-to-moderate irrigation volumes without compromising soil quality and peach orchard performance. According to a study by Rolbiecki et al. [76], it is estimated that due to climate change, there will be an increase in the water requirements by peach trees of about 26%. The irrigation volumes supplied in the peach orchard of this study were able to restore the evapotranspiration of the crop, so the plant was able to avoid water stress, even under high VPD conditions.

3.1.1. Soil Water Content

Figure 2a shows the daily soil water content in the A0, A1, and A2 treatments. A clear difference in SWC between A0 (0.28 m^3 m^{-3}) and the two conditioned soils, A1 (0.30 m^3 m^{-3}) and A2 (0.30 m^3 m^{-3}), at the start of monitoring season (May) is due to the time lapse between the soil amendment spreading (12 April 2021) and the beginning of the monitoring period (1 May 2021). SWC values of the three treatments ranged generally between the wilting point (before irrigation) and field capacity (after irrigation). Irrigation scheduling prevented the soil from exceeding field capacity and never allowed the soil to reach the wilting point, despite a weather pattern leading to high levels of evapotranspiration. The higher the evapotranspiration rate of the atmosphere, the earlier the soil moisture approaches the wilting point. Values of soil water content close to the wilting point were observed only in the case of the A0 treatment (without soil amendment). The value closest to the wilting point was approximately 0.25 m^3 m^{-3} in the A0 treatment at the end of the production cycle, after irrigation was stopped. The behavior of the two treatments that received the amendment (regardless of the amendment amount) differs from the treatment without soil amendment. The seasonal values of soil water content for A1 and A2 are systematically +13.8% and +11.4% higher than in A0, respectively.

Figure 2b summarizes, on a seasonal scale, the mean soil water content for the three treatments, with significant differences between soil with amendment (A1 and A2) and the control (A0). Figure 2b further shows that the variability of soil moisture data during the peach tree growing season is significantly higher in the treatment without soil amendments than in the two treatments that received soil amendments.

It is observed that adding amendment to the soil not only raises the measured soil moisture values but also reduces the fluctuations around the seasonal mean value. This means that where amendment has not been added to the soil, the crop is exposed to potentially dangerous fluctuations in soil moisture during the growing cycle. These moisture variations are reflected in the plant performance, which appears to be more exposed to the risks of water stress.

3.1.2. Relative Extractable Water

Considering the threshold of drought stress index is 0.4 [77,78], the seasonal REW values for A0 were close to the threshold (Figure 3a), which could indicate a risk of soil water stress. The two treatments that have benefited from the ACM (either complete dosage, A1; or half dosage, A2) had REW values far from the critical stress threshold during the whole peach tree growing season (see Figure 3a). The statistical analysis (Tukey's HSD) showed that REW for A1 (0.66) and A2 (0.61) were significantly different from A0 (0.50). At the seasonal level, the mean REW values showed that irrigation planned to avoid any water stress in the soil (irrigation performed by returning 100% of the ET0) was effective and that the use of soil conditioner improved the soil water condition in direct proportion to the quantity.

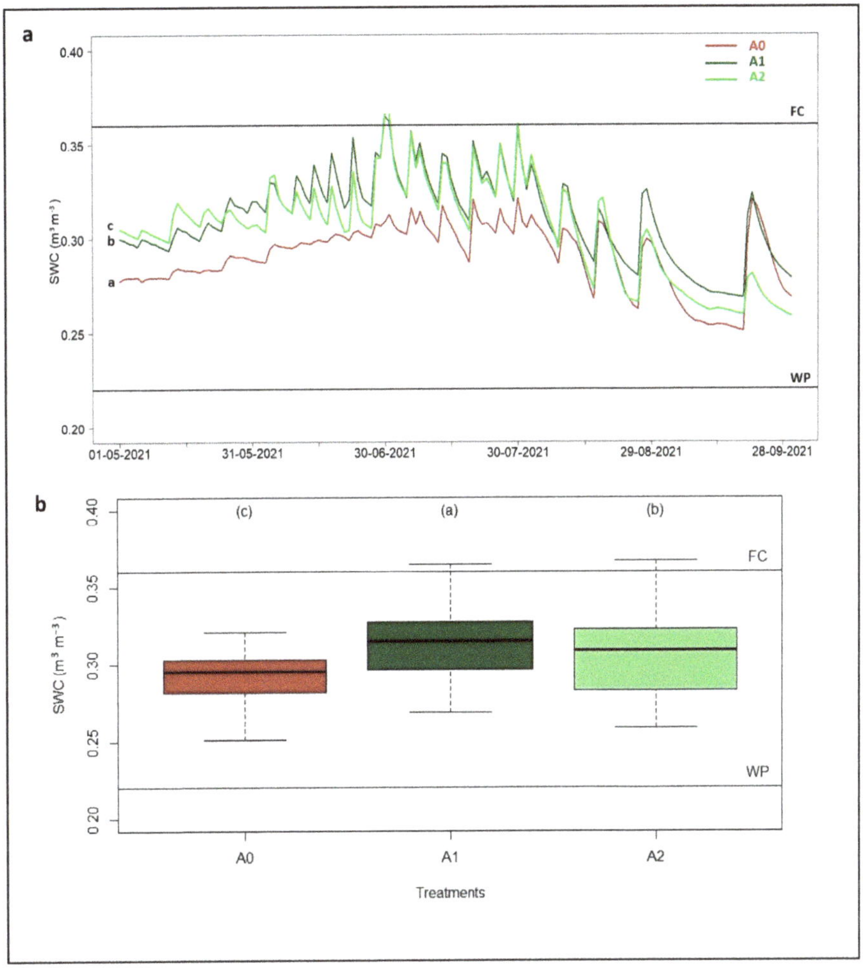

Figure 2. (**a**) Soil moisture values in the three treatments; (**b**) soil water content averaged on the peach tree growing season. Different letters indicate a significant difference (p-value < 0.05). A0 = control; A1 = complete dosage of soil amendment; A2 = half dosage of soil amendment; WP = wilting point, FC = field capacity.

The analysis of the REW values on a monthly scale (Figure 3b) reveals that the peach tree stand suffers soil water stress (particularly when the regular water supply is interrupted, e.g., in September) if the crop does not benefit from the ACM. The risk of soil stress does not occur when the soil receives a complete dose of soil amendment (A1) and only rises at the end of the cycle, i.e., in September (Figure 3b), in the case of a reduced supply of soil amendment (A2). The monthly analysis of the REW values indicates that soil water stress occurs in September in the treatments that did not receive amendment and in the treatment that had half a dose of amendment.

The average maximum and minimum values of REW were 0.72 and 0.22 for A0, 1.00 and 0.35 for A1, and 1.00 and 0.28 for A2, respectively. In particular, REW for A0 goes below the 0.4 threshold on 20% of the days in August and 68% of the days in September (Figure 3c), according to [47]. The A2 treatment experiences water stress on 16% of the days in August and 67% in September (Figure 3c); and soil water stress did not occur on any day except for in September (35% of the days in the month) in treatment A1. It was possible to better understand the contribution of soil amendment treatments, compared to A0: a full-dose organic matter supply, as in A1, guarantees better soil water retention such that no water stress is generated on any (or almost any) of the days of the season.

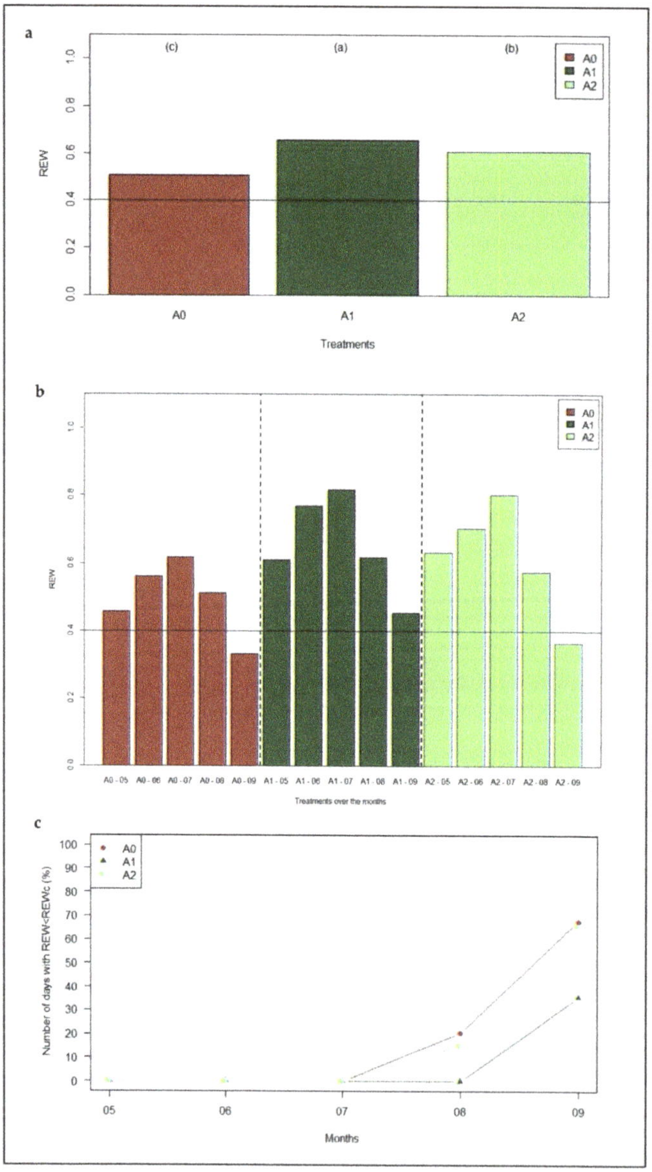

Figure 3. (**a**) Seasonal relative extractable water (REW) during the peach tree growing season. Different letters indicate a significant difference (p-value < 0.05); (**b**) REW values at monthly scale; (**c**) number of days with REW < critical REW value (REWc = 0.4). A0 = control; A1 = complete dosage of soil amendment; A2 = half dosage of soil amendment.

3.2. *Total Organic Carbon*

The TOC measurements in Figure 4 show how the total amount of organic carbon in the soil varies during the 2021 growing season. At the beginning of the season on 12 April, when the amendment had not been applied, all the samples measured the same amount of TOC in the soil. After amendment supply, the TOC levels in the soil of the three treatments changed considerably, especially in A1, while remaining relatively constant in the control (A0). The trend of TOC in A2, although higher than in A0 throughout the season, always remained lower than the treatment with A1. Note that on the 11th of November, all three treatments showed the same amount of TOC in the soil. The TOC variations measured in different periods, even beyond the growing season, have shown how the amendment increases the total amount of carbon in the soil relative to the greater amount of water in the soil [47]. The peak recorded in A1 on 31 May, as reported by Batiot et al. [79], is probably due to the high amount of rainfall [80], which caused a higher TOC concentration to be recorded. However, the TOC value for A1 showed a downward trend, probably due to lower soil moisture availability [81]. Furthermore, it can be seen that in the last measurement taken in November, the TOC in the three treatments is almost identical, probably also due to the low temperatures, as reported in a study by Lepistö et al. [82].

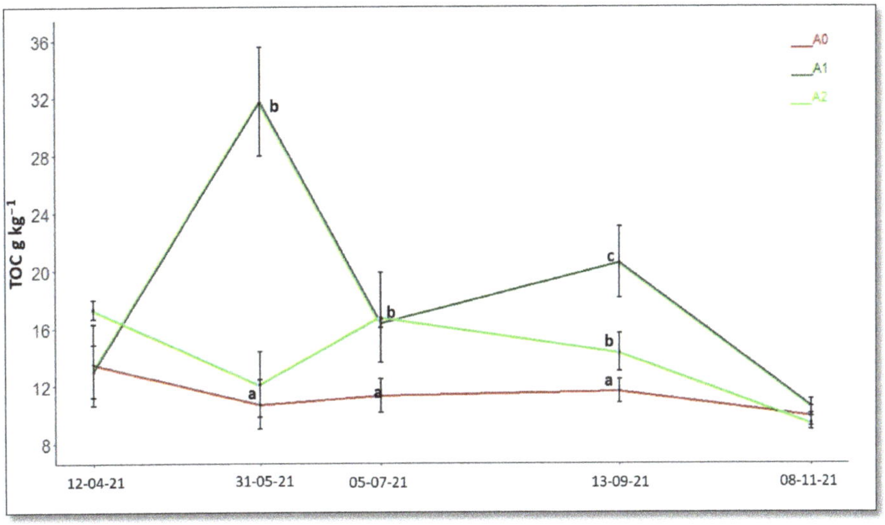

Figure 4. Total organic carbon (TOC) during the investigated season. Different letters indicate a significant difference (p-value < 0.05). A0 = control; A1 = complete dosage of soil amendment; A2 = half dosage of soil amendment.

3.3. *Plant Water Monitoring*

3.3.1. Stem Water Potentials and Stomatal Conductance

Figure 5a shows the evolution of the stem water potential during the peach growing season. The trends of SWC described in Section 3.1.1 are in agreement with the patterns of stem water potential. During the whole crop cycle, the highest stem potential values were observed in treatment A1, where the soil amendment was supplied in a complete dose. The lowest potential values were measured in the treatment without soil amendment.

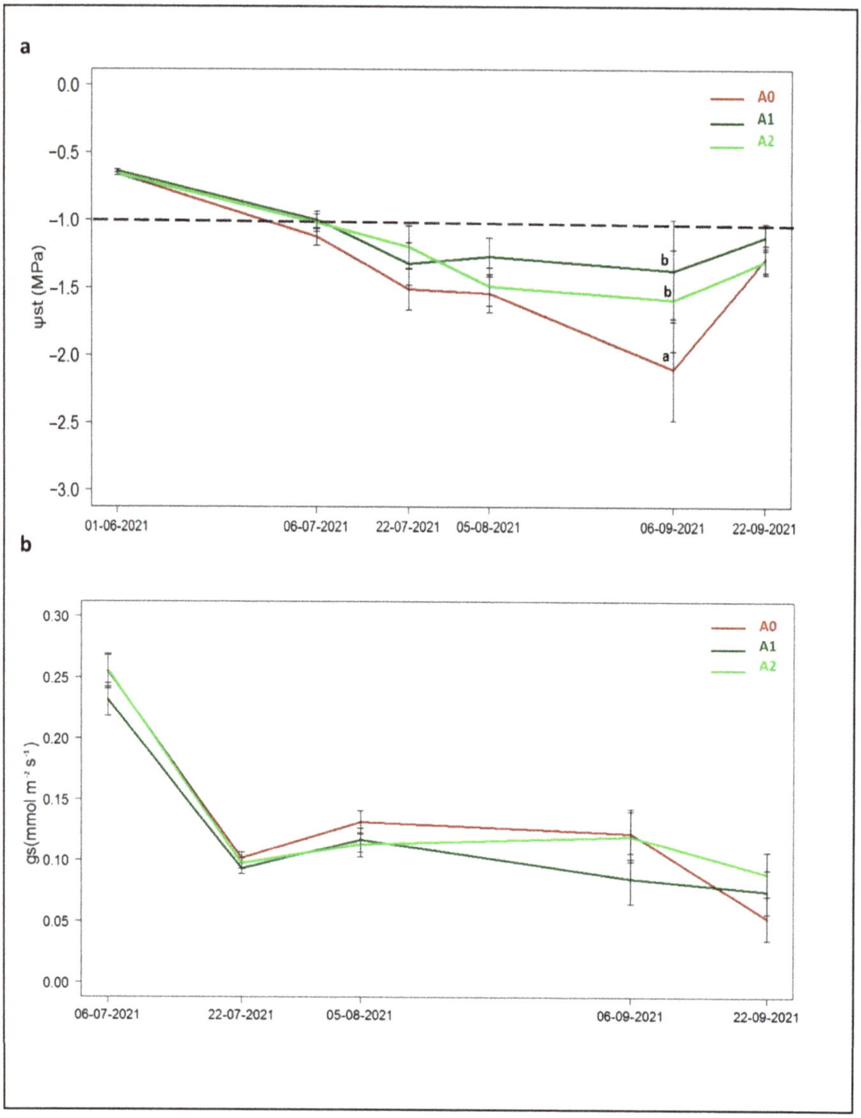

Figure 5. (**a**) Stem midday water potential (Ψ_{st}): different letters indicate a significant difference (p-value < 0.05); (**b**) stomal conductance (g_s) during the investigated season. There was no significant difference between the treatments (p-value < 0.05). A0 = control; A1 = complete dosage of soil amendment; A2 = half dosage of soil amendment.

The differences in stem water potential values between the treatments were not significant (Figure 5a) along the season, except on 9 September, when Ψ_{st} was equal to −1.37, −1.59, and −2.10 MPa for A1, A2, and A0, respectively. According to Rahmati et al. [78], a value of Ψ_{st} equal to −1.5 MPa could be considered as the threshold for peach water stress. According to these results, risks of water stress should arise at the beginning of September in treatment A0. The pressure chamber technique [83,84] measures leaf potential, i.e., expresses the force with which water is retained by leaves. This measurement makes it possible to assess the water status of the plant, to identify when the plant enters a water stress condition [85].

The stem water potential values confirm what was also observed for REW. Only in September were the stem water potential values measured in the treatment without modification statistically lower than those measured in the two treatments with modification.

The conductance values over time, shown in Figure 5b, did not statistically differ among the three treatments. The highest conductance values were measured at the beginning of the cycle from the first fully developed leaves.

The irrigation schedule set out by the experiment protocol ensured the stomatal opening and, as a consequence, the gas exchanges during the whole vegetative period of the peach tree. In our study, data on stem water potential and stomatal conductance (Figure 5a,b) showed no significant differences between treatments except at certain times; this is because stomata opening is not only determined by stem water potential, but also by PAR levels, evapotranspiration demand, and CO_2 concentration within the sub-stomatal chambers [86]. Stomatal behavior is also influenced by agronomic treatments [87], but these are seldom revealed in field trials [88].

3.3.2. Sensitivity of Plant Water Status Indicators

Figure 6a shows the relationship between REW and Ψ_{st} when the values of the different treatments during the season are combined. A good exponential increase of Ψ_{st} was obtained as REW increases, with maximum levels of REW at values above −1.00 MPa (r2 = +0.47). A similar exponential relationship was observed between REW and g_s, with the latter reaching a plateau of around 1.3 mmol m^{-2} s^{-1} (Figure 6b; r2 = +0.49). These two relationships seem to be in agreement, as reported by Alcaras et al. [89]. The correlation between stem water potential and REW was significant (Figure 6a). Regardless of the experimental treatment, the relationship confirms that in peach trees, the stem water potential follows an exponential function of the relative extractable water (REW) [89]. Since the scheduling irrigation was carried out in full irrigation conditions (100% ET0), the stomatal conductance seems to show no significant differences among the treatments studied, probably because the amendment resulted in an improved situation compared to the control treatment, which well exceeded the stress threshold of −1.50 MPa only in September. In a previous study [51], in moderate and severe water stress conditions (about 50% of the full irrigation), the xylematic potential results were not completely informative regarding plant water stress in late-ripening peach cultivars, and therefore should be used with caution as a plant water indicator; instead, the stomatal conductance could be a useful index.

In Figure 6c, the correlation between Ψ_{st} and g_s is reported and compared to previous figures, a lower correlation is noted (R^2 = +0.38). In addition, the gs measurements fall for all treatments in a range of 0.07 (mmol m^{-2} s^{-1}) to 0.12 (mmol m^{-2} s^{-1}), with Ψ_{st} from −1.5 (MPa) to −1.1 (MPa). It should be added that the measurements carried out on 1 June in all three treatments show a higher stomatal conductance compared to other days, which is linked to a higher value of stem water potential equal to about −0.66, −0.64, and −0.66, respectively, in A0, A1, and A2. These differences were found between the different dates between Ψ_{st}, and g_s, according to Ahumada et al. [90], can be influenced by agronomic factors and climatic conditions.

Figure 6b,c shows the relationship between the measurement of stomatal conductance with REW and stem water potential. The relationship confirms that irrespective of the experimental treatment, conductance in peach trees follows the REW, and the stem water potential, according to an exponential function and a quadratic function, is poor but significant ($p > 0.05$). Stomatal conductance is a direct function of the stem water potential [66] and is indirectly related to the soil water status.

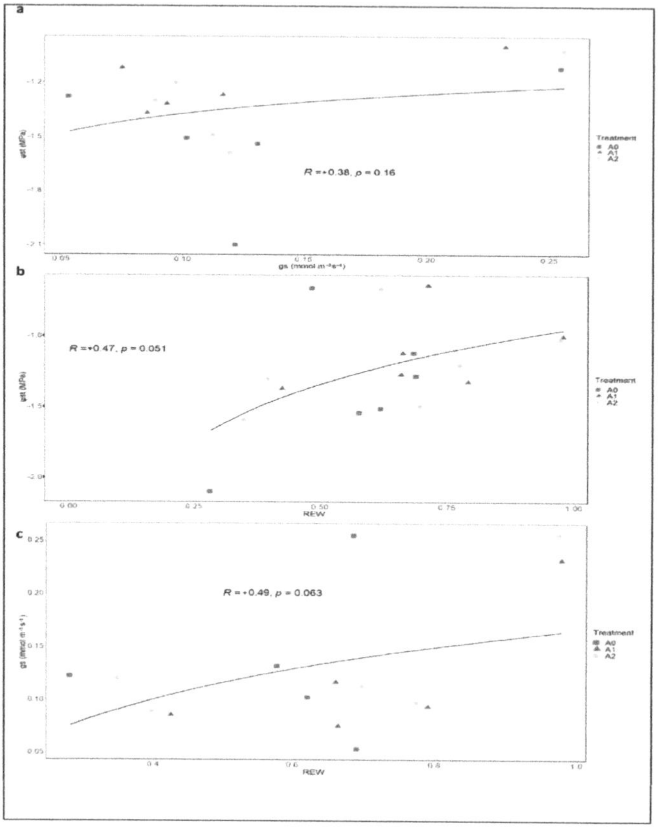

Figure 6. (**a**) Relationship between stem water potential and REW; (**b**) relationship between stomatal conductance (gs) and stem water potential (Ψst); (**c**) relationship between REW and stomatal conductance. A0 = control; A1 = complete dosage of soil amendment; A2 = half dosage of soil amendment.

3.4. Fruit and Shoot Growth

The *AGR* values for fruit (Figure 7a) show an increasing trend in July until reaching a plateau and then decreasing at the beginning of August. The season's fruit growth rate in A0 was lower than A1 and A2, except on 5 August, when the shoots reached their maximum length (Figure 7b). Shoots grow quickly from May to 5 August, showing differences between the treatments (Figure 7b). The fruits' volume shows a slow increase at the beginning, when the shoots are very active, then it increases considerably until the harvest, when there is no more competition with shoots (Figure 7c). The presence of soil amendment influences the three measured morphological parameters: *AGR*, shoot length, and fruit volume. Without soil amendment, growth rates are lower. Although the effect of the two soil amendment amounts is not significant for fruit growth, treatments A1 and A2 showed higher growth rates than A0 before harvest. In our study, we also noticed how the fruit growth rate and shoot length are influenced by the application of the amendment [91]. As reported by Nair and Ngouajio [92], the fruit growth measured during the season seems, in the smallest part, to be influenced by the application of the amendment because the growth of fruit is also influenced by different climate factors. The results discussed so far show that the addition of amendment does indeed affect the amount of water in the soil. This improvement in soil water content is also observed at the plant level when analyzing the behavior of stem water potential over time (and less clearly with stomatal conductance). The stem water potential indicates the improved hydration status of the plant tissue when adding amendment to the soil. Consequently, the analysis of fruit and shoot growth indicates greater growth in treatments with greater soil water availability, i.e., where amendment has been added to the soil.

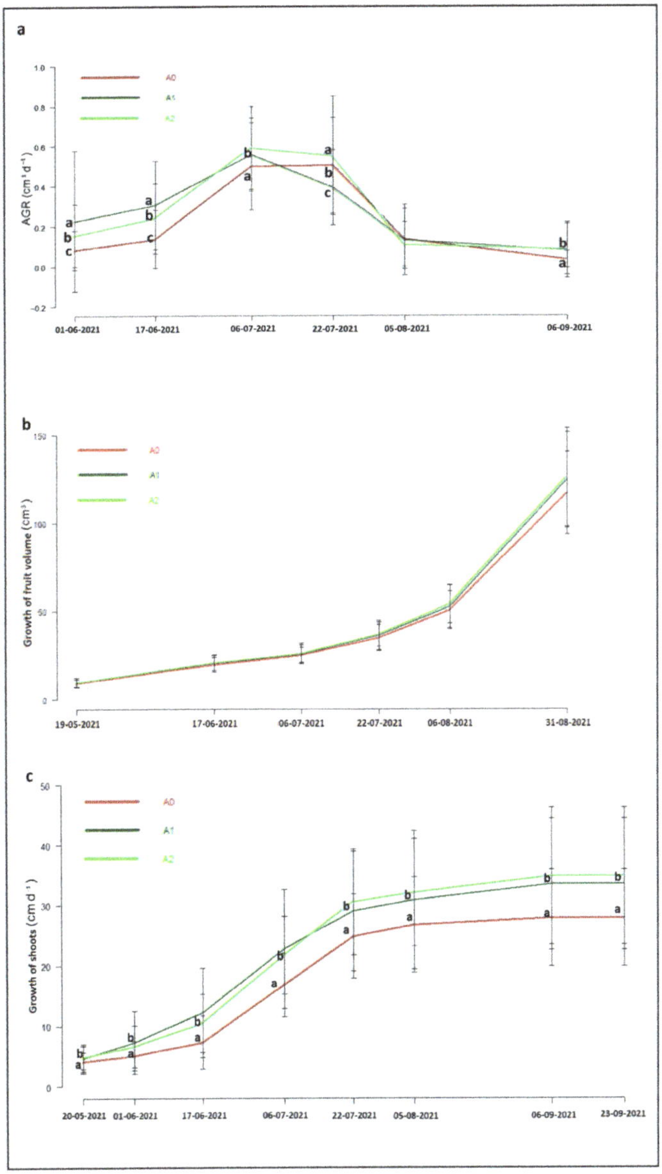

Figure 7. (**a**) Absolute growth rate (*AGR*) of fruits (cm^3 d^{-1}); (**b**) growth of shoots (cm d^{-1}); (**c**) growth of fruit volume (cm^3). Different letters indicate a significant difference (p-value < 0.05). A0 = control; A1 = complete dosage of soil amendment; A2 = half dosage of soil amendment.

4. Conclusions

This study shows that the addition of ACM to the soil at the beginning of the irrigation season increases the daily soil water content with the use of both complete (A1) and half (A2) dosages, with a slightly better performance for A1 in terms of soil water content. These results were better clarified by the use of the REW water stress index. The increase in soil water content led to an increase in the values of Ψs (stem water potential) above or near the water stress threshold (−1.5 MPa). Moreover, an improvement in the total organic carbon in the soil with the amendment supply was measured. The correlation between stem water potential and REW was significant. In late-ripening peach cultivars with conservative behavior, it is important to consider the right index when detecting the plant water status. In fact, in conditions of slight water stress, the stem water potential has confirmed reliability as a plant-based index. However, it is advisable to combine it with stomatal conductance in conditions of moderate and severe water stress. Further analyses are necessary to investigate the relationship between stomatal conductance and soil water storage. Improvements in soil water content also influence the plant with respect to increased fruit and shoot growth. Considering that the dose of the amendment did not affect the variability in soil and water parameters and plant performance in general, it would be desirable to use the halved dose to reduce management costs. Additional studies should deepen the soil amendment–soil–plant relationship, following annual applications, to highlight the medium-to-long-term effects of the amendment on soil water storage and the improved crop production. Providing sustainable methods by which to retain as much water as possible within the soil, while limiting its evaporation as much as possible, will be essential.

Author Contributions: Conceptualization, R.M.F. and P.C.; methodology, P.C.; software, O.C. and M.R.B.; validation, L.G. and A.F.M.; formal analysis, R.M.F. and P.C.; investigation, A.F.M.; resources, G.D.C.; data curation, O.C. and M.R.B.; writing—original draft preparation, O.C., M.R.B., and G.D.C.; writing—review and editing, R.M.F., P.C., and L.G.; visualization, L.G. and A.F.M.; supervision, R.M.F. and P.C.; project administration, P.C.; funding acquisition, P.C. All authors have read and agreed to the published version of the manuscript.

Funding: This study was supported by the project: "Water4AgriFood, Miglioramento delle produzioni agroalimentari mediterranee in condizioni di carenza di risorse idriche", PNR 2015–2020", Area Agrifood, funded by MIUR, PON ARS01_00825 "Ricerca e Innovazione" 2014–2020.

Data Availability Statement: Data available on request due to restrictions privacy. The data presented in this study are available on request from the corresponding author.

Acknowledgments: Special thanks to M. Mastrorilli for his contribution to this work and for inspiring us with his passion for scientific research.

Conflicts of Interest: The authors declare that there are no known competing financial interests or personal relationships that could have appeared to influence the work reported in this paper.

References

1. Bleu, U. UNEP/MAP-Plan Bleu: State of the Environment and Development in the Mediterranean. *Athens* **2009**. Available online: https://mcc.jrc.ec.europa.eu/documents/201607121602.pdf (accessed on 24 April 2023).
2. Change, Intergovernmental Panel on Climate. Ipcc. *Clim. Change* **2014**. Available online: https://www.ipcc.ch/site/assets/uploads/2018/02/SYR_AR5_FINAL_full.pdf (accessed on 24 April 2023).
3. Madsen, H.; Lawrence, D.; Lang, M.; Martinkova, M.; Kjeldsen, T. Review of trend analysis and climate change projections of extreme precipitation and floods in Europe. *J. Hydrol.* **2014**, *519*, 3634–3650. [CrossRef]
4. Fiori, E.; Comellas, A.; Molini, L.; Rebora, N.; Siccardi, F.; Gochis, D.; Tanelli, S.; Parodi, A. Analysis and hindcast simulations of an extreme rainfall event in the Mediterranean area: The Genoa 2011 case. *Atmos. Res.* **2014**, *138*, 13–29. [CrossRef]
5. Fahad, S.; Hussain, S.; Saud, S.; Tanveer, M.; Bajwa, A.A.; Hassan, S.; Shah, A.N.; Ullah, A.; Wu, C.; Khan, F.A. A biochar application protects rice pollen from high-temperature stress. *Plant Physiol. Biochem.* **2015**, *96*, 281–287. [CrossRef]
6. Trenberth, K.E. Atmospheric moisture residence times and cycling: Implications for rainfall rates and climate change. *Clim. Chang.* **1998**, *39*, 667–694. [CrossRef]

7. Findell, K.L.; Eltahir, E.A. An analysis of the soil moisture-rainfall feedback, based on direct observations from Illinois. *Water Resour. Res.* **1997**, *33*, 725–735. [CrossRef]
8. Wallace, J. Increasing agricultural water use efficiency to meet future food production. *Agric. Ecosyst. Environ.* **2000**, *82*, 105–119. [CrossRef]
9. Wang, S.; Wang, H.; Hafeez, M.B.; Zhang, Q.; Yu, Q.; Wang, R.; Wang, X.; Li, J. No-tillage and subsoiling increased maize yields and soil water storage under varied rainfall distribution: A 9-year site-specific study in a semi-arid environment. *Field Crops Res.* **2020**, *255*, 107867. [CrossRef]
10. Masri, Z.; Ryan, J. Soil organic matter and related physical properties in a Mediterranean wheat-based rotation trial. *Soil Tillage Res.* **2006**, *87*, 146–154. [CrossRef]
11. Stocker, T.F. *Climate Change 2013: The Physical Science Basis: Summary for Policymakers, a Report of Working Group I of the IPCC, Technical Summary, a Report Accepted by Working Group I of the IPCC But Not Approved in Detail and Frequently Asked Questions: Part of the Working Group I Contribution to the Fifth Assessment Report of the Intergovernmental Panel on Climate Change;* Intergovernmental Panel on Climate Change: Geneva, Switzerland, 2013.
12. González-Hidalgo, J.C.; Peña-Monné, J.L.; de Luis, M. A review of daily soil erosion in Western Mediterranean areas. *Catena* **2007**, *71*, 193–199. [CrossRef]
13. Philandras, C.; Nastos, P.; Kapsomenakis, J.; Douvis, K.; Tselioudis, G.; Zerefos, C. Long term precipitation trends and variability within the Mediterranean region. *Nat. Hazards Earth Syst. Sci.* **2011**, *11*, 3235–3250. [CrossRef]
14. Sumner, G.; Romero, R.; Homar, V.; Ramis, C.; Alonso, S.; Zorita, E. An estimate of the effects of climate change on the rainfall of Mediterranean Spain by the late twenty first century. *Clim. Dyn.* **2003**, *20*, 789–805. [CrossRef]
15. Bermúdez, F.L.; Díaz, M.A.R. Génesis y consecuencias erosivas de las lluvias de alta intensidad en la región mediterránea. *Cuad. Investig. Geográfica/Geogr. Res. Lett.* **1992**, *18*, 7–28. [CrossRef]
16. Poesen, J.W.; Hooke, J.M. Erosion, flooding and channel management in Mediterranean environments of southern Europe. *Prog. Phys. Geogr.* **1997**, *21*, 157–199. [CrossRef]
17. Zittis, G.; Bruggeman, A.; Lelieveld, J. Revisiting future extreme precipitation trends in the Mediterranean. *Weather Clim. Extrem.* **2021**, *34*, 100380. [CrossRef]
18. Tramblay, Y.; Koutroulis, A.; Samaniego, L.; Vicente-Serrano, S.M.; Volaire, F.; Boone, A.; Le Page, M.; Llasat, M.C.; Albergel, C.; Burak, S. Challenges for drought assessment in the Mediterranean region under future climate scenarios. *Earth-Sci. Rev.* **2020**, *210*, 103348. [CrossRef]
19. Rashid, A.; Ryan, J. Micronutrient constraints to crop production in soils with Mediterranean-type characteristics: A review. *J. Plant Nutr.* **2004**, *27*, 959–975. [CrossRef]
20. Brownrigg, S.; McLaughlin, M.J.; McBeath, T.; Vadakattu, G. Effect of acidifying amendments on P availability in calcareous soils. *Nutr. Cycl. Agroecosyst.* **2022**, *124*, 247–262. [CrossRef]
21. Umer, M.I.; Rajab, S.M.; Ismail, H.K. Effect of CaCO3 form on soil inherent quality properties of calcareous soils. In Proceedings of the Materials Science Forum; Trans Tech Publications Ltd.: Stafa-Zurich, Switzerland, 2020; pp. 459–467.
22. Mosley, L.; Biswas, T.; Cook, F.; Marschner, P.; Palmer, D.; Shand, P.; Yuan, C.; Fitzpatrick, R. Prolonged recovery of acid sulfate soils with sulfuric materials following severe drought: Causes and implications. *Geoderma* **2017**, *308*, 312–320. [CrossRef]
23. Bar-Ness, E.; Hadar, Y.; Chen, Y.; Romheld, V.; Marschner, H. Short-term effects of rhizosphere microorganisms on Fe uptake from microbial siderophores by maize and oat. *Plant Physiol.* **1992**, *100*, 451–456. [CrossRef] [PubMed]
24. Geng, S.; Yan, D.; Zhang, T.; Weng, B.; Zhang, Z.; Qin, T. Effects of drought stress on agriculture soil. *Nat. Hazards* **2015**, *75*, 1997–2011. [CrossRef]
25. Fernández, J.M.; Plaza, C.; García-Gil, J.C.; Polo, A. Biochemical properties and barley yield in a semiarid Mediterranean soil amended with two kinds of sewage sludge. *Appl. Soil Ecol.* **2009**, *42*, 18–24. [CrossRef]
26. Campi, P.; Gaeta, L.; Mastrorilli, M.; Losciale, P. Innovative soil management and micro-climate modulation for saving water in peach orchards. *Front. Plant Sci.* **2020**, *11*, 1052. [CrossRef]
27. Guzha, A. Effects of tillage on soil microrelief, surface depression storage and soil water storage. *Soil Tillage Res.* **2004**, *76*, 105–114. [CrossRef]
28. Aboudrare, A.; Debaeke, P.; Bouaziz, A.; Chekli, H. Effects of soil tillage and fallow management on soil water storage and sunflower production in a semi-arid Mediterranean climate. *Agric. Water Manag.* **2006**, *83*, 183–196. [CrossRef]
29. Zhang, S.; Li, P.; Yang, X.; Wang, Z.; Chen, X. Effects of tillage and plastic mulch on soil water, growth and yield of spring-sown maize. *Soil Tillage Res.* **2011**, *112*, 92–97. [CrossRef]
30. Rawls, W.; Nemes, A.; Pachepsky, Y. Effect of soil organic carbon on soil hydraulic properties. *Dev. Soil Sci.* **2004**, *30*, 95–114.
31. Wagner, S.; Cattle, S.R.; Scholten, T. Soil-aggregate formation as influenced by clay content and organic-matter amendment. *J. Plant Nutr. Soil Sci.* **2007**, *170*, 173–180. [CrossRef]
32. Dos Santos, I.; Bettiol, W. Effect of sewage sludge on the rot and seedling damping-off of bean plants caused by *Sclerotium rolfsii*. *Crop Prot.* **2003**, *22*, 1093–1097. [CrossRef]
33. Montemurro, F.; Fiore, A.; Campanelli, G.; Ciaccia, C.; Ferri, D.; Maiorana, M.; Diacono, M. Yield and performance and soil properties of organically fertilized fodder crops. *J. Plant Nutr.* **2015**, *38*, 1558–1572. [CrossRef]

34. Tisdell, S.E.; Breslin, V.T. *Characterization and Leaching of Elements from Municipal Solid Waste Compost*; American Society of Agronomy, Crop Science Society of America, and Soil Science Society of America: Madison, WI, USA, 1995; Volume 24, No. 5; pp. 827–833.
35. Pérez, D.V.; Alcantara, S.; Ribeiro, C.C.; Pereira, R.; Fontes, G.C.d.; Wasserman, M.; Venezuela, T.C.; Meneguelli, N.d.A.; De Macedo, J.; Barradas, C.A.A. Composted municipal waste effects on chemical properties of a Brazilian soil. *Bioresour. Technol.* **2007**, *98*, 525–533. [CrossRef] [PubMed]
36. Pigozzo, A.T.J.; Lenzi, E.; Luca Junior, J.d.; Scapim, C.A.; Costa, A.C.S.d. Transition metal rates in latosol twice treated with sewage sludge. *Braz. Arch. Biol. Technol.* **2006**, *49*, 515–526. [CrossRef]
37. Achiba, W.B.; Gabteni, N.; Lakhdar, A.; Du Laing, G.; Verloo, M.; Jedidi, N.; Gallali, T. Effects of 5-year application of municipal solid waste compost on the distribution and mobility of heavy metals in a Tunisian calcareous soil. *Agric. Ecosyst. Environ.* **2009**, *130*, 156–163. [CrossRef]
38. Lal, R. Soil organic matter and water retention. *Agron. J.* **2020**, *112*, 3265–3277. [CrossRef]
39. Jong, R.d.; Campbell, C.; Nicholaichuk, W. Water retention equations and their relationship to soil organic matter and particle size distribution for disturbed samples. *Can. J. Soil Sci.* **1983**, *63*, 291–302. [CrossRef]
40. Ellerbrock, R.; Gerke, H.; Bachmann, J.; Goebel, M.-O.J. Composition of organic matter fractions for explaining wettability of three forest soils. *Soil Sci. Soc. Am. J.* **2005**, *69*, 57–66. [CrossRef]
41. Bronick, C.J.; Lal, R. Soil structure and management: A review. *Geoderma* **2005**, *124*, 3–22. [CrossRef]
42. Liu, M.; Han, G.; Zhang, Q. Effects of soil aggregate stability on soil organic carbon and nitrogen under land use change in an erodible region in Southwest China. *Int. J. Environ. Res. Public Health* **2019**, *16*, 3809. [CrossRef]
43. Karami, A.; Homaee, M.; Afzalinia, S.; Ruhipour, H.; Basirat, S. Organic resource management: Impacts on soil aggregate stability and other soil physico-chemical properties. *Agric. Ecosyst. Environ.* **2012**, *148*, 22–28. [CrossRef]
44. Libohova, Z.; Seybold, C.; Wysocki, D.; Wills, S.; Schoeneberger, P.; Williams, C.; Lindbo, D.; Stott, D.; Owens, P.R. Reevaluating the effects of soil organic matter and other properties on available water-holding capacity using the National Cooperative Soil Survey Characterization Database. *J. Soil Water Conserv.* **2018**, *73*, 411–421. [CrossRef]
45. Taban, M.; Movahedi Naeini, S. Effect of aquasorb and organic compost amendments on soil water retention and evaporation with different evaporation potentials and soil textures. *Commun. Soil Sci. Plant Anal.* **2006**, *37*, 2031–2055. [CrossRef]
46. Azlan, A.; Aweng, E.; Ibrahim, C. The correlation between total organic carbon (TOC), organic matter and water content in soil collected from different land use of Kota Bharu, Kelantan. *Aust. J. Basic Appl. Sci.* **2011**, *5*, 915–922.
47. Díaz-Zorita, M.; Buschiazzo, D.E.; Peinemann, N. Soil organic matter and wheat productivity in the semiarid Argentine Pampas. *Agron. J.* **1999**, *91*, 276–279. [CrossRef]
48. Balogh, J.; Pintér, K.; Fóti, S.; Cserhalmi, D.; Papp, M.; Nagy, Z. Dependence of soil respiration on soil moisture, clay content, soil organic matter, and CO_2 uptake in dry grasslands. *Soil Biol. Biochem.* **2011**, *43*, 1006–1013. [CrossRef]
49. Zhang, Y.J.; Meinzer, F.C.; Qi, J.H.; Goldstein, G.; Cao, K.F. Midday stomatal conductance is more related to stem rather than leaf water status in subtropical deciduous and evergreen broadleaf trees. *Plant Cell Environ.* **2013**, *36*, 149–158. [CrossRef]
50. Naor, A. Midday stem water potential as a plant water stress indicator for irrigation scheduling in fruit trees. In Proceedings of the III International Symposium on Irrigation of Horticultural Crops 537, Lisbon, Portugal, 28 June–2 July 1999; pp. 447–454.
51. Gaeta, L.; Amendolagine, A.; Di Gennaro, D.; Navarro, A.; Tarricone, L.; Campi, P.; Stellacci, A.; Losciale, P. Managing orchard floor for saving water in a late ripening peach cultivar: A preliminary result. In Proceedings of the IX International Peach Symposium 1304, Bucureşti, Romania, 2–6 July 2017; pp. 207–214.
52. Lordan, J.; Pascual, M.; Fonseca, F.; Villar, J.; Rufat, J. Use of rice husk to enhance peach tree performance in soils with limiting physical properties. *Soil Tillage Res.* **2013**, *129*, 19–22. [CrossRef]
53. Baldi, E.; Toselli, M.; Marcolini, G.; Marangoni, B. Effect of mineral and organic fertilization on soil chemical, biological and physical fertility in a commercial peach orchard. In Proceedings of the V International Symposium on Mineral Nutrition of Fruit Plants 721, Talca, Chile, 16–21 January 2005; pp. 55–62.
54. Celano, G.; Dumontet, S.; Xiloyannis, C.; Nuzzo, V.; Dichio, B. Responses of peach-orchard system to green manuring and mineral fertilisation. In Proceedings of the III International Symposium on Mineral Nutrition of Deciduous Fruit Trees 448, Zaragoza, Spain, 27 May 1996; pp. 289–296.
55. Montanaro, G.; Dichio, B.; Bati, C.B.; Xiloyannis, C. Soil management affects carbon dynamics and yield in a Mediterranean peach orchard. *Agric. Ecosyst. Environ.* **2012**, *161*, 46–54. [CrossRef]
56. Cheng, Y.; Xie, W.; Huang, R.; Yan, X.; Wang, S. Extremely high N_2O but unexpectedly low NO emissions from a highly organic and chemical fertilized peach orchard system in China. *Agric. Ecosyst. Environ.* **2017**, *246*, 202–209. [CrossRef]
57. Xiao, Y.; Peng, Y.; Peng, F.; Zhang, Y.; Yu, W.; Sun, M.; Gao, X. Effects of concentrated application of soil conditioners on soil–air permeability and absorption of nitrogen by young peach trees. *Soil Sci. Plant Nutr.* **2018**, *64*, 423–432. [CrossRef]
58. Campi, P.; Palumbo, A.; Mastrorilli, M. Effects of tree windbreak on microclimate and wheat productivity in a Mediterranean environment. *Eur. J. Agron.* **2009**, *30*, 220–227. [CrossRef]
59. Katerji, N.; Mastrorilli, M.; Rana, G. Water use efficiency of crops cultivated in the Mediterranean region: Review and analysis. *Eur. J. Agron.* **2008**, *28*, 493–507. [CrossRef]
60. Soil Conservation Service, US Department of Agriculture. *Soil Taxonomy: A Basic System of Soil Classification for Making and Interpreting Soil Surveys*; Soil Conservation Service, US Department of Agriculture: Washington, DC, USA, 1975.

61. Allen, R.G.; Pereira, L.S.; Raes, D.; Smith, M. Crop evapotranspiration-Guidelines for computing crop water requirements-FAO Irrigation and drainage paper 56. *Fao Rome* **1998**, *300*, D05109.
62. Mastrorilli, M.; Katerji, N.; Rana, G.; Nouna, B.B. Daily actual evapotranspiration measured with TDR technique in Mediterranean conditions. *Agric. For. Meteorol.* **1998**, *90*, 81–89. [CrossRef]
63. Granier, A.; Breda, N.; Biron, P.; Villette, S. A lumped water balance model to evaluate duration and intensity of drought constraints in forest stands. *Ecol. Model.* **1999**, *116*, 269–283. [CrossRef]
64. Granier, A.; Loustau, D.; Bréda, N. A generic model of forest canopy conductance dependent on climate, soil water availability and leaf area index. *Ann. For. Sci.* **2000**, *57*, 755–765. [CrossRef]
65. Zhang, R.; Wang, D.; Sun, H.; Wei, C.; Wang, L. Comparison of transpiration of differently aged apple orchards on the Loess Plateau of China at multiple temporal scales. *Hydrol. Sci. J.* **2021**, *66*, 979–990. [CrossRef]
66. Tognetti, R.; Giovannelli, A.; Lavini, A.; Morelli, G.; Fragnito, F.; d'Andria, R. Assessing environmental controls over conductances through the soil–plant–atmosphere continuum in an experimental olive tree plantation of southern Italy. *Agric. For. Meteorol.* **2009**, *149*, 1229–1243. [CrossRef]
67. Ferrara, R.M.; Mazza, G.; Muschitiello, C.; Castellini, M.; Stellacci, A.M.; Navarro, A.; Lagomarsino, A.; Vitti, C.; Rossi, R.; Rana, G. Short-term effects of conversion to no-tillage on respiration and chemical-physical properties of the soil: A case study in a wheat cropping system in semi-dry environment. *Ital. J. Agrometeorol* **2017**, *1*, 47–58.
68. Ferrara, R.M.; Campi, P.; Muschitiello, C.; Leogrande, R.; Vittorio Vonella, A.; Ventrella, D.; Rana, G. Soil respiration during three cropping cycles of durum wheat under different tillage conditions in a Mediterranean environment. *Soil Use Manag.* **2022**, *38*, 1547–1563. [CrossRef]
69. Vitti, C.; Stellacci, A.M.; Leogrande, R.; Mastrangelo, M.; Cazzato, E.; Ventrella, D. Assessment of organic carbon in soils: A comparison between the Springer–Klee wet digestion and the dry combustion methods in Mediterranean soils (Southern Italy). *Catena* **2016**, *137*, 113–119. [CrossRef]
70. Naor, A.; Klein, I.; Doron, I. Stem water potential and apple size. *J. Am. Soc. Hortic. Sci.* **1995**, *120*, 577–582. [CrossRef]
71. Liu, Q.; Piao, S.; Janssens, I.A.; Fu, Y.; Peng, S.; Lian, X.; Ciais, P.; Myneni, R.B.; Peñuelas, J.; Wang, T. Extension of the growing season increases vegetation exposure to frost. *Nat. Commun.* **2018**, *9*, 426. [CrossRef]
72. Jones, H.G.; Serraj, R.; Loveys, B.R.; Xiong, L.; Wheaton, A.; Price, A.H. Thermal infrared imaging of crop canopies for the remote diagnosis and quantification of plant responses to water stress in the field. *Funct. Plant Biol.* **2009**, *36*, 978–989. [CrossRef] [PubMed]
73. Aragüés, R.; Medina, E.; Martínez-Cob, A.; Faci, J. Effects of deficit irrigation strategies on soil salinization and sodification in a semiarid drip-irrigated peach orchard. *Agric. Water Manag.* **2014**, *142*, 1–9. [CrossRef]
74. Ruiz-Sanchez, M.C.; Domingo, R.; Castel, J.R. Deficit irrigation in fruit trees and vines in Spain. A review. *Span. J. Agric. Res.* **2010**, *8*, S5–S20. [CrossRef]
75. Pedrero, F.; Camposeo, S.; Pace, B.; Cefola, M.; Vivaldi, G.A. Use of reclaimed wastewater on fruit quality of nectarine in Southern Italy. *Agric. Water Manag.* **2018**, *203*, 186–192. [CrossRef]
76. Rolbiecki, S.; Piszczek, P. Effect of the forecast climate change on the peach tree water requirements in the Bydgoszcz region. *Infrastrukt. I Ekol. Teren. Wiej.* **2016**, *IV/3*, 1499–1508.
77. Sadras, V.; Milroy, S. Soil-water thresholds for the responses of leaf expansion and gas exchange: A review. *Field Crops Res.* **1996**, *47*, 253–266. [CrossRef]
78. Rahmati, M.; Davarynejad, G.H.; Génard, M.; Bannayan, M.; Azizi, M.; Vercambre, G. Peach water relations, gas exchange, growth and shoot mortality under water deficit in semi-arid weather conditions. *PLoS ONE* **2015**, *10*, e0120246. [CrossRef]
79. Batiot, C.; Liñán, C.; Andreo, B.; Emblanch, C.; Carrasco, F.; Blavoux, B. Use of Total Organic Carbon (TOC) as tracer of diffuse infiltration in a dolomitic karstic system: The Nerja Cave (Andalusia, southern Spain). *Geophys. Res. Lett.* **2003**, *30*. [CrossRef]
80. Volk, C.; Wood, L.; Johnson, B.; Robinson, J.; Zhu, H.W.; Kaplan, L. Monitoring dissolved organic carbon in surface and drinking waters. *J. Environ. Monit.* **2002**, *4*, 43–47. [CrossRef] [PubMed]
81. Yang, C.; Yang, L.; Ouyang, Z. Organic carbon and its fractions in paddy soil as affected by different nutrient and water regimes. *Geoderma* **2005**, *124*, 133–142. [CrossRef]
82. Lepistö, A.; Räike, A.; Sallantaus, T.; Finér, L. Increases in organic carbon and nitrogen concentrations in boreal forested catchments—Changes driven by climate and deposition. *Sci. Total Environ.* **2021**, *780*, 146627. [CrossRef] [PubMed]
83. Scholander, P.F.; Hammel, H.; Hemmingsen, E.; Bradstreet, E. Hydrostatic pressure and osmotic potential in leaves of mangroves and some other plants. *Proc. Natl. Acad. Sci. USA* **1964**, *52*, 119–125. [CrossRef]
84. Scholander, P.; Hammel, H.T.; Bradstreet, E.D.; Henningson, E.A. Sap Pressure in Vascular Plants. *Science* **1965**, *148*, 339–346. [CrossRef] [PubMed]
85. Turner, N.C. Measurement of plant water status by the pressure chamber technique. *Irrig. Sci.* **1988**, *9*, 289–308. [CrossRef]
86. Barillot, R.; Frak, E.; Combes, D.; Durand, J.-L.; Escobar-Gutiérrez, A.J. What determines the complex kinetics of stomatal conductance under blueless PAR in Festuca arundinacea? Subsequent effects on leaf transpiration. *J. Exp. Bot.* **2010**, *61*, 2795–2806. [CrossRef]
87. Chadha, A.; Florentine, S.K.; Chauhan, B.S.; Long, B.; Jayasundera, M. Influence of soil moisture regimes on growth, photosynthetic capacity, leaf biochemistry and reproductive capabilities of the invasive agronomic weed; *Lactuca serriola*. *PLoS ONE* **2019**, *14*, e0218191. [CrossRef]

88. Yu, C.-L.; Hui, D.; Deng, Q.; Wang, J.; Reddy, K.C.; Dennis, S. Responses of corn physiology and yield to six agricultural practices over three years in middle Tennessee. *Sci. Rep.* **2016**, *6*, 1–9. [CrossRef]
89. Alcaras, L.M.A.; Rousseaux, M.C.; Searles, P.S. Responses of several soil and plant indicators to post-harvest regulated deficit irrigation in olive trees and their potential for irrigation scheduling. *Agric. Water Manag.* **2016**, *171*, 10–20. [CrossRef]
90. Ahumada-Orellana, L.; Ortega-Farías, S.; Poblete-Echeverría, C.; Searles, P.S. Estimation of stomatal conductance and stem water potential threshold values for water stress in olive trees (cv. Arbequina). *Irrig. Sci.* **2019**, *37*, 461–467. [CrossRef]
91. Papafilippaki, A.; Paranychianakis, N.; Nikolaidis, N.P. Effects of soil type and municipal solid waste compost as soil amendment on *Cichorium spinosum* (spiny chicory) growth. *Sci. Hortic.* **2015**, *195*, 195–205. [CrossRef]
92. Nair, A.; Ngouajio, M. Integrating rowcovers and soil amendments for organic cucumber production: Implications on crop growth, yield, and microclimate. *HortScience* **2010**, *45*, 566–574. [CrossRef]

water

MDPI

Article

Cost-Effectiveness of Sustainable Agricultural Water Policies: Source Switching versus Irrigation Buyout Auctions in Georgia's Lower Flint River Basin

Jeffrey D. Mullen * and Yizhou Niu

Department of Agricultural and Applied Economics, University of Georgia, Athens, GA 30605, USA; yizhouniu@uga.edu

* Correspondence: jmullen@uga.edu; Tel.: +1-706-542-0767

Abstract: In this paper, a new methodology for comparing the cost-effectiveness of sustainable agricultural water policies during times of drought is developed. The methodology explicitly accounts for regional economic impacts from policy implementation and uncertainty related to drought frequency. The methodology is applied to two policy options being considered by the state of Georgia in the lower Flint River basin: irrigation buyout auctions and source switching. The results demonstrate the following: (1) the importance of modeling uncertainty associated with both the frequency and timing of drought, and the hydrologic effects of source switching; (2) as the frequency of drought increases, the cost-effectiveness of irrigation buyout auctions decreases. Failure to incorporate the regional economic impacts of each policy significantly underestimates the costs of both, but more so for irrigation buyout auctions than source switching. The ability to proactively manage the uncertainty associated with source switching through research and the judicious site selection of new irrigation wells increases its cost-effectiveness.

Keywords: water policy; drought; irrigation; economics; cost-effectiveness

1. Introduction

Water supplies were long considered abundant in Georgia, but a growing population, increased agricultural water use, and a changing environment and climate have highlighted the need for effective water management strategies, especially during drought. And droughts are occurring more frequently—drought conditions have been recorded in the state during 10 of the last 25 years [1]. Over those same 25 years, statewide, irrigated agriculture has accounted for 30–45% of total withdrawals [2]. Irrigated agriculture, however, accounts for well over 90% of withdrawals in the ecologically sensitive lower Flint River Basin (FRB). Most withdrawals within the lower FRB come either from the Floridan aquifer or surface water [3].

The FRB covers nearly 8500 square miles, with the Flint River stretching 349 miles from the southern edge of the Atlanta metropolitan area in the upper Piedmont region to the wetlands of the Coastal Plain in the southwest corner of Georgia [4]. South of Dooly County, in the lower FRB, the Flint River and many of its tributaries are in hydrologic connection with the Floridan aquifer and either receive water from the aquifer or lose water to it depending on the head difference between the streams and the aquifer [4]. That connectivity has implications for in-stream flows, especially during periods of heavy pumping, and even more acutely during drought. Interestingly, the Floridan is not the only aquifer lying below the lower FRB. This area of the Coastal Plain actually has a stratified groundwater system that also includes the Claiborne, Clayton, and Cretaceous aquifers [5] (see Figure 1).

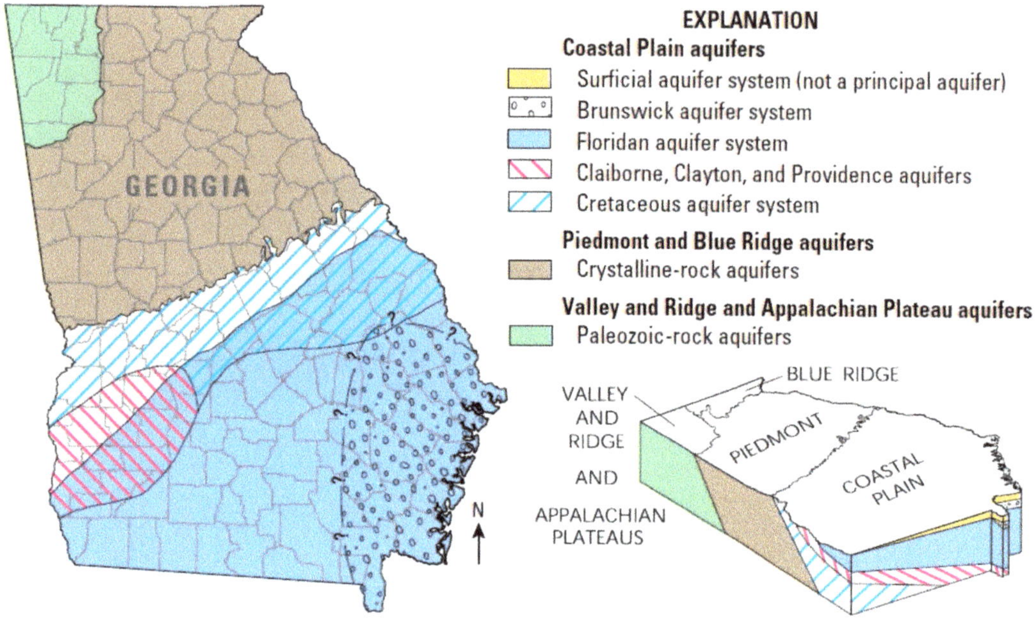

Figure 1. Areas of mMajor aAquifers in Georgia, modified from [5].

In addition to supporting irrigated agriculture, the Flint River and its tributaries in the lower FRB are also home to six freshwater mussel species protected by the U.S. Fish and Wildlife Service [6]. Under the rules of the Endangered Species Act, the state of Georgia is obligated to ensure minimum flows in streams that support these species. During the prolonged drought of 1998–2002, the state executed irrigation buyout auctions in which farmers with surface water withdrawal permits were paid not to irrigate their fields during the 2001 and 2002 growing seasons [4,7]. These auctions were executed under the auspices of the 2001 Flint River Drought Protection Act (FRDPA).

The FRDPA was amended in 2006 to allow groundwater permit holders to participate in future auctions. Mullen [7] examined the cost-effectiveness of different auction rules for protecting instream flows when both surface water and groundwater permit holders were allowed to participate. The results of that study emphasized the need to incorporate the hydrologic connectivity between surface and groundwater at the point of the groundwater withdrawal into the auction rules to maximize the cost-effectiveness of the auction. With the stratified nature of the aquifer system in the lower FRB, there is another policy option available beyond irrigation buyout auctions, namely, source switching.

Source switching is the act of changing the source of an irrigation withdrawal and usually refers to switching from a surface water withdrawal to a groundwater withdrawal. In the lower FRB, there is the possibility of switching surface water withdrawals to any of the four underlying aquifers and of switching from the Floridan—an aquifer with relatively high hydrologic connectivity to the streams—to one of the underlying aquifers that are hydrologically disconnected.

The primary objective of this study is to develop a general methodology to compare the cost-effectiveness of source switching to an irrigation buyout auction as a policy for managing stream flow and/or sustainable yields from an aquifer. Two source switching approaches are considered in this paper: "standard" source switching, in which the new water source is developed and the old source is no longer used, and "emergency" source switching, in which the new source is developed but the old source continues to be used and withdrawals are only switched to the new source during periods when conditions require it. The secondary objective is to apply that methodology to the lower Flint River Basin and determine the conditions under which each policy is more cost-effective. Finally, we discuss the suitability of the methodology for other systems.

Importantly, other studies have examined the cost-effectiveness of paying farmers not to irrigate [8–10], but none of them have considered the policy of source switching. Additionally, while those studies have alluded to regional economic impacts from converting irrigated land to dryland production or fallowing a field, they have not explicitly incorporated those impacts into their cost-effectiveness assessments. The methodology developed below does explicitly account for regional economic impacts and includes them in the policy analysis.

2. Materials and Methods

The methodology developed here entails specifying the cost components of the two policies under consideration over a given time horizon. Because some of those cost components are only realized under certain circumstances, the methodology actually calculates the expected present value of costs for each policy over time. Assuming the policies both generate the same benefits—in our empirical example, ensuring ecologically sufficient stream flow is the benefit—then comparing the expected present value of costs would also represent a comparison of their relative cost-effectiveness.

2.1. Expected Present Value of Policy Costs

For each policy (z) considered, the expected present value of costs over a time horizon of length T is shown in Equation (1):

$$E[PV_{z,T}] = \sum_{t=1}^{T} Pr(C_{z,t}) \times \frac{C_{z,t}}{(1+r)^{t-1}} \quad (1)$$

where $E[PV_{z,\,T}]$: expected present value of policy z from years 1 to T;

$C_{p,t}$: cost of policy z in year t;

$Pr(C_{z,t})$: probability of incurring cost of policy z in year t;

r: discount rate.

2.1.1. Costs of Irrigation Buyout Auctions

One of the critical differences between an irrigation buyout auction and source switching is that the auction adversely affects agricultural production either by fallowing the field or by producing under rainfed conditions during drought. As such, the costs of irrigation reduction auctions include both direct payments to farmers and regional economic impacts resulting from reduced agricultural production. In fact, as we will see in our empirical example, the regional economic impacts can dwarf the direct payments, so it is critical to account for them. Because the regional economic impacts can vary based on the location of the field, cost estimates for the irrigation reduction auctions are expressed here at the county level, as shown in Equation (2):

$$C_{Auction,c,t} = Pay_t + EI_{Auction,c,t} \quad (2)$$

where $C_{Auction,c,\,t}$: the cost of an auction in county c, year t;

Pay_t: direct auction payments in year t;

$EI_{Auction,c,t}$: regional economic impact of lost agricultural production in county c, year t.

While the auction payments are a distinct cost, the regional economic impact has several components. First, there is the reduction in the value of agricultural production, also known as the direct economic impact ($EI_{Ag\ Direct}$), resulting from fallowing or not irrigating a field during a drought. Next, the lost economic activity in the agricultural sector affects the purveyors of goods and services needed to (a) prepare the field; (b) sow, grow, protect, and harvest the crop; and (c) process, store, market, and distribute the harvested product. These impacts are referred to as the indirect economic impact ($EI_{Ag\ Indirect}$). However, the firms and employees that support agricultural production also purchase goods and services outside the agricultural sector (e.g., gasoline, electricity, accounting services, restaurants, etc.), so when the agricultural sector expands or contracts, there are also effects on the larger economy. These are referred to as the induced economic impact ($EI_{Ag\ Induced}$). Furthermore, the contraction (expansion) of economic activity in the agricultural sector subsequently leads to a reduction (increase) in tax revenues (ΔTR). Note that from a policy cost perspective, a reduction in tax revenues is a positive cost. The present value of these costs is accounted for in Equation (3):

$$EI_{Auction,c,t} = \frac{\left(EI_{AgDirect,c,t} + EI_{AgIndirect,c,t} + EI_{AgInduced,c,t} + \Delta TR_{Ag,c,t}\right)}{(1+r)^{t-1}} \quad (3)$$

where $EI_{Ag\ Direct,\ c,t}$: direct economic impact due to the change in the value of agricultural production in county c, year t;

$EI_{Ag\ Indirect,\ c,t}$: indirect economic impact due to the change in value of agricultural production in county c, year t;
$EI_{Ag\ Induced,\ c,t}$: induced economic impact due to the change in value of agricultural production in county c, year t;
$\Delta TR_{Ag,c,t}$: change in tax revenue due to the change in value of agricultural production in county c, year t;
r: discount rate.

It is important to remember that the costs of the auction are realized only if the auction is actually held. So, when we consider the cost of the auction as a policy, we need to consider the expected county-level cost of the auction, i.e., the sum of the yearly auction cost multiplied by the likelihood of the auction being held in any given year. An auction would only be held if a drought was severe enough to require the suspension of irrigation. Throughout the remainder of this article, when we refer to "drought", we are referring to a drought of that severity. The likelihood of an auction in year t, then, is equal to the probability of a drought in that year (PrD_t). The expected present value of county-level costs of the auction over T years is represented by Equation (4):

$$E[C_{Auction,c}] = \sum_{t=1}^{T} \frac{PrD_t \times (Pay_t + EI_{Auction,c,t})}{(1+r)^{t-1}} \quad (4)$$

2.1.2. Costs of Source Switching

In this study, source switching refers to switching either from a surface water source to a groundwater source, or from a more hydrologically connected groundwater source to a less hydrologically connected groundwater source. (It is also possible to switch from one surface source to another in order to preserve stream flow. The methodology can handle that situation as well, simply by accounting for the additional conveyance infrastructure costs in place of the well construction costs defined above.) In general, the closer the water table of an aquifer is to the surface, the greater the hydrologic connectivity of the system is likely to be. In other words, source switching often entails digging a well to switch from a surface water source or digging a deeper well to switch from one groundwater source to another.

Digging wells generates both fixed and variable costs. The fixed costs (*FC*) are the costs of drilling, lining, and capping the well. The variable costs (*VC*) of source switching are the extra energy costs required to pump water from a greater depth. Both *FC* and *VC* are functions of well depth, although the *FC* is a function of the actual depth of the well (*Depth*) while the *VC* is a function of the depth to the new water table compared to the old water table ($\Delta Depth$) and the amount of water pumped. For standard source switching (*SSS*), because the original source is no longer used, the extra pumping costs are realized every year. For emergency source switching (ESS), however, the extra pumping costs are only realized during drought. Therefore, when evaluating the variable costs of emergency source switching in a given year, we need to multiply them by the probability of a drought occurring that year. Furthermore, the water used for emergency source switching will be the water needed during times of drought. The water used for standard source switching will vary during wet years, typical years, and drought years.

There are additional costs that may or may not be incurred by owners of a deeper well. As a well gets deeper, there are more opportunities for breakages or malfunctions. More importantly, the deeper aquifers may have lower yields, slower recharge rates, and/or their hydrology may be less well understood. As such, wells in those aquifers could have a higher likelihood of running dry as withdrawals increase, especially if the wells are concentrated in a relatively small area. We refer to a well that is inoperable, either through over-drafting or due to breakage or malfunction, as "well failure." When well failure occurs, the value of agricultural production in that field is affected, leading to adverse regional economic impacts.

We can write the expected present value of the costs of standard source switching as in Equation (5), and the emergency source switching as in Equation (6).

$$E[C_{SSS,c,t}] = \sum_{t=1}^{T} \frac{FC_{c,t}(Depth_c) + VC_{SSS,c,t}(\Delta Depth_c, Water_{SSS,c,t}) + PrF \times EI_{F,c,t}}{(1+r)^{t-1}} \quad (5)$$

$$E[C_{ESS,c,t}] = \sum_{t=1}^{T} \frac{FC_{c,t}(Depth_c) + PrD_t \times VC_{ESS,c,t}(\Delta Depth_c, Water_{ESS,c,t}) + PrF \times EI_{F,c,t}}{(1+r)^{t-1}} \quad (6)$$

In Equations (5) and (6), $FC_{c,t}$ is defined as above, $Water_{SSS,c,t}$ and $Water_{ESS,c,t}$ are the amount of water applied in county c in year t, $VC_{SSS,c,t}$ and $VC_{ESS,c,t}$ are the extra pumping costs in county c in year t, PrF is the probability of well failure, and $EI_{F,c,t}$ is the regional economic impact of well failure in county c at time t. In Equation (6), PrD_t is the probability of drought in year t.

As the fixed and variable costs are increasing functions of well depth, the costs of standard and emergency source switching also strictly increase with well depth in a given county. It is important to note, however, that the costs of both types of source switching are also a function of the economic impact of well failure. The economic impact of well failure reflects both the productivity of the land in the county and the strength of the economic linkages between agricultural production and other sectors of the economy. As such, a shallower well in one county could have higher expected costs of source switching than a deeper well in another county.

Another important point is that the expected present value of standard and emergency source switching will only be equal if a drought occurs every year, i.e., $PrD_t = 1$. When the likelihood of drought is less than one, the variable costs of ESS will be less than the variable costs of SSS. Additionally, because SSS will draw water out of the new source each year and ESS will not, the probability of well failure from SSS is likely to be greater than that of ESS. As a result of these factors, the expected present value of standard source switching will be greater than that of emergency source switching.

2.2. Comparing Policy Costs

Cost-effectiveness is an economic measure used to compare alternative options for achieving a given objective [11,12]. Cost-effectiveness is, essentially, the cost of implementing the option divided by the units of desired outcome generated by the option. For example, a business firm can calculate the cost-effectiveness of a marketing strategy by dividing the cost of the strategy by the number of sales that strategy is likely to generate. The firm could do the same for alternative marketing strategies, and then, determine the most cost-effective among them.

In this study, we compare three water management policies that are assumed to have the same outcome—namely, the avoidance of stream-flow impacts associated with irrigation withdrawals during drought. The irrigation auction accomplishes this goal by prohibiting water withdrawals, whereas both standard and emergency source switching accomplish it by diverting irrigation withdrawals into aquifers that are not hydrologically connected to the streams. Because the policies have the same outcome, the denominator of their respective cost-effectiveness measure is the same and can, therefore, be ignored. The relative cost-effectiveness of the policies is determined entirely by the relative cost of each.

Equations (4)–(6) represent the expected present values of the costs of each policy. The challenge is to find the conditions under which one policy is unambiguously more cost-effective than the others. As noted above, the costs of emergency source switching are less than or equal to the costs of standard source switching, so we focus on *ESS* here. We begin by equating the present value of the expected costs of the policies, as shown in Equation (7).

$$\sum_{t=1}^{T} PrD_t \times (Pay_t + EI_{Auction,c,t})/(1+r)^{t-1} = \sum_{t=1}^{T}[FC_{c,t}(Depth_c) + PrD_t \times VC_{ESSc,t}(\Delta Depth_c, Water_{ESSc,t}) + PrD_t \times PrF_t \times EI_{F,c,t}]/(1+r)^{t-1} \quad (7)$$

For a given probability of drought, Equation (7) can be rearranged to find the probability of well failure for which the present value of expected costs over a horizon of T years is the same. We refer to this as the threshold probability of well failure (*PrF**). If the actual probability of well failure is greater than *PrF**, then the auction has a lower present value of expected costs than emergency source switching—in other words, the auction is more cost-effective. Emergency source switching is more cost-effective when the actual probability of well failure is less than the threshold probability, *PrF**.

Alternatively, we can use Equation (7) to identify, for a given *PrF*, the probability of drought that equates the present value of expected costs for the two policies (*PrD**). If the likelihood of drought in any given year is greater than *PrD**, then source switching is more cost-effective, and vice versa.

We can also use Equation (7), with a slight modification, to investigate a different question. Imagine that the water manager (e.g., the state) decides to pay for source switching in Year 1. By doing this, the state has avoided the costs of an auction in the future. But the present value of the costs of the auction depends critically on when in the future the auction is held. If the auction is held in Year 1, the present value of the cost is much higher than if the auction is held in Year 20, due to discounting. We can calculate a unique *PrF** that equates the present value of expected costs of emergency source switching implemented in Year 1 to the present value of the costs of a single auction held in any given year of the *T*-year horizon.

Finally, Equation (7) can identify, for any *PrD* and *PrF* pair, the unique auction payment (*Pay**) that equates the two policies. Actual payment levels below *Pay** would make the auction more cost-effective; emergency source switching would be more cost-effective if auction participants required payments greater than *Pay**.

2.3. Comparing Policies in the Lower Flint River Basin

In this section, we demonstrate the application of the methodology in the context of Georgia's lower Flint River Basin. Eleven counties in the lower FRB have more than one aquifer beneath them: Baker, Calhoun, Decatur, Dougherty, Early, Lee, Miller, Mitchel, Randolph, Terrell, and Worth. The components of the expected present value of costs of an irrigation buyout auction and emergency source switching were estimated for each of these counties. The analysis was performed for a 150-acre field over a 25-year time horizon using 2020 as the baseline year. The study area is shown in Figure 2.

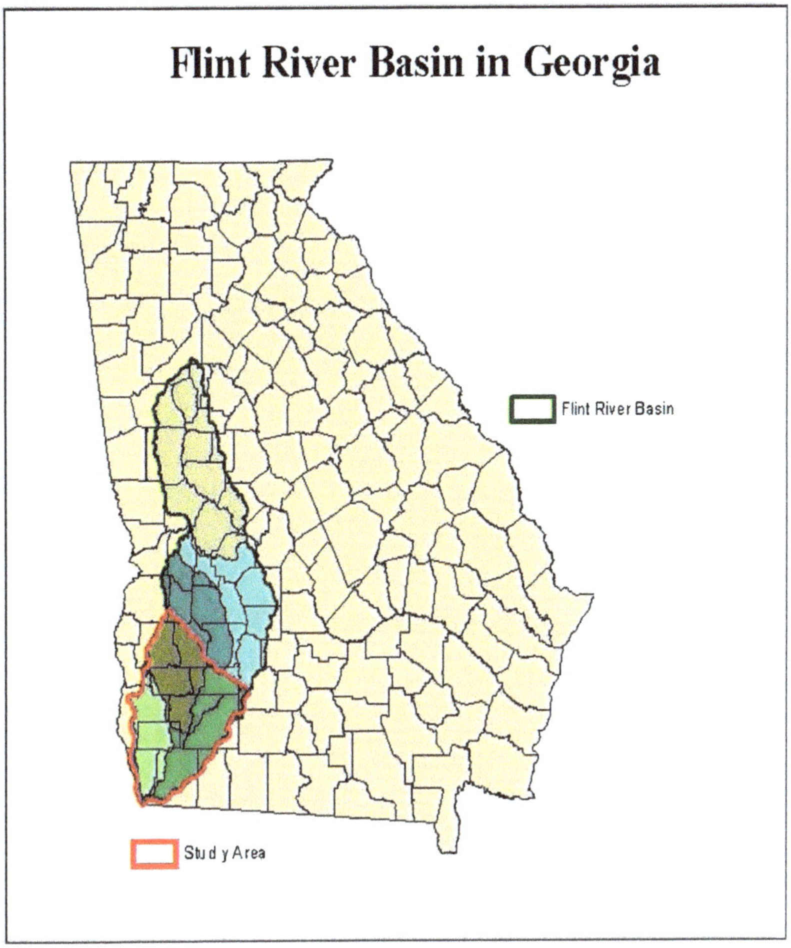

Figure 2. Study area [4].

2.3.1. Calculating Auction Costs

As shown in Equation (2), the auction has two major cost components: direct payments ($P_{auction,t}$) and regional economic impacts ($EI_{auction,c,t}$). The regional economic impacts can be further broken down into direct, indirect, induced, and tax revenue effects associated with lost agricultural production. These regional economic impacts are estimated using IMPLAN (version 6.9), an input–output model of the linkages across economic sectors. IMPLAN is widely used to study how changes in economic activity in one or more sectors ripple through an economy [13,14]. Country-specific IMPLAN models have been developed for use in 66 countries around the world [15]. These analyses can be conducted at varying spatial scales, including at the U.S county level.

Direct Auction Payments

For our analysis, the direct auction payments (Pay_t) are straightforward to calculate. We simply inflate the average payment per acre from the 2002 auction (USD 135/acre) (2022 auction USD 483/acre) to 2020 dollars (USD 195/acre) using the U.S. Bureau of Labor Statistics' inflation calculator [16] and multiply by 150 acres. This means the auctioning agency, e.g., the state, would incur and a farmer would receive a USD 29,250 direct payment.

Alternatively, if a reference auction value were not available, the prevailing rental rate of irrigated land in the study area could be used. Even without that information, *Pay** could be calculated.

Regional Economic Impacts of Lost Agricultural Production

Our IMPLAN analysis was conducted at the county level using version 6.9. For each of the 11 counties in the lower FRB, the task at hand was to determine the direct economic impacts of taking a standard 150-acre irrigated field out of production ($EI_{Ag\ Direct,c,t}$).

Irrigated land in the lower FRB is dominated by four major row crops: cotton, peanuts, corn, and soybeans. To estimate the lost value of production from a standard 150-acre irrigated field in each county, we calculated the share of harvested irrigated acres for each of these crops in the county using Equation (8). We then multiplied each share by 150 acres, the crop price, and the yield, as in Equation (9), to obtain the direct economic impact associated with the crop. Adding up the crop-specific direct economic impacts (Equation (10)) gives the total direct economic impact of taking the field out of production.

$$S_{c,y} = \frac{HA_{c,y}}{\sum_Y HA_{c,y}} \tag{8}$$

$$EI_{Ag\ Direct,c,y,t} = S_{c,y} \times 150 \times P_{y,t} \times Q_{y,t} \tag{9}$$

$$EI_{Ag\ Direct,c,t} = \sum_Y EI_{Ag\ Direct,c,y,t} \tag{10}$$

For the IMPLAN (version 6.9) analyses, the loss of the value of agricultural production must be specified. That specification includes identifying the impacted region (county), the impacted industries, and the change in the value of output for each industry (i.e., the direct impact). Each crop was assigned an industry: corn was assigned "grain farming"; cotton was assigned "cotton farming"; peanut was assigned "all other crop farming"; and soybean was assigned "oil seed farming." The direct, indirect, and induced impacts of taking a 150-acre irrigated field out of production in each county, as well as the state and local tax impacts, are reported in Table 1. Details regarding the county-level data used to calculate $EI_{Ag\ Direct,c,t}$ are available in the Supplemental Materials.

Table 1. Direct, indirect, and induced impacts and state and local tax revenue change (2020 USD) from taking a 150-acre irrigated field out of production.

County	Direct	Indirect	Induced	State and Local Tax	Total
Baker	USD (119,648)	USD (62,127)	USD (24,526)	USD (2629)	USD (208,931)
Calhoun	USD (136,718)	USD (42,699)	USD (23,140)	USD (2180)	USD (204,737)
Decatur	USD (118,184)	USD (49,077)	USD (28,469)	USD (2439)	USD (198,169)
Dougherty	USD (113,111)	USD (54,891)	USD (71,437)	USD (3176)	USD (242,616)
Early	USD (112,829)	USD (30,053)	USD (20,998)	USD (1955)	USD (165,835)
Lee	USD (105,655)	USD (66,128)	USD (39,732)	USD (3108)	USD (214,624)
Miller	USD (131,375)	USD (42,079)	USD (5287)	USD (889)	USD (179,630)
Mitchell	USD (110,705)	USD (44,765)	USD (32,197)	USD (2726)	USD (190,392)
Randolph	USD (121,118)	USD (57,221)	USD (9191)	USD (1384)	USD (188,913)
Terrell	USD (124,101)	USD (61,676)	USD (27,078)	USD (2407)	USD (215,261)
Worth	USD (108,130)	USD (32,299)	USD (11,633)	USD (1158)	USD (153,219)

There are a few critical points to understand about Table 1. First, the direct impacts of taking an irrigated field out of production vary across counties (from USD 105 k to USD 136 k) due to variation in the share of crops in each county and agricultural productivity. Second, due to differences in economic diversity across counties, the indirect and induced effects of lost agricultural production vary significantly. For example, Dougherty county is home to a university and many of the region's retail, restaurant, and entertainment establishments. As a result, the indirect and induced effects of lost agricultural production are more pronounced than in other counties. Third, counties with the largest direct effects do not necessarily have the largest total effects. And finally, the combined economic impacts of lost agricultural production from a 150-acre field are many times larger than the USD 29,250 auction payment to the farmer. All of these points emphasize the need for an economic impact analysis when evaluating the cost-effectiveness of alternative agricultural water management policies.

2.3.2. Calculating Emergency Source Switching Costs

As shown in Equation (6), the costs of emergency source switching are a function of the fixed costs (*FC*), variable costs (*VC*), likelihood of drought, likelihood of well failure, and regional economic impact of well failure. The regional economic impact of well failure is also estimated using IMPLAN (version 6.9).

Fixed Cost

The *FC* of source switching depends on the costs per foot of drilling, lining, and capping the well ($C_{Drilling,c,t}$) in county *c*, at year *t*, and the well depth ($Depth_c$), as shown in Equation (11). $Depth_c$ is the average depth (feet) to the aquifer in county *c*. The Claiborne aquifer that underlies the Upper Floridan in the study area is a viable alternative source of irrigation water. However, there is less information about its depth, thickness, water quality, and water-bearing characteristics [17].

$$FC_{c,t} = C_{Drilling,c,t} \times Depth_c \tag{11}$$

Variable Costs

The *VC* is a function of the depth to the water table and the amount of water pumped. To estimate the marginal cost of pumping water from different depths, we modify the engineering relationships among depth, pressure, and total dynamic head (*TDH*) in [18], to reflect the change in pumping costs due to source switching:

$$\Delta TDH_c = psi \times 2.31 + \Delta Depth_c \tag{12}$$

where *psi* is the pumping pressure and $\Delta Depth_c$ is the difference between the depth to the water table of the new source and the original source. If the original source is surface water, $\Delta Depth_c$ is simply the depth to the water table of the aquifer. The value of *psi* is taken from the literature [18].

Equation (13) calculates the amount of water pumped, in acre-feet, to irrigate a standard 150-acre field growing our four selected crops.

$$Water_{ESS,c,t} = \sum_{y=1}^{4} WD_{ESS,c,y,t} \times S_{c,y} \times 150 \tag{13}$$

In Equation (13), $WD_{ESS,c,\,y,t}$ is the water application rate (acre-feet/acre) for crop *y* in county *c* in year *t* during a drought year.

Equation (14) is used to derive the extra fuel consumed due to source switching, where Fuel usage is the number of units of fuel needed to lift one acre-foot of water by one foot (units/acre-foot/foot). Fuel usage depends on the type of fuel used.

$$\Delta Total\ fuel\ consumed_{ESS,c,t} = Fuel\ usage \times \Delta TDH_c \times Water_{ESS,c,t} \tag{14}$$

$$VC_{ESS,c,t} = \Delta Total\ fuel\ consumed_{ESSc,t} \times P_{Electricity,t} \tag{15}$$

The *VC* of emergency source switching is the extra pumping cost, which equals the change in total fuel consumed times the fuel price. This estimate of pumping cost is imperfect, as it does not contain possible changes in the cost of distribution once the water has been raised to surface level. Here, we assume those distribution costs to be the same regardless of the water source.

For each county, the fixed costs of a new well tapping into the Claiborne aquifer are reported in Table 2. Also reported in Table 2 are the extra pumping costs during a drought year from a well in the Claiborne aquifer when the original source was either surface water or the Floridian aquifer. The extra pumping costs from an average year are also included in Table 2 as a point of comparison. Details regarding the county-level data used to calculate the fixed and variable costs are available in the Supplemental Materials.

Table 2. Fixed and variable costs of installing a new well into the Claiborne aquifer, by county (2020 USD).

County	Fixed Costs (Drilling)	Annual Variable Costs (Extra Pumping)			
		Switch Surface to Claiborne		Switch Floridan to Claiborne	
		Median Year	Drought Year	Median Year	Drought Year
Baker	USD 100,151	USD 4957	USD 7850	USD 3190	USD 5052
Calhoun	USD 86,660	USD 4063	USD 6707	USD 2425	USD 4003
Decatur	USD 191,184	USD 6492	USD 12,520	USD 4873	USD 9399
Dougherty	USD 149,465	USD 5928	USD 9852	USD 4569	USD 7594
Early	USD 73,002	USD 2641	USD 4731	USD 1303	USD 2334
Lee	USD 69,768	USD 2646	USD 4632	USD 1183	USD 2070
Miller	USD 102,432	USD 4229	USD 7163	USD 2682	USD 4544
Mitchell	USD 160,492	USD 7138	USD 11,935	USD 3495	USD 5843
Randolph	USD 23,717	USD 726	USD 1658	USD 263	USD 600
Terrell	USD 59,985	USD 2595	USD 4489	USD 1003	USD 1734
Worth	USD 124,315	USD 3465	USD 6944	USD 1192	USD 2388

Economic Impacts of Well Failure

When a well failure occurs, farmers are unable to irrigate. Of course, the direct economic impacts of well failure depend on when the well fails. If the well fails prior to planting, then the direct economic impacts are the same as the irrigation buyout auction. If the well fails after planting, however, the impacts would be reduced as the farmer has already spent money to purchase inputs. In this study, we only consider well failure that occurs before planting. As such, the inability to irrigate a field due to well failure leads to the same direct, indirect, and induced impacts as the irrigation reduction auction. The regional economic impacts of well failure used in our analysis, then, are equal to the regional economic effects of lost agricultural production presented in Table 1.

3. Results: Relative Cost-Effectiveness under Different Scenarios

The costs of each policy option depend on the timing of the policy implementation and the likelihood of unknown events occurring. In this section, we investigate the relative cost-effectiveness of emergency source switching versus an irrigation buyout auction under a variety of scenarios. In these scenarios, a drought refers to an event in the study area severe enough that either emergency source switching or a buyout auction must be executed to preserve stream flows for federally protected, endangered aquatic species.

3.1. Imminent Drought in Current Year, No Well Failure

To begin, we consider a situation where a drought is imminent and policy makers need to decide whether to hold an auction or pay the fixed and variable costs of emergency source switching. In the simplest case, we compare, for the current year alone, estimates of the cost of an irrigation buyout auction to the cost of ESS without well failure. This is carried out through Equation (7) with $T = 1$, $P_{Auction}$ = USD 29,250, the economic impact estimates in Table 1, the fixed cost and drought variable cost estimates in Table 2, and setting $Pr(Drought_1) = 1$ and $PrF = 0$. The results for each county are shown in Table 3.

Table 3. Policy costs, by county, for a single year, with *Pr(Drought)* = 1 and *PrF* = 0.

County	Auction Payments	Total Auction Cost	Total *ESS* Cost Surface to Claiborne	Total *ESS* Cost Floridan to Claiborne
Baker	USD 29,250	USD 238,181	USD 108,001	USD 105,203
Calhoun	USD 29,250	USD 233,987	USD 93,367	USD 90,663
Decatur	USD 29,250	USD 227,419	USD 203,804	USD 200,683
Dougherty	USD 29,250	USD 271,866	USD 159,317	USD 157,059
Early	USD 29,250	USD 195,085	USD 77,733	USD 75,336
Lee	USD 29,250	USD 243,874	USD 74,400	USD 71,839
Miller	USD 29,250	USD 208,880	USD 109,595	USD 106,976
Mitchell	USD 29,250	USD 219,642	USD 172,427	USD 166,335
Randolph	USD 29,250	USD 218,163	USD 25,374	USD 24,317
Terrell	USD 29,250	USD 244,511	USD 64,474	USD 61,719
Worth	USD 29,250	USD 182,469	USD 131,259	USD 126,703

In this scenario, in every county except Randolph, the auction payments made to farmers are much lower than the costs of emergency source switching, regardless of the original source. When the economic impacts of taking an irrigated field out of production are included, however, the total costs of the auction far exceed those of ESS. In other words, if the probability of well failure is zero and a drought occurs in the year the new well is dug, source switching is more cost-effective in every county when the full economic costs of the auction are considered. If policy makers focus only on the auction payments, they would erroneously conclude that the auction is more cost-effective. This reinforces the imperative of including a regional economic impact analysis when comparing costs across different water management policies.

Another point of interest in Table 3 is the fact that the cost differential between an auction and emergency source switching varies by county. Furthermore, due to the variation in both the total auction costs and emergency source switching costs, the county with the lowest cost differential is not necessarily the one with the lowest auction cost.

The next question is as follows: how high does the probability of well failure need to be for the auction to become the cost-effective policy option?

3.2. One Drought, Known Drought Year, Non-Zero Likelihood of Well Failure

When the probability of well failure is non-zero, the expected value of the costs of emergency source switching in a given year increases by the product of the regional economic impacts of lost agricultural production and the probability of well failure in that year. We can calculate county *c*'s threshold probability of well failure (PrF_c*), the likelihood of well failure above which the auction is more cost-effective, by rearranging Equation (7), as shown in Equations (16) and (17), below.

$$PrF_c^* = \frac{\frac{(Pay + EI_{Auction,c}) - VC_{ESS,c})}{(1+r)^{t^*-1}} - FC_c}{EI_{Failure,c} / (1+r)^{t^*-1}} \quad (16)$$

$$PrF_c^* = 1 + \frac{(Pay - VC_{ESS,c})}{EI_{Fail,c}} - \frac{FC_c}{EI_{Fail,c} / (1+r)^{t^*-1}} \quad (17)$$

In this section, we examine the threshold probability of well failure for a specific scenario, namely, when a well is dug for a source in Year 1 and there is only one drought over a 25-year time horizon, and it occurs in year t*. Looking at Equation (16), we are solving for PrF_c* by setting PrD_t equal to 1 in the drought year and equal to zero in all other years. The numerator of Equation (16) comprises the expected present value of the auction costs minus the expected present value of the fixed and variable costs of emergency source switching. The fixed costs of ESS occur only in year 1 when the well is dug. The denominator of Equation (16) is the present value of the regional economic impacts of well failure in year t^*. Figure 3 plots each county's PrF_c^* against the year of the drought. Looking at Year 1, we see some variation in PrF_c^* across counties, with a high of 1 for Randolph and a low of 0.12 for Decatur. This means that source switching in Year 1 would be more cost-effective than an auction in Year 1 in Randolph county no matter the chance of the new well failing in the drought year. In Decatur county, on the other hand, the Year 1 auction is more cost-effective when the chance of well failure in the auction year is 12% or higher.

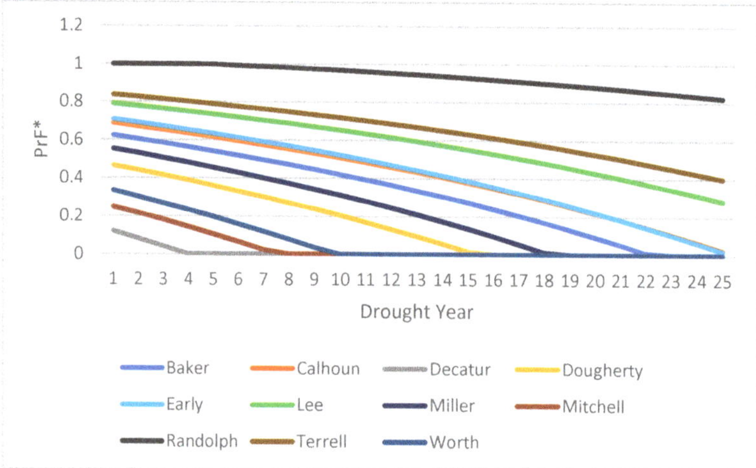

Figure 3. Probability of well failure above which an auction is cost-effective, by drought-year and county, when only 1 drought occurs in 25 years.

For all counties, as the year of the auction moves further into the future, the first two components of Equation (17) are unaffected while the denominator of the fixed cost component falls. As a result, PrF_c^* also decreases when the auction occurs further in the future. Once the threshold probability of well failure goes to zero, the auction is always more cost-effective than source switching. This occurs in Year 4 for Decatur county, Year 8 for Mitchell county, Year 10 for Worth county, etc. It never occurs in Randolph, Terrell, and Lee counties over a 25-year time horizon.

If there is more than one drought during the 25-year time horizon, the expected present value of auction costs would increase. The timing of the additional drought(s) would dictate the change in the expected present value of auction costs as well as the threshold probability of well failure. When looking over a 25-year time horizon, however, we do not know when drought(s) will occur. We examine this in the next section.

3.3. Uncertain Drought Timing and Likelihood, Non-Zero Probability of Well Failure

In this scenario, the timing of drought(s) over a 25-year time horizon is unknown, and the probability of well failure is assumed to be non-zero. Here, policy makers are faced with the decision of whether to invest in source switching now (Year 1) to avoid an auction(s) in an unknown year(s) in the future (in our analysis, we assume the probability of a drought is the same every year). This is the situation policy makers in the lower Flint River Basin currently face each year.

To gain insight into this issue, we vary the probability of drought and use Equation (7) to find the associated threshold probability of well failure for each county. Figure 4 plots PrF^* over a range of drought probability from 0.01 to 0.40. The first thing to notice in Figure 4 is that when the probability of drought is one in one hundred, the threshold probability of well failure is zero in every county except Randolph. This means that, in the other ten counties, if we only expect one drought per century, it is more cost-effective to hold an auction to address low stream flows; in Randolph county, emergency source switching would still be more cost-effective than an auction up to the point where the likelihood of well failure is over 37%. The second thing to notice in Figure 4 is how steeply *PrF** rises in every county as the likelihood of drought increases.

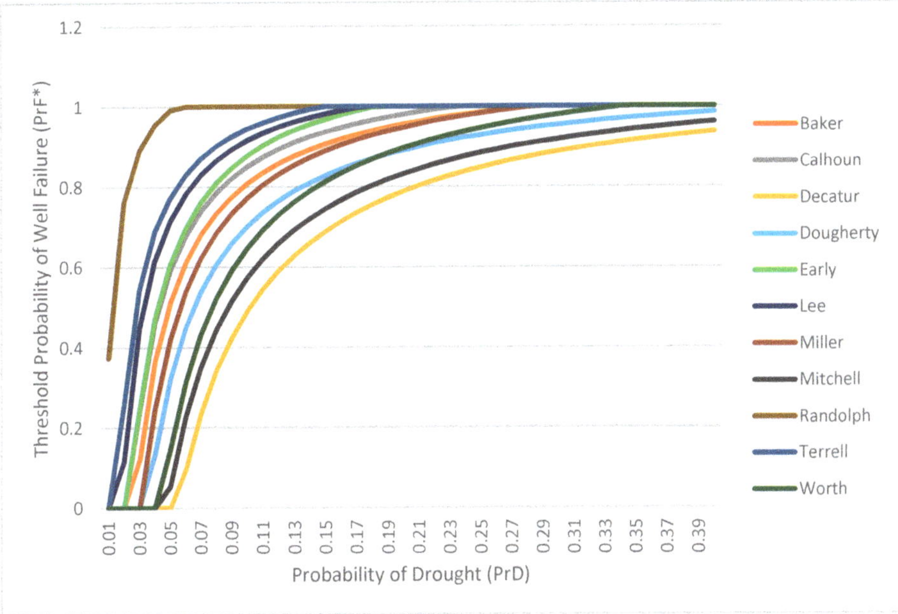

Figure 4. Threshold probability of well failure as a function of probability of drought. Note: If the actual likelihood of well failure is greater than *PrF**, then an irrigation buyout auction is more cost-effective than emergency source switching, and vice versa.

To put the probability of drought into perspective, there were two auctions in the lower FRB from 2001 to 2020, i.e., a 10% probability of an auction. However, as mentioned in the introduction, over the last 25 years, drought conditions occurred in 10 years (40% of the time) in the lower FRB, although no funds were budgeted for an irrigation buyout auction in eight of those years. If the drought probability were 10%, emergency source switching would always be more cost-effective in Randolph county, regardless of the likelihood of well failure; even in Decatur county, the county with the highest source switching costs, the likelihood of well failure would have to exceed 49% for the auction to be more cost-effective.

As mentioned in the introduction, over the last 25 years, drought conditions occurred in 10 years (40% of the time) in the lower FRB, although no funds were budgeted for an irrigation buyout auction in eight of those years. At that frequency of drought, for the auction to be more cost-effective, the likelihood of well failure would need be to over 93% in every county. A well failing 93% of the time would only happen if the well construction was exceptionally poor, or if the water table of the aquifer was extremely sensitive to irrigation pumping. However remote it is, the latter is more likely to occur as more withdrawals are switched from the original source to the new source.

Identifying the Threshold Value of Economic Impacts

In this section, rather than relying on the IMPLAN (version 6.9) results, we allow the regional economic impacts of taking a field out of production to vary. By doing so, we can identify, for a given *PrF* and *PrD* pair, how high the regional economic impacts would have to be for source switching to become more cost-effective than an auction. We use Equation (7) to solve for the threshold regional economic impacts (EI^*). Table 4 presents the results for two scenarios: (1) when the probability of drought is 0.1 and the probability of well failure is 0.01, and (2) when the probability of drought is 0.1 and the probability of well failure is 0.25. Also included for reference are the regional economic damages estimated by IMPLAN (version 6.9). The results in Table 4 show that emergency source switching would be more cost-effective than an irrigation buyout auction even if the regional economic impacts were significantly lower than the IMPLAN (version 6.9) estimates. This is the case even when there is a one-in-four chance the new well will fail each year.

Table 4. Threshold level of regional economic impacts (EI^*) from taking a 150-acre irrigated field out of production, given drought and well failure probabilities.

County	*EI* from IMPLAN [a]	EI^* [b]	
		$PrD = 0.1, PrF = 0.01$	$PrD = 0.1, PrF = 0.25$
Baker	USD 208,931	USD 40,649	USD 53,657
Calhoun	USD 204,737	USD 31,107	USD 41,062
Decatur	USD 198,169	USD 102,026	USD 134,674
Dougherty	USD 242,616	USD 73,331	USD 96,797
Early	USD 165,835	USD 20,620	USD 27,218
Lee	USD 214,624	USD 18,509	USD 24,432
Miller	USD 179,630	USD 41,374	USD 54,613
Mitchell	USD 190,392	USD 82,291	USD 108,624
Randolph	USD 188,913	USD 0	USD 0
Terrell	USD 215,261	USD 12,282.24	USD 16,212.56
Worth	USD 153,219	USD 54,757.60	USD 72,280.03

Notes: [a]: Values in this column are the total of the direct, indirect, and induced economic impacts and the change in tax revenue reported in Table 1, i.e., the total regional economic impact. [b]: EI^* is the amount of total regional economic impact above which emergency source switching is more cost-effective than an irrigation buyout auction, given the probability of drought and the probability of well failure.

Figure 5 plots EI^* as a proportion of the IMPLAN (version 6.9) estimates across a range of well failure probabilities. In panel (a), the probability of drought is set to 0.1 (the proportion of times an auction was held between 2001 and 2020); in panel (b), $PrD = 0.4$ (the proportion of times a drought occurred in the study area between 2001 and 2020). When the likelihood of drought is 0.4, emergency source switching is strictly more cost-effective than an irrigation buyout auction in eight of the eleven counties in the study area. In the three other counties (Decatur, Mitchell, and Dougherty) emergency source switching would be the cost-effective policy when the actual regional economic impacts are a fraction of the IMPLAN (version 6.9) estimates.

In 2006, when the Georgia Environmental Protection Division evaluated the costs of the 2001 and 2002 irrigation buyout auctions, the costs of the auction focused exclusively on the payments to farmers [4]. This is analogous to setting the regional economic impacts to zero. Our final scenario examines this situation. Here, we set the regional economic impacts to zero and calculate the threshold probability of drought ($PrD^*_{EI=0}$) above which emergency source switching is more cost-effective. The results of doing so are presented in Table 5. The value of $PrD^*_{EI=0}$ ranges from a low of 0.05 in Randolph county to a high of 0.7 in Decatur county. In other words, if the regional economic impacts are ignored, in Decatur county, emergency source switching would only appear to be an attractive alternative to an irrigation buyout if a drought was expected to occur in 7 out every 10 years, or more frequently. From Figure 5b, we can see that the regional economic impacts do not have to be very high to make emergency source switching more cost-effective when the likelihood of drought is 0.4. This is true even in Decatur county.

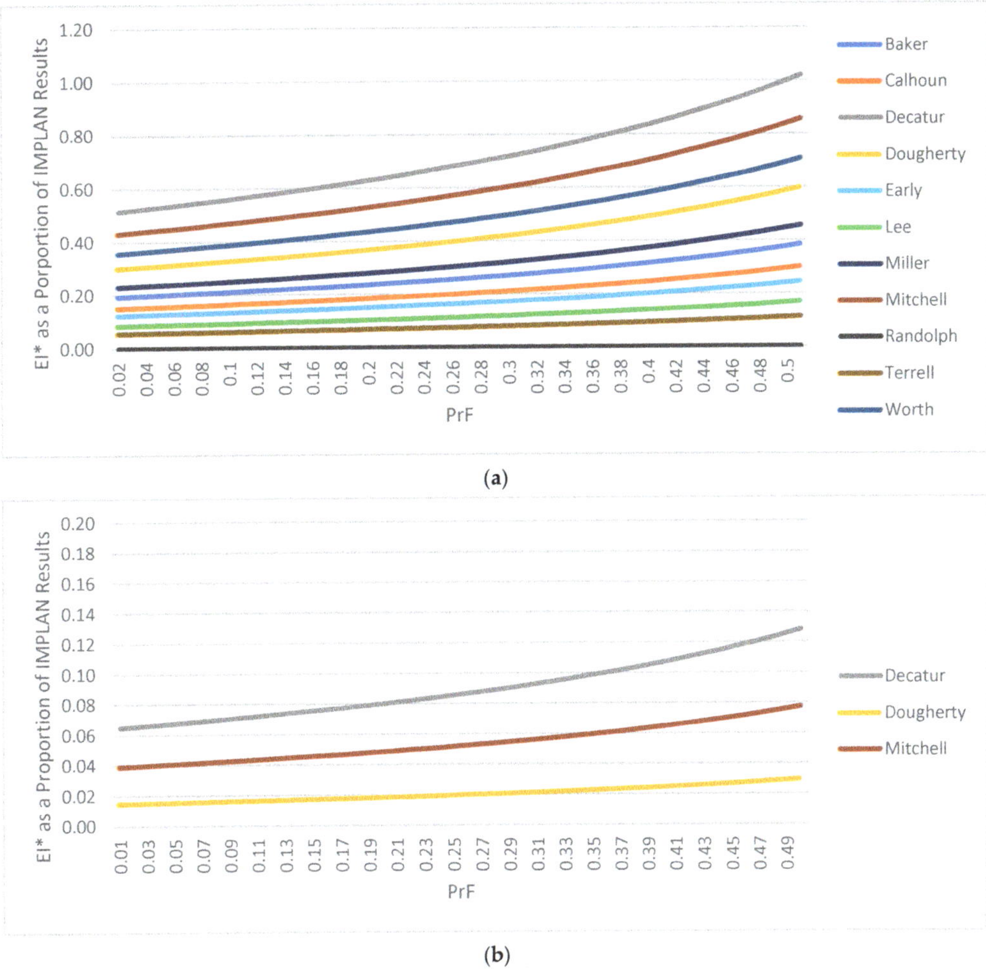

Figure 5. Threshold level of regional economic impacts (*EI**) as a function of probability of well failure (*PrF*). (**a**) Probability of drought = 0.1. (**b**) Probability of drought = 0.4.

Table 5. Threshold probability of drought when regional economic impacts are zero (*EI* = USD 0).

County	$PrD^*_{EI=0}$
Baker	0.29
Calhoun	0.24
Decatur	0.70
Dougherty	0.47
Early	0.18
Lee	0.17
Miller	0.29
Mitchell	0.57
Randolph	0.05
Terrell	0.15
Worth	0.34

4. Discussion

From an ecological perspective, the lower Flint River Basin in southwest Georgia supports a diverse array of freshwater aquatic species, including several mussel populations protected by the Endangered Species Act. From an economic perspective, agricultural irrigation withdrawals from the region's surface water and groundwater resources are critical to its economic vitality. To ensure the long-term ecological and economic sustainability of the region, water policy options need to be evaluated and deployed with respect to their cost-effectiveness.

The hydrologic connectivity between surface water and the Floridan aquifer in the lower FRB is well documented [19]. Over the past 20 years, the state of Georgia has considered paying to switch surface water and Floridan aquifer withdrawals to deeper aquifers with less hydrologic connectivity to the streams. The state has also funded irrigation buyout auctions to actively protect instream flows during times of drought. When funds are limited for water management, it is critical to select a policy that can achieve sustainability goals—both economic and ecological—in a cost-effective manner. The methodology developed in this study identifies the essential components and key parameters for evaluating the cost-effectiveness of these two policies. These include accounting for the full range of regional economic impacts, including the indirect and induced effects of lost agricultural production, and understanding the capacity and reliability of the new source to accommodate withdrawals.

A state agency with a limited budget can use the information in Figure 4 to prioritize source switching investments by county. That same information can also prioritize investments in research related to the hydrologic features of the new source. For example, because Randolph county has the highest threshold probability of well failure, it should be the first county to focus funds on source switching. Source switching should be considered in the other counties in descending order of their PrF_c*. On the other hand, because Decatur county has the lowest PrF_c*, it is especially important to understand the hydrology of the Claiborne aquifer in that area, so research funds should be directed there first.

There are two sources of uncertainty related to the cost-effectiveness of these water management policies: the likelihood of drought, and likelihood of well failure. Given recent climate trends and the anticipated acceleration of those trends in the future, the probability of drought is likely to increase, thereby increasing the expected present value of irrigation buyout auctions. Importantly, the probability of drought is something that cannot be managed at the state level. There is, however, some ability to manage the probability of well failure through a better understanding of the spatial variability of maximum sustainable yields from the Claiborne aquifer. That information is essential to the judicious selection of where and how much to invest in source switching to protect the economic and ecological health of the lower Flint River Basin.

The cost-effectiveness of irrigation management policies to address water shortages has been investigated in multiple settings. Ding and Peterson [8] compared the relative cost-effectiveness of subsidizing improved irrigation technology to paying farmers for not irrigating. They found that crop prices and aquifer water levels were important parameters in determining the relative cost-effectiveness of those policies. The regional economic impacts of converting irrigated fields to dryland production, however, were not evaluated.

Similarly, Luitel et al. [9] modeled irrigation water restrictions coupled with irrigation water trading in the Ogallala aquifer in Texas. In that paper, the authors state that both the restrictions and water trading would have definite regional economic impacts, but those impacts are not assessed in their analysis.

The impacts of irrigation efficiency gains have been examined in multiple countries. Scott et al. found that gains at the intensive margin (water use per hectare) would likely be negated through an expansion of the extensive margin (irrigated hectare) in Chile, the US, and Spain [20]. The implications for water scarcity are discussed but the regional economic impacts are not estimated.

Mulligan et al. parameterize an optimal control model to compare a water quota to a water tax in the Republican River basin in the US [21]. That model was optimized with respect to the agents' profits, not regional economic performance. The linkages across economic sectors were not incorporated into the impacts of groundwater policy in the optimization model.

In China, Pang et al. evaluated the costs of achieving stream-flow targets through irrigation restrictions by compensating farmers for their lost revenues [22], although the regional economic impacts of those lost revenues were not taken into consideration. Our results suggest that this is a mistake that would likely significantly underestimate the true costs of the policy. Also in China, Zou et al. compare the cost-effectiveness of alternative irrigation technologies for mitigating climate change's impacts on agriculture [23].

In Spain, Ballesteros-Olza et al. [24] investigated the effects of using reclaimed water as a substitute for groundwater; Perni and Martinez-Paz used stakeholder surveys to estimate the cost-effectiveness of policies to restore waterways [25]; and Blanco-Gutierrez et al. [10] examined the cost-effectiveness of irrigation water price structures and irrigation water markets. None of those studies included the regional economic impacts of these policies in their analyses. Blanco-Gutierrez et al. do, however, conclude that "Additional studies on net social costs are highly recommended" [10].

Aulong, Bouzit, and Dorfliger also stress the importance of including social costs in their case studies of river basin management in Lebanon and Jordan [26]. In particular, they acknowledge the need to account for environmental costs and resource scarcity costs, but do not attempt to trace the sectoral linkages and resulting economic impacts of water management policy through their study regions.

The key features of the methodology developed in this paper are that it explicitly accounts for (1) the regional economic impacts of irrigation water management policies and (2) the uncertainty of drought frequency. Furthermore, the methodology is applicable to other locations weighing multiple policy options to address water scarcity. The state of California, for example, is considering expanding their forced groundwater recharge program in which surface water is injected into the ground as a storage mechanism. This is analogous to source switching, with the injection costs taking the place of drilling costs. It is an expensive proposition at face value. There are, however, likely to be significant regional economic impacts of fallowing irrigated land when surface water is unavailable. The lesson from our analysis is that considering the regional economic impacts is critical to conducting an accurate assessment of the cost-effectiveness of such policies.

Supplementary Materials: The following supporting information can be downloaded at: https://www.mdpi.com/article/10.3390/w15193381/s1, Table S1: 2020 Harvested irrigated acres by crop and county; Table S2: 2020 Share of irrigated acreage, by crop and county; Table S3: 2020 Crop price and yield, by county; Table S4: Well depth data; Table S5: Water application rates (acre-feet/acre) by county and crop; Equations (S1)–(S3): Calculating $EI_{\text{Ag Direct}, c, t}$; Equation (S4): Calculating county-level fixed costs of source switching; Equations (S5)–(S8): Calculating county-level extra pumping costs of source switching.

Author Contributions: J.D.M.: conceptualization, methodology, writing—original draft preparation and review and editing; Y.N.: data curation, writing—original draft preparation. All authors have read and agreed to the published version of the manuscript.

Funding: The authors would like to thank the United States Department of Agriculture for providing partial funding for this project through Multistate Project W4190.

Data Availability Statement: All data used in this study are presented in the main text of the article, above, and in the Supplementary Mterials accompanying the article.

Acknowledgments: The authors would like to thank the College of Agricultural and Environmental Sciences for its support during this project.

Conflicts of Interest: The authors declare no conflict of interest.

References

1. National Drought Mitigation Center at the University of Nebraska-Lincoln (NDMC), the United States Department of Agriculture, and the National Oceanic and Atmospheric Administration. United States Drought Monitor: Time Series. 2019. Available online: https://drought.unl.edu/ (accessed on 5 June 2020).
2. U.S. Geological Survey. National Water Information System: USGS Water Data for the Nation. 2021. Available online: https://waterdata.usgs.gov/nwis (accessed on 8 August 2022).
3. Georgia Department of Natural Resources. The Flint River Basin: Technical Summary of Hydrogeology, Farm Water Use, and Ecology. 2004. Available online: https://www1.gadnr.org/frbp/Assets/Documents/SAC_Executive_Summary.pdf (accessed on 7 June 2020).
4. Couch, C.A.; McDowell, R.J. *Flint River Basin Regional Water Development and Conservation Plan*; Georgia Department of Natural Resources Environmental Protection Division: Atlanta, GA, USA, 2006; pp. 18–56.
5. Gordon, D.W.; Painter, J.A. Groundwater conditions in Georgia, 2015–2016. *US Geol. Surv. Sci. Investig. Rep.* **2018**, *59*, 2017–5142. [CrossRef]
6. U.S. Fish & Wildlife Service. Species Reports. Environmental Conservation Online System. 2019. Available online: https://ecos.fws.gov/ecp/ (accessed on 15 August 2022).
7. Mullen, J.D. Agricultural Water Policy During Drought: A Strategy for Including Groundwater Permits in Future Irrigation Buyout Auctions in the Flint River Basin. *Water* **2019**, *11*, 151. [CrossRef]
8. Ding, Y.; Peterson, J. Comparing the Cost-Effectiveness of Water Conservation Policies in a Depleting Aquifer: A Dynamic Analysis of the Kansas High Plains. *J. Agric. Appl. Econ.* **2012**, *44*, 223–234. [CrossRef]
9. Luitel, K.P.; Tewari, R.; Mitchell, D.M.; Benson, A.; Bancorporation, Z.; Johnson, P. Evaluating the Possibility of a Water Trading Scenario Coupled With Water Use Restriction in the Texas Southern High Plains. *Environ. Manag. Sustain. Dev.* **2015**, *4*, 194. [CrossRef]
10. Blanco-Gutiérrez, I.; Varela-Ortega, C.; Flichman, G. Cost-effectiveness of groundwater conservation measures: A multi-level analysis with policy implications. *Agric. Water Manag.* **2011**, *98*, 639–652. [CrossRef]
11. Balana, B.B.; Vinten, A.; Slee, B. A Review on Cost-Effectiveness Analysis of Agri-Environmental Measures Related to the EU WFD: Key Issues, Methods, and Applications. *Ecol. Econ.* **2011**, *70*, 1021–1031. [CrossRef]
12. Convery, F.J. Reflections—Shaping Water Policy: What Does Economics Have to Offer? *Rev. Environ. Econ. Policy* **2013**, *7*, 156–174. [CrossRef]
13. English, B.C.; Menard, R.J.; Wilson, B. The Economic Impact of a Renewable Biofuels/Energy Industry Supply Chain Using the Renewable Energy Economics Analysis Layers Model System. *Front. Energy Res.* **2022**, *10*, 780795. [CrossRef]
14. Howitt, R.; MacEwan, D.; Medelin-Azuara, J.; Lund, J.; Sumner, D. *Economic Analysis of the 2015 Drought for California Agriculture*; UC Davis Center for Watershed Sciences: Davis, CA, USA, 2015.
15. Li, Y.; Mei, B.; Linhares-Juvenal, T. The Economic Contribution of the World's Forest Sector. *For. Policy Econ.* **2019**, *100*, 236–253. [CrossRef]
16. U.S. Bureau of Labor Statistics. Available online: https://www.bls.gov/data/inflation_calculator.htm (accessed on 22 August 2021).
17. Gordon, D.W.; Gonthier, G. *Hydrology of the Claiborne Aquifer and Interconnection with the Upper Floridan Aquifer in Southwest Georgia*; USGS Report; US Geological Survey: Reston, VA, USA, 2017. [CrossRef]
18. Rogers, D.H.; Alam, M. Comparing irrigation energy costs. In *Irrigation Management Series MF-2360, Kansas State University Agricultural Experiment Station and Cooperative Extension Service*; Extension Service Publication: Manhattan, KS, USA, 1999; 4p.
19. Rugel, K.; Golladay, S.W.; Jackson, C.R.; Rasmussen, T.C. Delineating groundwater/surface water interaction in a karst watershed: Lower Flint River Basin, southwestern Georgia, USA. *J. Hydrol. Reg. Stud.* **2016**, *5*, 1–19. [CrossRef]
20. Scott, C.A.; Vicuña, S.; Blanco-Gutiérrez, I.; Meza, F.; Varela-Ortega, C. Irrigation efficiency and water-policy implications for river basin resilience. *Hydrol. Earth Syst. Sci.* **2014**, *18*, 1339–1348. [CrossRef]
21. Mulligan, K.B.; Brown, C.; Yang, Y.-C.E.; Ahlfeld, D.P. Assessing groundwater policy with coupled economic-groundwater hydrologic modeling. *Water Resour. Res.* **2014**, *50*, 2257–2275. [CrossRef]
22. Pang, A.; Sun, T.; Yang, Z. Economic compensation standard for irrigation processes to safeguard environmental flows in the Yellow River Estuary, China. *J. Hydrol.* **2013**, *482*, 129–138. [CrossRef]
23. Zou, X.; Li, Y.; Cremades, R.; Gao, Q.; Wan, Y.; Qin, X. Cost-effectiveness analysis of water-saving irrigation technologies based on climate change response: A case study of China. *Agric. Water Manag.* **2013**, *129*, 9–20. [CrossRef]
24. Ballesteros-Olza, M.; Blanco-Guttierrez, I.; Esteve, P.; Gomez-Ramos, A.; Bolinches, A. Using reclaimed water to cope with water scarcity: An alternative for agricultural irrigation in Spain. *Environ. Res. Lett.* **2022**, *17*, 125002. [CrossRef]
25. Perni, A.; Martínez-Paz, J.M. A participatory approach for selecting cost-effective measures in the WFD context: The Mar Menor (SE Spain). *Sci. Total Environ.* **2013**, *458–460*, 303–311. [CrossRef] [PubMed]
26. Aulong, S.; Bouzit, M.; Dörfliger, N. Cost–Effectiveness Analysis of Water Management Measures in Two River Basins of Jordan and Lebanon. *Water Resour. Manag.* **2009**, *23*, 731–753. [CrossRef]

water

MDPI

Article

Management Techniques of Ancestral Hydraulic Systems, Nasca, Peru; Marrakech, Morocco; and Tabriz, Iran in Different Civilizations with Arid Climates

Doris Esenarro [1,2,*], Jesica Vilchez [1,2], Marie Adrianzen [1], Vanessa Raymundo [1,2], Alejandro Gómez [1] and Pablo Cobeñas [1]

1 Faculty of Architecture and Urbanism, Ricardo Palma University (URP), Lima 15039, Peru; jesica.vilchez@urp.edu.pe (J.V.); 201310161@urp.edu.pe (M.A.); 202112586@urp.edu.pe (V.R.); alejandro.gomez@urp.edu.pe (A.G.); pablo.cobenas@urp.edu.pe (P.C.)

2 Research Laboratory for Formative Investigation and Architectural Innovation (LABIFIARQ), Lima 15039, Peru

* Correspondence: doris.esenarro@urp.edu.pe

Abstract: The research aims to evaluate various management techniques of Ancient Hydraulic Systems (AHS) in different civilizations in arid climates, in cities located in Nasca in Perú (Puquio), Marrakech in Marruecos (Khettara) andTabriz in Irán (Qanat). The scarcity of water resources in these areas compelled the inhabitants to seek water management solutions to meet the necessary water supply for the population at the time. The methodology employed was a case study in which climatic data, supply, and operation of AHS were analyzed. The different indicators studied resulted in findings that, in the case of Nasca, the system relied on lintels, utilizing robust materials such as stone. They employed geometry to control water flow velocity, inclined walls to prevent collapses, and terraces to facilitate access to underground galleries. In the cases of Tabriz and Marrakech, their systems were based on excavations and reinforcements primarily using clay and earth as materials. In conclusion, the techniques employed in different civilizations are responses to contextual realities, offering an adaptive solution to environmental and physical challenges with a sustainable focus within their immediate surroundings.

Keywords: hydraulic systems; ancestral hydraulic systems; Puquio; Khettara; Qanat

1. Introduction

Since ancient times, water resources have been crucial for both domestic use and agriculture. Approximately 20% of the world's water supply is used for irrigating cultivable land, which generates about 40% of the global food supply and 60% of cereals [1]. Moreover, 80% of the water is on the surface, while the rest is underground or in the form of vapor in the atmosphere [2]. These facts are intrinsically related to global challenges outlined in the 6th Sustainable Development Goal (SDG), which promotes sustainable water and sanitation management, which takes special significance in arid or semi-arid regions [2]. Due to its limited availability, water requires careful management to meet human needs, protect ecosystems, and mitigate climate change [3,4].

T5% of the world's population lives in the driest half of the planet, which means that millions of people do not have access to this resource [1]. Therefore, ancient techniques were developed for water collection, such as Khettaras or Qanats [5]. In the Middle East, they are called Falaj or Qanat; in the Mediterranean, they are known as Foggara or Khettara; in Central Asia, Karez; in Peru, there is the Puquio; and in Japan, Manbo [6]. Furthermore, throughout the history of different cultures, they have been called various names such as quanat, Canant, moonlight, kanat, khanate, khad, kanaye, or ghannat in Iran; kahrez, kariz, kah-riz, karaz, or kakoriz in Southeast Asia; mayon, iffeli, mgoula, khottara, or rhettara in North Africa; and falaj, aflaj, or felled in Arabia [7,8].

It is generally assumed that they originated under Persian influence (Iran and/or the Arabian Peninsula) since these regions have a significant number of Qanats today. They then spread westward and southward into Europe during the reign of Alexander III of Macedon and the New World [9]. As for the Nasca Puquios, authors view them as a byproduct of the Nasca culture [9–12].

The application of these ancestral water collection and distribution techniques in agriculture has proven beneficial in arid and semiarid regions where water resources are limited [13–16]. These techniques first involve identifying available underground water sources and then constructing and maintaining channels and vertical wells that connect a point on the surface with groundwater, creating a slope [12,16,17]. The extracted water is distributed to the fields through channels and gravity-based irrigation systems [13,17]. These traditional techniques, adapted to local conditions, enable efficient and sustainable water utilization in agriculture [16,17].

Throughout history, human communities have faced difficulties in accessing water due to geographic, climatic, and demographic factors; likewise, AHS has encountered an increasing number of issues [18–20]. These challenges include rapid population growth, irregular rainfall, droughts, depletion of groundwater and aquifers, lack of maintenance, new technologies, the disorderly growth and urban expansion in rural areas, disappearance of construction techniques, disconnect between governance-population, and water pollution [1,3,5,13,21,22].

The growth of the global population is occurring at a significant pace, with an annual increase of approximately 80 million individuals. This situation entails a requirement of about 64 billion cubic meters of freshwater each year [1]. This scenario indicates that the demand for water will continue to escalate in the coming years, exerting increasing pressure on the available land resources [1,23,24].

Similarly, water scarcity is common in arid regions due to the irregularity of rainfall, which varies in time and intensity [21]. Since heavy rainfall events are characteristic of drylands and desert margins, a large portion of the water is lost through surface runoff and does not get stored in AHS [21]. On the other hand, droughts are a problem for communities as they can last for extended periods and cause water shortages, as the water is quickly absorbed by the soil or evaporates before it can be utilized [5,18,19,21]. This has a negative impact on agricultural production, which in turn can lead to food insecurity and economic problems, especially in communities whose main source of income is agriculture.

The extraction of groundwater often exceeds natural recharge, resulting in persistent depletion of groundwater and aquifers, which subsequently impacts environmental flows in rivers and places associated ecosystems in a vulnerable situation [5,25,26]. This excessive extraction, coupled with intensified water exploitation, reduces the use of AHS as they do not have sufficient supplies to maintain their flow [5,22].

Similarly, in these arid regions, there is a tendency for disconnection between their governments and the population regarding collaborative and controlled management, as there have been cases of Khettaras drying up due to the absence of proper management [13]. Additionally, they face indifference from the population towards their maintenance and preservation, partly due to the high economic cost of drilling for access to deeper groundwater but also due to the lack of knowledge among the population about their existence and importance for sustainable development and cultural heritage [3,5]. An example of this lack of care from the population and their governments, as well as their disinterest, can be seen in Figure 1.

With technological advancements, new modern hydraulic systems such as deep wells and underground water pumps have been developed, among others. These systems deplete water resources, impact the surrounding ecosystems, and tend to degrade the soil [25,27]. Furthermore, the disorderly growth and urban expansion in rural areas irrigated and supplied by these systems, such as the pumping system, have caused deterioration and destruction of some of AHS, severing their ties with traditional social structures and the community, leading to abandonment by the population [13]. Likewise, construction techniques that were once used (AHS) have vanished due to the decline of specialized social groups in construction, as subsequent generations have lost this knowledge, thus limiting the capacity to develop new AHS [3].

Figure 1. Photograph of the deteriorated Khettara in the study area, Mechouar-Kasbah city, Marrakech Prefecture.

In addition to these issues, a significant problem arises related to pollution, which intensifies due to constant interactions between groundwater and surface waters, creating a profound effect on their chemical composition, which in turn can result in a decrease in the availability of clean and safe water for use [26,28]. In other words, this interaction has practical consequences on the quantity and quality of water in any of the systems [25,26].

Great hydraulic engineering works were carried out that were significant in their time and continue to have a great impact on engineering, such as in Peru (city of Nasca), Morocco (city of Mechouar-Kasbah), and Iran (city of Tabriz) [4].

Therefore, the present investigation aims to evaluate the various management approaches of Ancient Hydraulic Systems (Puquio, Khettara, Qanat) in different existing civilizations of arid climates.

2. Materials and Method

In this section, a detailed description of the study areas is provided, followed by a comprehensive explanation of the methodology employed in the present study.

2.1. Places of Study

Three cities belonging to arid climates were analyzed: Nasca City in Ica, Peru (Figure 2A), which is located near the equator in South America; while Mechouar-Kasbah City in Marrakech-Safi, Morocco (Figure 2B) is situated in the northern region near the Greenwich line in Africa, and finally Tabriz City in Tabriz County, Iran (Figure 2C), which is located to the east of the Greenwich line in the Middle East [29–31].

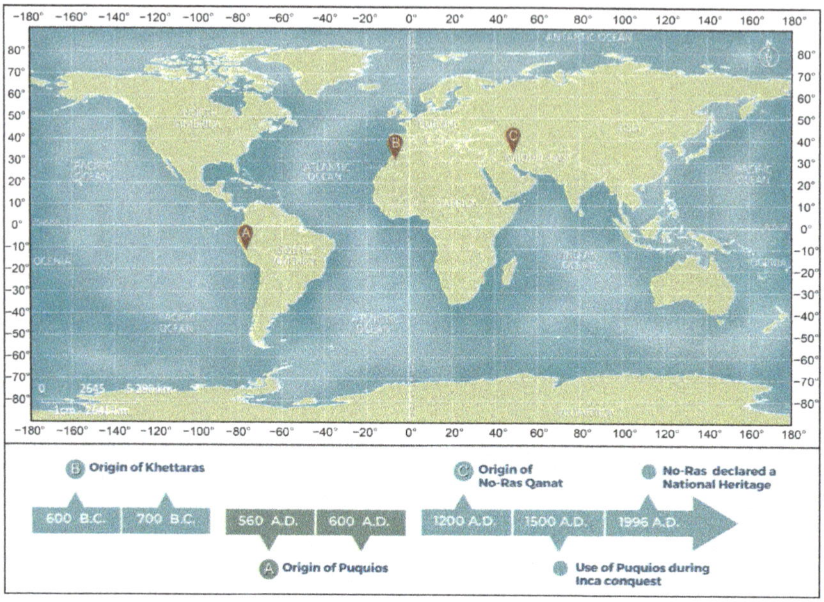

Figure 2. Study sites of the investigation where Ancient Hydraulic Systems (**A**) Puquio, (**B**) Khettara, (**C**) Qanat are found.

2.1.1. Cantalloc Aqueduct, Puquio Hydraulic System—Nasca City

The Puquio is a hydraulic system that means "fuente" or "manantial" [9,32]. However, in the context of the Nasca culture, it is known as "huncólpi," which consists of a network of channels, both surface and underground, that allow the extraction of subterranean waters [9]. In Ica, Nasca, the Cantalloc aqueduct is supplied by the lower Tierras Blanca's River, and this river serves 135 users and 908.18 hectares through irrigation [33]. These were constructed by the indigenous populations before the arrival of the Spanish and represented a valuable legacy of hydraulic engineering [12]. In Ica, alongside traditional underground aqueducts, modern drip and sprinkler irrigation systems have also been introduced in certain commercial crops [34]. The archaeological studies show that puquios were absent during the Early Nasca period (Early Intermediate Period) and were used during the Inca conquest in the 16th century [11]. Furthermore, radiocarbon analyses by accelerator mass spectrometry reveal that puquio structures have existed since cal AD 560–650 and cal AD 600–660 [35].

2.1.2. Khettara Hydraulic System—Mechouar-Kasbah City

The Khettara is a hydraulic system that, through the use of gravity, transports water from the water table to the surface, allowing for the irrigation of fields in oases [3,36]. These structures are characterized by their resilience, adaptability, and sustainability, as they leverage ancestral knowledge to address environmental challenges and ensure a sustainable water supply for agriculture; with management rules in place, they ensure maintenance and a fair distribution of water among farmers, thus avoiding overexploitation of water resources and maximizing benefits for communities [37]. In Morocco, sprinkler irrigation projects and drip irrigation systems have been developed to optimize water use in agriculture, sometimes used in conjunction with khettaras [38].

In Morocco, khettaras are known for their significance in water capture and distribution within a context where 80% to 95% of water resources are allocated to agriculture, with around 40% sourced from groundwater [5]. These structures are recognized to have

Persian influence, as seen in Iran and the Arabian Peninsula. It is considered that they were invented between 600 and 700 BC [13].

2.1.3. Qanat Hydraulic System—Tabriz City

Qanats are an ancient solution for water supply in arid and semi-arid regions, playing a vital role in the history of irrigation and human settlement [39,40]. In different regions, Qanats are constructed in a traditional manner and are essential for irrigation and domestic water supply in numerous settlements [17]. Their use has a significant impact on the socioeconomic organization and patterns of land ownership and tenure [14]. Furthermore, Qanats are not just a technique for accessing and managing groundwater; they have also influenced the landscape and prosperity of Iran, making their importance go beyond the technical aspects and being key to understanding the culture and civilization in the Iranian plateau [14]. Moreover, dams and drip irrigation systems have been constructed in Iran to store and enhance water use efficiency, respectively [41,42].

The use of qanats in Tabriz, such as No-Ras, dates back to the 13th century AD, having been a part of various hydraulic activities during the Mongol rule in the region [39,43]. In 1996, the No-Ras Qanat, particularly its final stretch encompassing the Fath Abad garden, point of outlet, and the historic Garden Mansion, was recognized as a historical heritage [39]. The majority of the qanat's route traverses the village of Chavan, which houses an average of 2000 inhabitants partially interconnected through the use of No-Ras, extending to the suburbs of Tabriz [39].

2.2. Locations of Ancestral Hydraulic Systems

2.2.1. Case of Nasca City: Cantalloc Aqueduct Puquio

Figure 3C shows the route of the Cantalloc Aqueduct (Puquio), which has a length of 371.8 m. It originates from the Tierra Blanca River at point (1), located at an elevation of 640 m.a.s.l., with geographic coordinates 14°49′33.36″ south latitude and 74°54′38.23″ west longitude [29]. The aqueduct extends to a midpoint where it branches into two sections: One is the endpoint (2), which is located at an elevation of 638 m.a.s.l., with coordinates 14°49′35.87″ south latitude and 74°54′41.23″ west longitude; and the other end reaches a point (3) situated at 647 m.a.s.l., with coordinates 14°49′36.94″ south latitude and 74°54′31.38″ west longitude [29].

Figure 3. (**A**) Map of the South American continent, country Peru, Ica department; (**B**) Map of Nasca province and city of Nasca; (**C**) Map of the intervention area corresponding to the Puquio of the Cantalloc Aqueduct in the city of Nasca.

2.2.2. Case of Mechouar-Kasbah City: Khettara

Figure 4C shows the route of the Khettara in the city of Mechouar-Kasbah, which has a length of 660 m. It has different points, one being point (1) located at an altitude of 476 m.a.s.l., with geographic coordinates of 31°36′16.13″ north latitude and 7°59′44.48″ west longitude, and it extends to point (2) which is located at an altitude of 479 m.a.s.l., with coordinates of 31°36′5.47″ north latitude and 7°59′42.76″ west longitude [30]. Another point (3) is situated at an altitude of 477 m.a.s.l., with coordinates of 31°36′13.81″ north latitude and 7°59′42.11″ west longitude, and it extends to point (4), which is located at an altitude of 479 m.a.s.l., with coordinates of 31°36′4.94″ north latitude and 7°59′38.02″ west longitude [30].

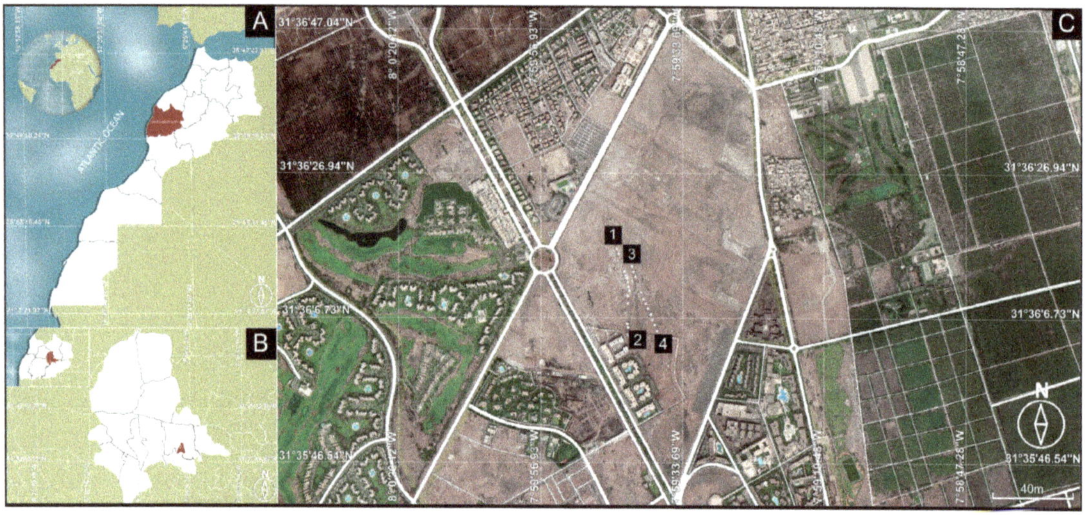

Figure 4. (**A**) Map of the North Africa continent, country Morocco, Marrakech-Safi region; (**B**) Map of Marrakech Prefecture and Mechouar-Kasbah city; (**C**) Map of the intervention area, corresponding to the Khettara of Mechouar-Kasbah city.

2.2.3. Case of Tabriz City: No-Ras Qanat

The study site is located on the outskirts of the city of Tabriz, which is situated in the southern region of the Asian continent in the country of Iran. It serves as the capital of East Azerbaijan province and is part of Tabriz County [31,39]. The site forms part of a complex environmental and urban system that revolves around the city of Tabriz, the Zagros Mountains, and Lake Urmia [40].

Figure 5C shows the route of the No-Ras Qanat of Tabriz, which has an approximate length of 3.5 km [31]. It extends from the slopes of Sahand Mountain, with its starting point (1) at an altitude of 1780 m.a.s.l., with geographical coordinates of 37°58′52″ latitude north and 46°23′60″ longitude east, near the village of Chavan, to the drainage located in the Fath Abad garden, with its endpoint (2) at an altitude of 1683 m.a.s.l., and coordinates of 38°00′26″ latitude north and 46°23′19″ longitude east [31]. There is an interruption at the midpoint (3) due to the presence of a sand and gravel mine [39].

2.3. Methodology

The study was divided into three phases, as illustrated in Figure 6, and was conducted according to the sections detailed below.

Figure 5. (**A**) Map of Middle East, Iran country, East Azerbaijan province, (**B**) Map of Tabriz County and Tabriz city (**C**) Map of intervention area, No-Ras Qanat de Tabriz city.

Phase	1 Climate Analysis	2 Supply and Operation Analysis	3 Analysis and interpretation of results
Goal	Climate analysis and its influence on the studied ancestral hydraulic systems.	Analysis of supply and operation of the three studied ancestral hydraulic systems.	Analysis of distribution, sizing, soil typology, construction system, and materials for the three ancient hydraulic systems.
Data Sources	(A) Weather Spark is a platform for collecting climate data through EPW files.	(A) Government data (B) Scientific articles (C) Autocad (D) Google Earth	(A) Historical and Archaeological sources (B) Scientific articles (C) Autocad (D) Google Earth
Criteria for Selection and/or Classification of Information	(A) Climatology	(A) Supply and operation	(A) Distribution and Sizing (B) Soil Typology (C) Construction Systems and Materials
Result	Assess the influence of climate in the study locations.	Understand the supply and operation of the rivers that impact each study location.	Understanding how physical and spatial factors influence the three studied systems.

Figure 6. The methodology applied in the study.

2.3.1. Climate Analysis

Climatic analysis is a fundamental tool for the evaluation of groundwater resource quality in different regions [44]. This process involves a meticulous climatological study that encompasses essential aspects such as temperature, wind speed, humidity, and precipitation in the various intervention areas of arid climates. This study allows for an understanding of how these factors impact ancestral hydraulic systems [44,45]. The process is detailed as follows:

1. Data compilation of meteorological information from the Weather Spark EPW for the year 2022, which includes temperature (°C), wind speed (km/h, hours per year), humidity (%), and monthly and annual precipitation (mm);
2. Rigorous statistical processing of the collected data;
3. Generation of graphs that present monthly data for parameters such as maximum and minimum temperature, maximum and minimum humidity, maximum and average annual precipitation, as well as monthly wind speed;
4. Analysis of the results obtained from each region and their influence on the studied hydraulic systems.

2.3.2. Supply and Operation Analysis

The initial phase of hydrographic analysis begins with accurate geolocation of each study region, supported by detailed cartographic representations that provide a comprehensive view of the geographical arrangement of rivers in each area and their intrinsic relationship with the ancestral hydraulic system. Next, a thorough characterization of the river basins in each region is carried out, investigating key aspects such as their length and flow direction and delving into the geological peculiarities that define them. Furthermore, a meticulous analysis of the river discharge is conducted, contributing to a more comprehensive understanding of its impact on local water supply and, ultimately, on the operation and relevance of the ancestral hydraulic system in its respective geographical context.

2.3.3. Analysis and Interpretation of Results

The final phase of the research analyzes the three hydraulic systems based on the following criteria:

- Analysis of Distribution Area and Sizing: In this stage, a meticulous collection of information is conducted, including the location of each ancestral hydraulic system, a description and sizing of its underground components, a specific distribution of its elements, a detailed schematic section, and an evaluation of the topography of the study area.
- Soil Typology: This analysis is based on the collection of relevant data regarding the types of soil present in each intervention region, which includes a detailed description of soil characteristics such as texture, water retention, infiltration capacity, and erosion. Additionally, it assesses how these characteristics may influence the implementation and durability of ancestral hydraulic systems in each specific area.
- Construction System and Materials: This analysis is based on multidisciplinary data collection, documentary research, and geological and historical analysis of the construction systems and materials used in the three study areas. Furthermore, a thorough review of historical and scientific sources is carried out to provide a comprehensive understanding of the evolution and uniqueness of these ancestral hydraulic systems, considering geological peculiarities and specific needs of each region.

3. Climate Analysis

According to the Köppen classification, the study area in Nasca corresponds to a desert climate (BWh). Similarly, the study area in Marrakech corresponds to an arid, steppe, and hot climate (BSh), while the study area in Tabriz corresponds to an arid, steppe, and cold climate (BSk). All three study areas belong to the arid climate, yet they exhibit distinct climatic features that differentiate and uniquely influence the behavior of groundwater throughout various months of the year, including factors such as temperatures, winds, humidity, and precipitation in 2022. It is crucial to highlight that any climatic variations among these study areas are pivotal as they can impact the recharge, discharge, and quality of groundwater and its behavior across diverse contexts [46–48].

3.1. Temperature

It is noteworthy that these climatic variations exert a direct impact on the management and operation of subterranean hydraulic systems [44]. In Figure 7, Nasca reaches average maximum temperatures during the months of February and March, at 30 °C, and its average minimum temperatures range around 15 °C from June to August [49]. Additionally, the annual temperature fluctuation is 15 °C, categorized as a moderate amplitude [49].

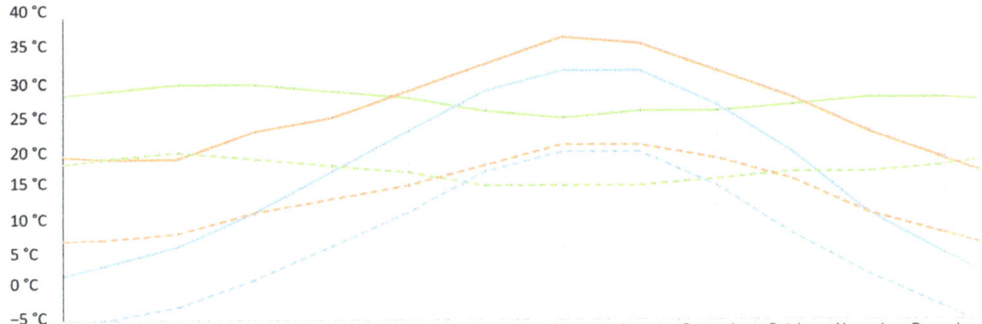

		January	February	March	April	May	June	July	August	September	October	November	December
Max. temp.	Nazca	29 °C	30 °C	30 °C	29 °C	28 °C	26 °C	25 °C	26 °C	26 °C	27 °C	28 °C	28 °C
	Marrakech	19 °C	19 °C	23 °C	25 °C	29 °C	33 °C	37 °C	36 °C	32 °C	28 °C	23 °C	19 °C
	Tabriz	3 °C	6 °C	11 °C	17 °C	23 °C	29 °C	32 °C	32 °C	27 °C	20 °C	11 °C	5 °C
Min. temp.	Nazca	19 °C	20 °C	19 °C	18 °C	17 °C	15 °C	15 °C	15 °C	16 °C	17 °C	17 °C	18 °C
	Marrakech	7 °C	8 °C	11 °C	13 °C	15 °C	18 °C	21 °C	21 °C	19 °C	16 °C	11 °C	8 °C
	Tabriz	–5 °C	-3 °C	1 °C	6 °C	11 °C	17 °C	20 °C	20 °C	15 °C	8 °C	2 °C	-3 °C

Figure 7. Comparison of the average maximum and minimum temperatures of Nasca, Marrakech, and Tabriz.

In areas with higher temperatures, such as Marrakech, where average maximum temperatures of 37 °C are reached in July, and average minimum temperatures are around 7 °C in January, with a temperature fluctuation between 15 °C and 16 °C [49]. During the summer period, there is an increase in temperature that leads to higher evaporation rates and water demand, which impacts the availability of storage systems, resulting in more frequent and prolonged droughts [49,50]. These conditions contribute to the formation of soil crusting and hydrophobic soils [50]. Consequently, during precipitation events, there is an elevated surface runoff and a decrease in groundwater recharge [50].

In the case of Tabriz, it experiences average maximum temperatures of 32 °C in the months of July and August and average minimum temperatures in January, reaching as low as −5 °C [49]. Challenges arise in locations with sub-zero temperatures like Tabriz, particularly concerning water freezing in the systems during the cold months [50]. On the other hand, in winter, temperatures vary slightly, ranging between −3 °C and 5 °C at the beginning of the season and between −3 °C and 6 °C towards the end, with an average temperature fluctuation of 8 °C during this period. The annual temperature fluctuation is 37 °C, which is considered a high amplitude [49].

3.2. Wind Rose and Wind Speed

Figure 8A shows that in the Nasca study area, there are two notable wind directions: northeast and southeast, with the former being predominant [51]. Northeast to southeast winds are the most frequent and strong, with a total duration of 4192 h/year and speeds that can reach up to 38 km/h during a period of 103 h/year [51]. These winds contribute to the flow of water in the puquio by means of wind force [52]. Likewise, Figure 8A′ shows that the months with the highest wind speeds are March and from May to October, with speeds ranging from 12 km/h to a maximum of 38 km/h. However, the months of March, May, and October have speeds starting from 5 km/h, and July records a maximum speed of up to 50 km/h during a period of 0.1 days. In contrast, the months from November to February and April have the lowest speeds, ranging from 5 km/h to a maximum of 28 km/h, except for January and December, which reach a maximum of 19 km/h. It is important to note that February is characterized by having the longest duration of minimum speed, lasting for 2.7 days [51].

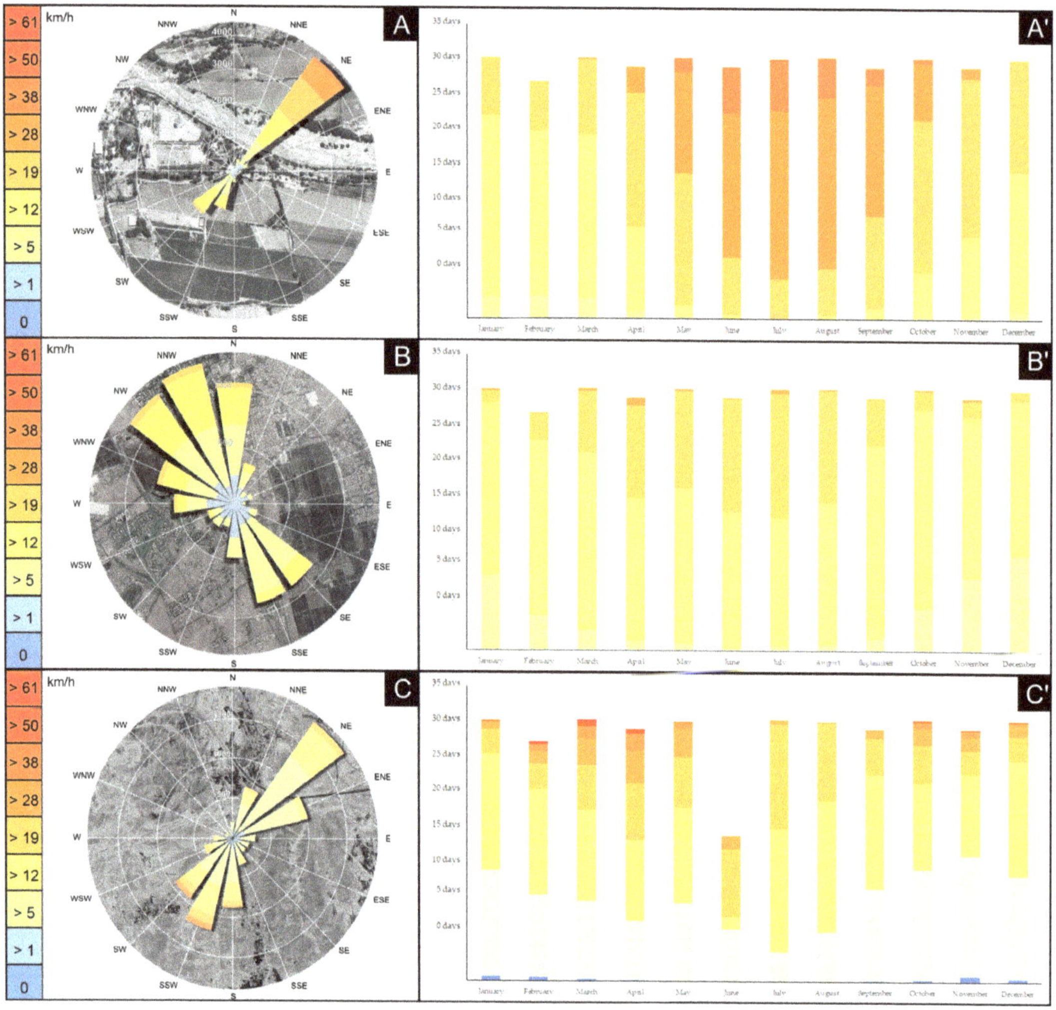

Figure 8. Wind conditions in the cities of Nasca (**A**,**A′**), Marrakech (**B**,**B′**), and Tabriz (**C**,**C′**).

In Figure 8B, Marrakech also exhibits notable winds, such as north-northwest, north-east, and in the opposite direction, southeast and south-southeast, with the predominant direction being north-northwest [53]. North-northwest to south-southeast winds are the most frequent and strong, with a total duration of 1262 h/year and speeds that can reach up to 19 km/h, with a maximum duration of 61 h/year at their maximum speed [53]. In Figure 8B′, it can be observed that the wind speed in all months ranges from 5 km/h to a maximum of 28 km/h, except for December, where it reaches up to 19 km/h for 1.2 days [53]. It is important to note that December has the longest duration of low winds, with 11.3 days, while June has the shortest duration of low winds, with 0.3 days. On the other hand, April has the longest duration of strong winds, with 0.9 days, while the months from August to October have the lowest durations, with only 0.1 days [53].

In Tabriz (Figure 8C), there are two notable wind directions: northeast to southwest and in the opposite direction, south-southeast to north-northeast, with the predominant direction being northwest due to its duration [54]. Northeast to southwest winds are the most frequent, with a total duration of 1814 h/year and speeds that can reach up to 28 km/h, with a maximum duration of 2 h/year [54]. Additionally, Figure 8C′ shows that the months with the highest wind speeds range from February to April, with speeds ranging from 1 km/h to a maximum of 61 km/h, with February and March having the longest durations of maximum speeds, at 0.2 days [54]. On the other hand, the months of May and from October to January are characterized by relatively low wind speeds, with a maximum of 50 km/h and a minimum of 1 km/h, except for May, which starts at 5 km/h. It is worth noting that November has the longest duration of low winds, with a duration of 0.7 days. Similarly, June and September have slightly lower speeds, with a maximum of 38 km/h, while in June, the minimum speed is 5 km/h, and in September, it is 1 km/h. On the other hand, the months of July and August have the lowest speeds, ranging from 5 km/h to a maximum of 28 km/h. Among these months, July stands out as the longest duration of minimum speed, with a duration of 0.5 days [54].

3.3. Relative Humidity

Figure 9 shows that, in Nasca, the month with the highest relative humidity is February (79%). The month with the lowest relative humidity is October (56%). In summer, the average relative humidity is 70.6%, which in turn benefits soil moisture retention and aquifer recharge, thereby increasing water availability [50,55]. In contrast, in winter, it is 66.3%, which reduces the soil's water retention capacity, which might necessitate additional measures to ensure sufficient water supply during those periods [50]. Additionally, Nasca has an average relative humidity of 65.5% [55].

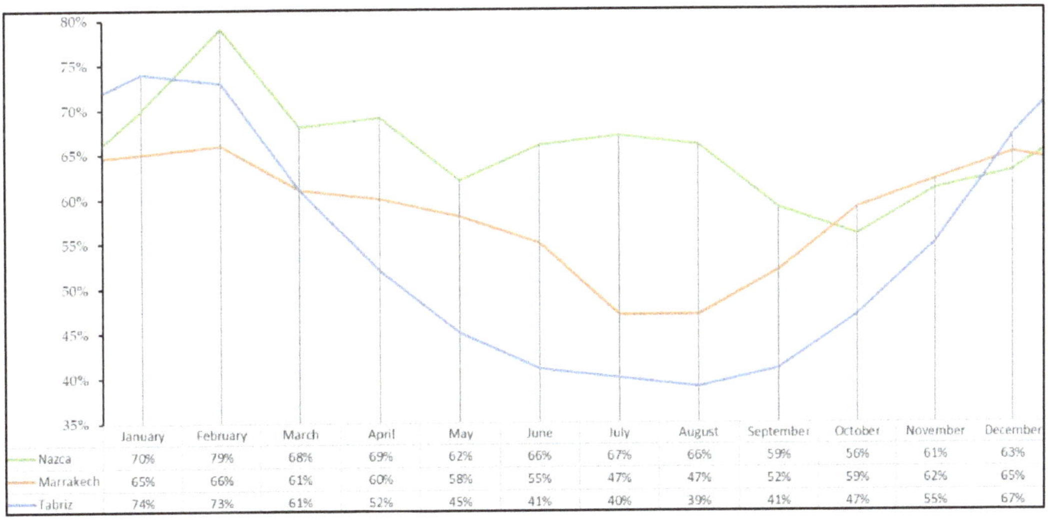

Figure 9. Comparison of average relative humidity in Nasca, Marrakech, and Tabriz.

In Marrakech, a lower relative humidity is observed compared to Nasca. In this locality, the month with the highest relative humidity is February (66%). Furthermore, the months with the lowest relative humidity are July and August (47%). In the summer months, the average relative humidity is 66.3%, whereas in winter, it rises to 70.6% [56]. These climatic conditions can lead to increased evaporation and impact the availability of water in the hydraulic systems [50]. Marrakech has an annual average relative humidity of 58.1% [56].

In Tabriz, the month with the highest relative humidity is January (74%), and the month with the lowest relative humidity is August (39%). On the other hand, the average relative humidity in summer is 40%, and during winter it is 71.3% [57]. These climatic patterns generate benefits for water retention and recharge during winter, but they also present challenges in summer due to low humidity levels, which result in higher water demand and can impact the efficiency of water capture and storage systems [50]. The annual average relative humidity in Tabriz is 52.9% [57].

3.4. Precipitation

Figure 10 shows that in Nasca, the month with the highest amount of precipitation is February, with an average of 8.6 mm [49], leading to increased aquifer recharge [58,59]. Conversely, the lowest amount of precipitation occurs in July and August, with an average of 0.5 mm [49]. This scarcity necessitates crucial water management and conservation, resulting in soil crust formation and hydrophobic soils [60]. During precipitation events, there is heightened surface runoff and reduced groundwater recharge [60]. The average precipitation is 40.7 mm per year [49].

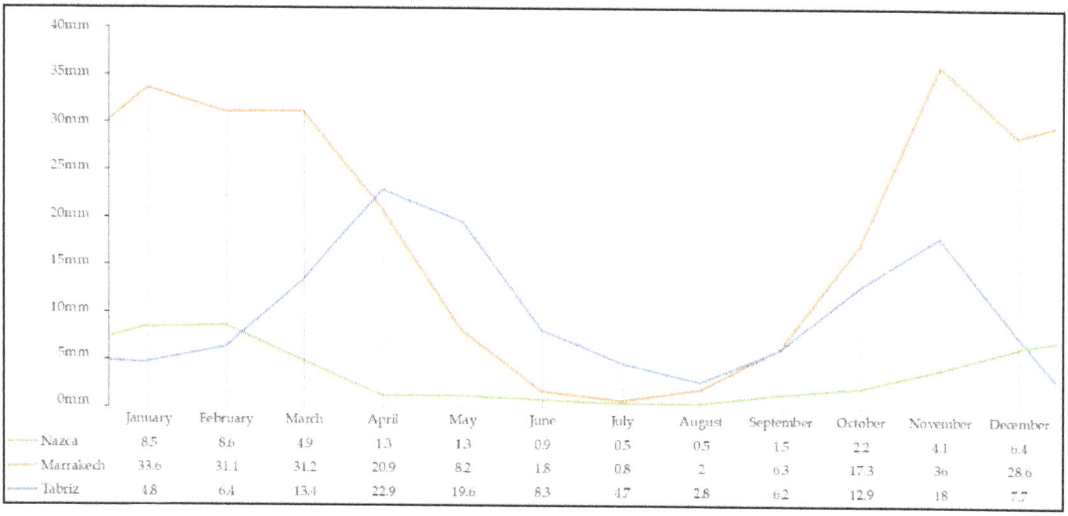

Figure 10. Comparison of average annual precipitation in Nasca, Marrakech, and Tabriz.

In Marrakech, water management poses fewer challenges due to the consistent recharge of the hydraulic system. November has the highest precipitation, with an average of 36.0 mm, which could exceed soil infiltration capacity, causing increased surface runoff and consequently reducing groundwater infiltration and recharge [60,61]. In contrast, July has the lowest, with an average of 0.8 mm [49]. The average precipitation is 217.8 mm per year [49].

Finally, in Tabriz, April has the highest amount of precipitation, with an average of 22.9 mm, leading to increased aquifer recharge [58,59]. Conversely, the lowest amount is recorded in August, with an average of 2.8 mm [49]. The average precipitation is 127.7 mm per year [49].

4. Supply and Operation Systems

4.1. Case of Nasca City: Cantalloc Aqueduct-Puquio

Figure 11A provides the location map of rivers in Nasca City, where the study site of the Cantalloc Aqueducts is situated. These rivers exhibit different directions, lengths, and altitudes along their course. The journey begins on the high plateau at an altitude of 4000 m.a.s.l., where the water flows into the Tierra Blanca River, also known as the Tambo Quemado River [62]. This river follows an east-west direction and spans a length of 75 km, with a maximum width of 700 m. It has an average slope of 7%, and its basin covers an area of 508 square kilometers, with 230 square kilometers corresponding to the active basin. This river plays a significant role in supplying the study area, although its flow is irregular, ranging from 2.43 m^3/s in May to 0 m^3/s in September, with peak discharges occurring from January to April and a sharp decline in flow from April to May, leading to complete drying of the channels [33,62].

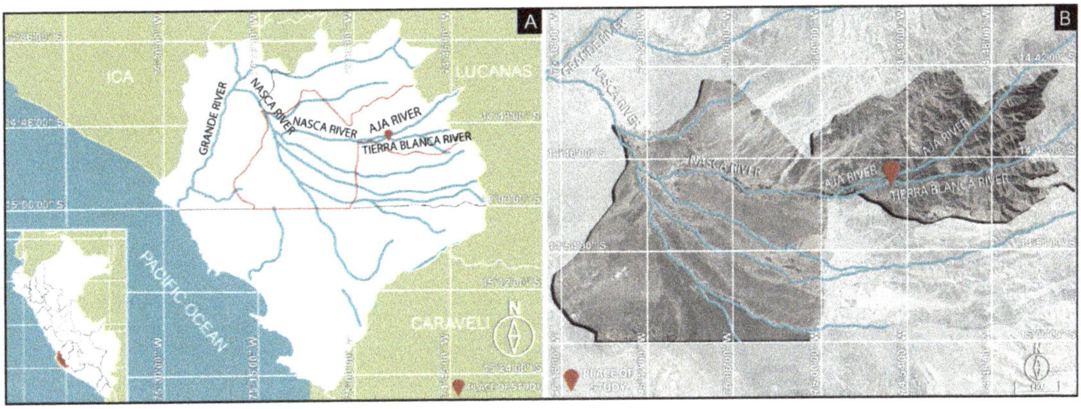

Figure 11. (**A**) Map of rivers in Nasca, Ica; (**B**) Map of rivers in Nasca city.

Subsequently, the water continues its course and meets the Aja River at an altitude of 400 m. The Aja River is formed by the convergence of three streams, the Hospicio River, Chillhua Stream, and Tototumi Stream, at an altitude of 4200 m [62]. The Aja River has a length of 1000 m and flows in an east-west direction. The altitude at the confluence point is 462 m, with a height difference of 317 m. The average slope is 5%, decreasing to 2% and even less before joining the Tierra Blanca River [62]. The Aja River basin covers an area of 513 square kilometers, with 295 square kilometers attributed to the active basin, while its flow varies from 3.65 m^3/s in March to 0 m^3/s in September [33,62]. Afterward, the water flows into the Nasca River, formed by the confluence of the Aja River and the Tierra Blanca River, at an altitude of approximately 460 m. The river has a length of 42 km and empties into the Pacific Ocean when it connects with the Grande River at an altitude of 155 m [62]. The Grande River has an approximate area of 4584.0 square kilometers, with an average elevation calculated at 50% of the area of 1736.62 m above sea level [33]. It is also intermittent in nature, flowing during the rainy months from December to April, while the rest of the year remains dry [33].

On the other way, Figure 11B provides a closer view of the water bodies in the city of Nasca, specifically the Cantalloc Aqueduct. It can be observed that the Tierra Blanca River directly connects to it. This basin is of great importance as its functioning contributes to the water supply of the region, both for agriculture and for domestic use by the population of Nasca City [52].

4.2. Case of Mechouar-Kasbah City: Khettara

Figure 12A presents the location map of rivers in the Marrakech region, where the study site of the Khettara is situated. It illustrates the various directions and branches of the rivers as they traverse the region. The journey begins in the High Atlas, a sub-range of the Moroccan Atlas, at an elevation of 4167 m above sea level. From the northern flank of the High Atlas, the rivers divide into five basins belonging to the Tensift River: Nfiss, Gheraya, Ourika, Zat, and Rdat, which connect to the main river from west to east [63]. In this sequence, the Rdat River spans 50 km, and its significant erosion has contributed to the sedimentation of the river [64]. The next notable river is the Issyl, whose basin originates from the High Atlas and stretches for 87 km, flowing in a south-north direction [65]. These two aforementioned rivers merge into the Tensift River, which flows from east to west and extends for 250 km, making it the most important river as it reaches the Atlantic Ocean [63]. The Tensift Basin receives a flow of 6224 Mm^3 and generates an estimated runoff of 1117 Mm^3; it flows through the basin via the Tensift River and its tributaries [66].

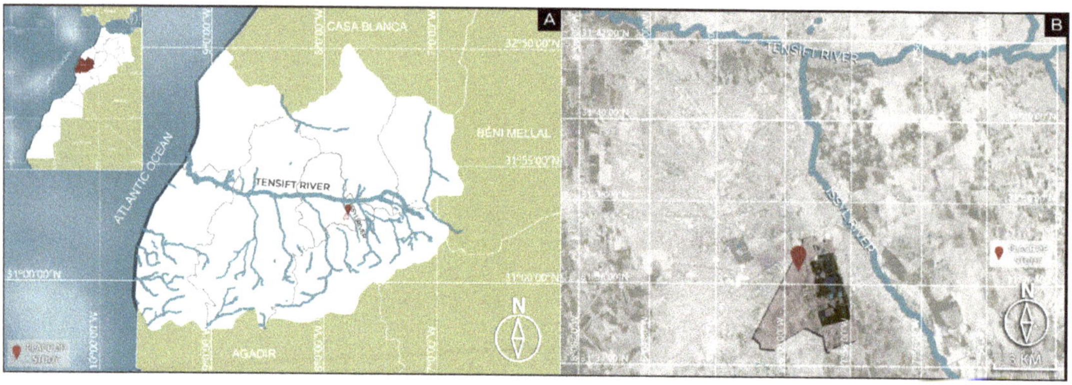

Figure 12. (**A**) Map of rivers in Marrakech, Morocco; (**B**) Map of rivers in Mechouar-Kasbah city.

Furthermore, Figure 12B provides a closer view of the water bodies in the city of Mechouar-Kasbah. It shows that the Issyl River is the closest river to the Khettara system in the study area. Its significance lies in its role of supplying water to the region, and it is considered a river with easy collection of rainwater for domestic use by the population of Mechouar-Kasbah [65]. However, the Khettaras in the Tensift Basin sector are nearly all out of service, and the amount extracted from these tunnels is now negligible [66].

4.3. Case of Tabriz City: No-Ras Qanat

In Figure 13A, the location of the No-Ras Qanat study area is depicted, showing the rivers in Tabriz County. These rivers have varying lengths and flow directions. The journey begins at the Sahand Mountain, situated at an elevation of 3707 m.a.s.l., where the Mehran-Rood River originates. This river flows through the city of Tabriz in a southeast-to-northwest direction. Ultimately, it joins the Aji-Chay River on the western side of the city, which is considered the most significant river in the region. With a length of 265 km and an average elevation of 1481 m.a.s.l., the Aji-Chay River flows from east to west. The waters of the Aji-Chay River have relatively high salinity, and it discharges into Lake Urmia, located between East Azerbaijan and West Azerbaijan [39,67].

Figure 13B provides a closer view of the aquifers surrounding the No-Ras Qanat in the Tabriz study area. Two river branches intersect with the studied Qanat system; however, they are seasonal and only have water flow during specific periods. The Mehran-Rood River, which significantly contributes to water transportation from the highlands to Tabriz, is also identified. It plays a vital role in irrigation for agricultural areas and domestic use [67].

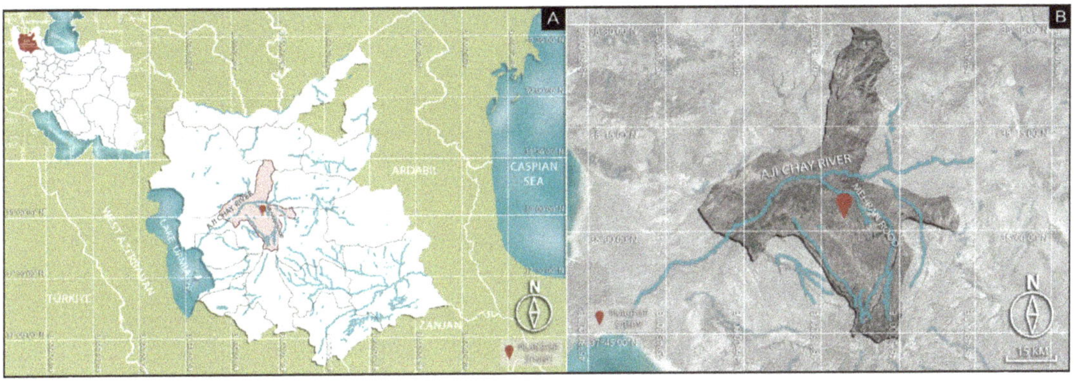

Figure 13. (**A**) Map of rivers in Tabriz, East Azerbaijan, East Azerbaijan; (**B**) Map of rivers in Tabriz city.

5. Results

5.1. Area Distribution and Sizing

5.1.1. Case of Nasca City: Cantalloc Aqueduct Puquio

In Figure 14A, the location of the Puquio near the southern bank of the Tierra Blanca River is depicted. Additionally, the figure provides valuable information about the direction of the underground flow, which moves from east to west, following the natural slope of the terrain [52]. The Puquio consists of two main zones: the underground zone and the open trench [52].

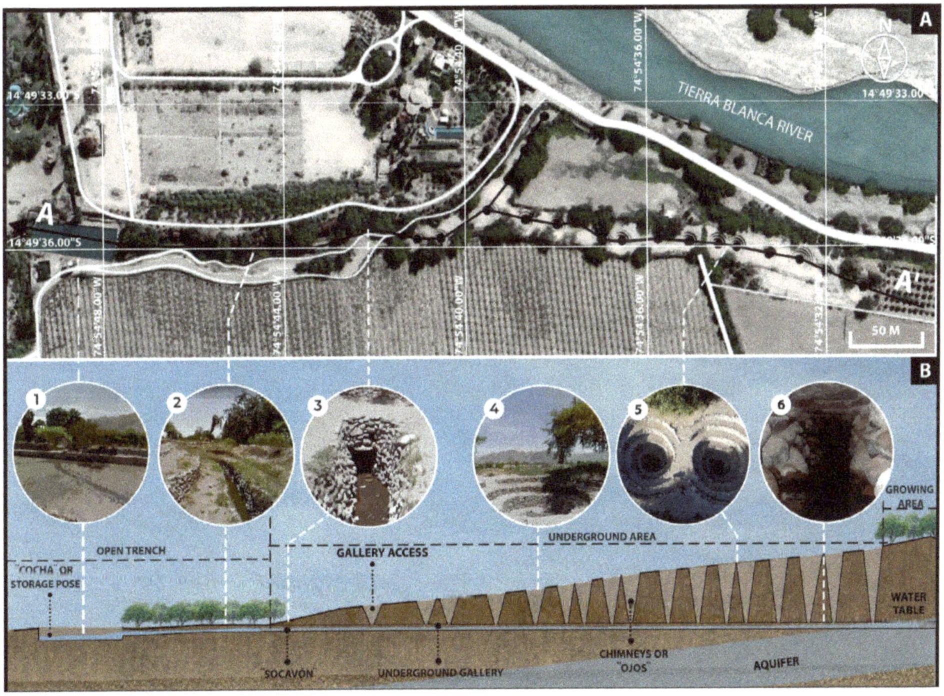

Figure 14. (**A**) Location plan of Puquio; (**B**) Sectional Cut A–A′: Schematic representation of the underground water system's operation.

In Figure 14B (4–6), the parts of the underground zone, such as the chimneys, also known as "ayes", and the underground galleries, are shown [68]. The funnel-shaped "ojos" facilitate the flow of water through the force of the wind and provide access for the maintenance of the underground galleries, where the water travels towards the surface [52]. As shown in Figure 14B (3), before entering the open trench zone, the water passes through the "socavón", which serves as the transition point between the two zones, to reach the open trench, which is the surface part of the Puquio [52,69]. Afterward, as depicted in Figure 14B (2), the water follows a meandering path with curves and sometimes abrupt changes in direction to control its speed, eventually reaching the reservoir called "cocha" Figure 14B (1) located at the end of the system for subsequent use in irrigation or to flow directly through the irrigation canals [16,52,69].

Regarding the specific distribution, the starting point of the Puquio is the first branch in the northeast, near the river. This branch extends for 71 m and has three chimneys [52]. As it progresses, the first branch divides into two: one on the east side, with a length of 265 m and 13 chimneys, continuing below the bed of the Tierra Blanca River for an unknown distance, and another on the west side, with 104 m and six chimneys [52]. This latter branch reaches the "socavón", where the underground gallery transitions into an open trench and culminates its flow in the cocha [52].

In Figure 14B, a schematic section A–A′ of the Puquio is presented, and a total of 15 vertical wells "ayes" are identified in the cross-section, with spacing ranging from 0.80 to 1.00 m, corresponding to the slope of the graph [70]. The conical "ayes" can have an opening as wide as 15 m in diameter at the surface of the ground, narrowing down to one or two meters at the bottom, with a depth of a few meters up to approximately 10 m [52]. Likewise, the underground water gallery has a width ranging from 50 cm to 80 cm and a height of 90 cm to 150 cm, allowing for the entry of maintenance personnel [52,70]. They have a length ranging from 300 to 1500 m and exhibit a rectangular or slightly trapezoidal section [68,70]. These galleries are excavated at depths ranging from 3.00 to 8.00 m below the surface of the ground.

It exhibits a flat, undulating, and strongly undulating topography, indicating a significant variation in altitude within the study area, ranging from 636 m.a.s.l. to 641 m.a.s.l. [29]. It covers a length of 208 m with a maximum slope of 10.5%, resulting in sectors prone to channel erosion, and the average gradient is 1.4%, which facilitates the continuous and controlled flow of water at the underground level [29,71].

5.1.2. Case of Mechouar-Kasbah City: Khettara

In Figure 15A, the location of the Khettara in the study area is depicted, indicating the direction of the underground flow according to the slope of the terrain (from south to north). It is also shown in relation to the surrounding environment, with the Raid Garden Hotel situated in the southern area Figure 14B (4), while the northern direction consists of vacant land Figure 15B (1). The Khettara exhibits minimal vegetation along its course and surroundings, primarily due to its state of abandonment and deterioration, as evident in Figure 15B (2,3), which depicts blocked wells filled with waste. In this sector, a total of 24 vertical wells are identified [30].

Figure 15B presents a schematic section A–A′ of the Khettara, where a total of 13 vertical wells are identified in the cut, spaced at intervals of 18 to 50 m, corresponding to the slope of the graph [3]. These wells have varying widths between 50 cm and 100 cm, with depths ranging from a few meters to approximately 10 m. Furthermore, the underground water channel has a width between 50 cm and 80 cm and a height ranging from 90 cm to 150 cm [3], allowing access for maintenance personnel.

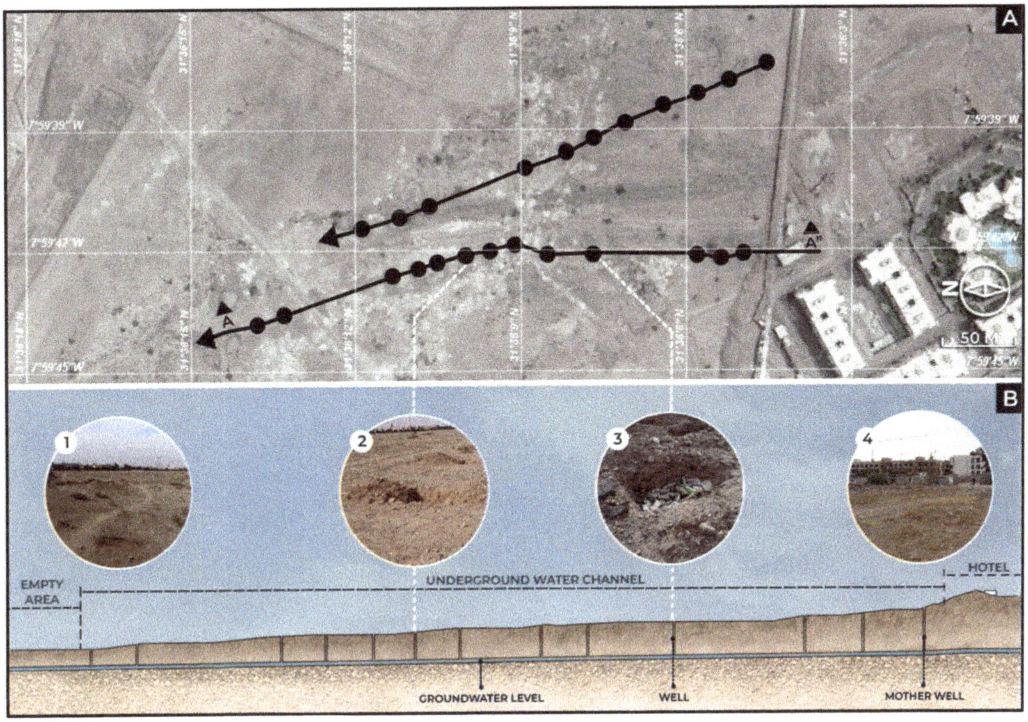

Figure 15. (**A**) Location plan of Khettara; (**B**) Sectional Cut A–A'': Schematic representation of the underground water system's operation.

The study area presents an undulating and rugged topography, with an altitude ranging from 477 m.a.s.l. to 480 m.a.s.l. [30]. It spans a distance of 280 m, with a maximum slope of 13.3%, leading to sectors of the channel that are susceptible to erosion due to increased flow velocity, and it has an average gradient of 1.1%, ensuring controlled water flow within the khettara [30,71].

5.1.3. Case of Tabriz City: No-Ras Qanat

Figure 16A, the location of the Qanat in the study area is depicted, indicating the direction of the underground flow according to the slope of the terrain (from south to north). It is also shown in relation to the surrounding environment, with agricultural areas to the south and the historical Fath-Abad Garden to the north Figure 16B (1) [39,40]. Some parts of its course and the surrounding areas exhibit moderate vegetation, supporting the agricultural practices of the Chavan community.

Figure 16B presents a schematic section A–A′ of the Qanat, where a total of 54 vertical shafts are identified in the cut, spaced at intervals of 18 to 50 m. However, only four of them are accessible and in operation, while the rest are either unidentified or blocked, as seen in Figure 16B (3), which shows the access to one of the blocked shafts due to a rock obstruction [39]. Moreover, the Qanat has a depth ranging from a few meters to approximately 49 m, as observed in the Mother Well. Additionally, the underground water level, also known as "dehliz," extends with a slight slope until reaching the surface outlet. It has an average width of 1 m and a height of approximately 1.5 m, providing ventilation, preventing water stagnation, and facilitating maintenance access [14,39].

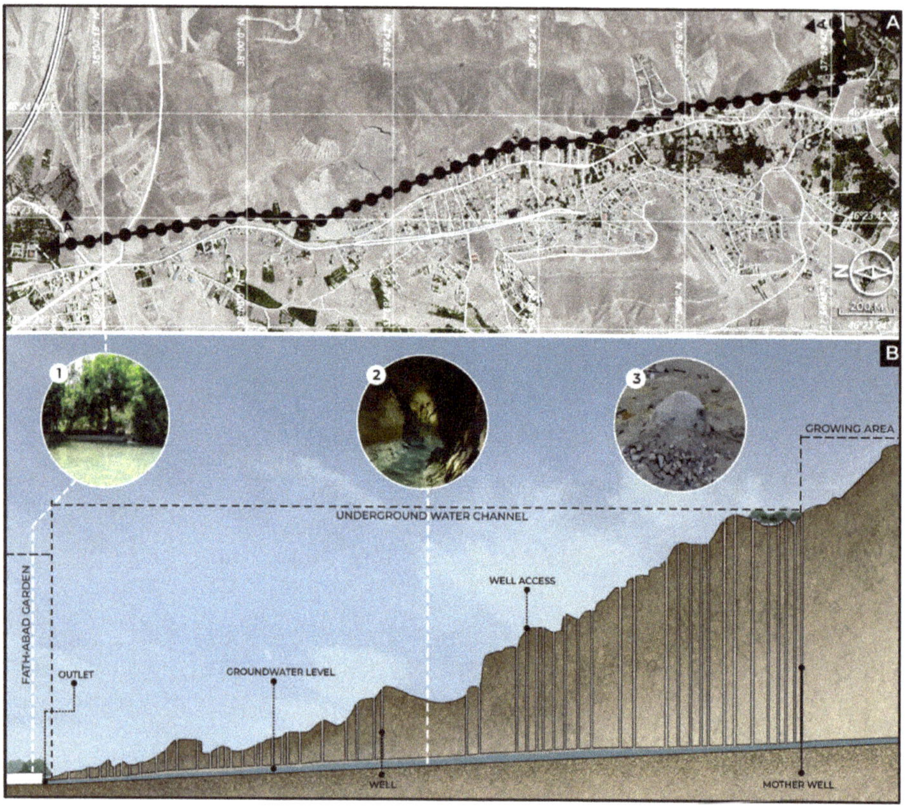

Figure 16. (**A**) Location of the No-Ras Qanat; (**B**) Sectional Cut A–A: Schematic representation of the underground water system's operation.

The study area presents a highly undulating topography surrounded by mountainous features, with altitudes ranging between 1670 m.a.s.l. and 1782 m.a.s.l., covering an extent of 3.50 km [31]. The maximum slope is 12.6%, and the average gradient is 2.7%, which significantly influences water flow and distribution [31,71].

5.2. Soil Typology

In Table 1, the comparison of each intervention site, Nasca, Marrakech, and Tabriz, shows that although they share a semi-arid climate, the soils in Nasca, Marrakech, and Tabriz have different characteristics and behaviors.

Table 1. The soil typology of the study locations: Nasca City, Nasca; Mechouar-Kasbah City, Marrakech; and Tabriz City, Tabriz.

	Nasca	Marrakech	Tabriz
Region	Semi-arid	Semi-arid	Semi-arid
Type	Sandy clay (Sc) [72,73]	Silt loam (Sl) [15,72]	Clay loam (Cl) [74]
Texture	Medium [75,76]	Medium [75,76]	Fine soil [74]
Water Retention	Medium-High [75]	Medium-High [75]	High [75]
Water Infiltration	Low	High	Low
Erosion	High	Medium-Low	High [74]

The soil in Nasca is mainly sandy clay (Sc), giving it a medium texture and a medium to high water retention capacity [72,73]. However, its infiltration capacity is low and prone to erosion. On the other hand, the soil in Marrakech is loamy, sandy clay, providing it with a medium texture and a medium to high water retention capacity [15,72]. However, it has a high infiltration capacity and a medium to low erosion potential. In the case of Tabriz, the soil is loamy clay, classified as fine-textured alluvial soil [74], with a high water retention capacity. However, it has a low infiltration capacity, making it prone to erosion, similar to Nasca.

In summary, we can affirm that the soils in Nasca and Marrakech share similarities in terms of water retention capacity and texture, while the soil in Tabriz is characterized by a finer texture and slightly higher water retention capacity [75,76]. However, Marrakech has an important characteristic: a high water infiltration capacity, making it more resistant to erosion.

This attribute potentially enhances the overall stability and durability of these systems, as the higher resistance to erosion minimizes the potential adverse effects of sedimentation and material wear. It also contributes to a prolonged lifespan of the subterranean structures and a reduced need for frequent maintenance or repairs compared to areas with lower infiltration capacity, such as Nasca and Tabriz.

Each region has its own geological significance and may require different soil management and conservation practices. The soils in Nasca and Tabriz are suitable for agriculture, while the soils in Marrakech may require careful water management for agriculture and soil protection. In conclusion, understanding the properties and behavior of the soil is essential for sustainable development and proper management of natural resources in any region.

5.3. Construction System and Materials

Table 2 provides detailed information regarding the construction system and materials used in the three study locations. In Nasca, Pre-Incas employed a construction system based on stable geometric figures to protect the structures from erosion caused by groundwater. They used a lintel system to resist the pressure of the soil, coating them with river stones to ensure proper settlement and improve adherence [52].

Table 2. Comparison of construction systems and materials in different study locations.

Ancestral Hydraulic Systems	Construction System	Roof	Wall	Floor
Puquio	Lintel	Flat stones [68,70] Wood (Huarango tree) [68,70]	River Stone Slate soil Clay [16]	Clay
Khettara	Excavation [3] and/or masonry [58]	Clay Stone	Clay Stone	Clay -
Qanat	Conventional excavation [40]	Soil Clay Pumice stone	Soil Clay Pumice stone	Soil Clay -

These lintels are made from flat stones extracted from nearby quarries [52]. Additionally, wooden beams made from the trunk of the Huarango tree were used in certain roof sections. The Huarango tree, scientifically known as "Prosopis pallida" and commonly referred to as "Algarrobo" in the Northern and Central Coast of Peru and "Huarango" in the Department of Ica, is abundant in the dry region of Peru and exhibits high water resistance [52,68,70,77]. In terms of the walls, inclined planes were used, and terraces were constructed to facilitate vertical access. The walls had an appropriate incline and were coated with river pebbles to guarantee stability [52]. River cobblestones, dry-stacked stones, stone slabs, as well as filling with earth and clay were employed [52].

In the case of the upper part, the open trench was lined with river cobblestones, and the aqueduct was covered with stone slabs or beams made from Huarango trunks. The "ojos" or openings were constructed using stones and Huarango logs, covered with excavated material [52,62]. Finally, in the present day, the walls are reinforced with concrete, similar to the "socavón" [52].

For Khettaras and Qanats, prior to construction, an exploratory well is excavated to verify the presence of the water table in the subsurface [3,78]. The construction process starts with the excavation of the first two wells, which are connected through a lower channel. Subsequently, a third well is excavated, following the same connection with the channel, and this process is repeated until reaching the final well, where the water table is reached. This forms the underground horizontal tunnel called "dehliz", which directs the water flow [3,14,78]. Additionally, the vertical wells serve not only for water access but also for determining the correct direction and suitable slope for the dehliz, as well as facilitating the removal of excavated materials [14].

During construction, natural elements of the terrain and subsurface are utilized, and in some cases, clay rings are used as support. In the Khettara system, it has been documented that, depending on the soil stability, stone masonry is employed as an additional structure to prevent collapse [78].

The Puquio was constructed using a complex process with stronger structural support based on a lintel system and careful management of inclinations. Additionally, in the construction of some Khettaras, stone masonry walls were used as additional support, providing greater stability and containment. On the other hand, in Tabriz, a system consisting of an excavation with clay linings for containment was employed, making use of natural elements of the terrain, such as stones, rocks, and earth.

6. Discussion

Ancestral hydraulic systems have played a vital role in water supply, both for human consumption and for the agriculture that characterizes these regions. These systems have been characterized by their ability to efficiently harness available water resources, capturing and storing groundwater and distributing it equitably within the community. The existence of structures such as the Khettara in the city of Mechouar-Kasbah, Morocco, the Qanat in the city of Tabriz, Iran, and the Puquio in the city of Nasca, Peru, showcases the ability of these societies to adapt to their environment and ensure a reliable water supply even in adverse climatic conditions. Additionally, in Roman civilizations, renowned for their aqueducts, these systems have proven to be an effective solution for meeting basic water needs. Ambitious engineering projects, decorative fountains, and private villas have contributed to the development of communities in these areas despite geographical challenges, difficult terrain, and adverse climates [79].

The analysis of the locations reveals clear climatic differences among them. Marrakech and Tabriz, situated in the northern hemisphere, experience extreme temperatures, while Nasca, located in the southern hemisphere, presents distinctive climatic conditions. These climatic disparities, including relative humidity and precipitation, play a crucial role in water scarcity in these regions and their hydraulic systems, as they share arid climate classifications. For instance, in the southern regions of Morocco, an arid Saharan climate is observed, hosting valuable oasis ecosystems threatened by rapid desertification, irrational exploitation of water resources, and inadequate agricultural practices, which exert significant pressure on groundwater [3,5,80,81].

Through an exhaustive analysis of rivers in the different studied cities, which supply water for agriculture, various forms of distribution along the basins have been observed, ranging from high to transition zones. The close relationship between rivers and the feeding of groundwater bodies has been highlighted, demonstrating the importance of this connection for water supply in these areas. These three systems exhibit significant connections with the rivers that supply the nearby city in the study area. This connection is also evident in the filtrating galleries of Alto Lerma, Mexico, where groundwater in the Lerma River basin plays a fundamental role in the region's development, being part of the exploited aquifer for potable water supply in the local populations and industrial establishments in the area [82]. Thus, an important connection is established among these three elements: rivers, groundwater bodies, and ancestral hydraulic systems, all of which contribute to the water supply in the city adjacent to the study area.

7. Conclusions

The importance of the use of ancestral water resources in extreme conditions and in desert places was evaluated using efficient techniques and deep knowledge of the population in the behavior of groundwater.

The design of the puquios shows the use of winds with the objective of taking advantage of the slope of the underground flow in the galleries, contributing to the efficiency and functionality of the hydraulic system. Also, in the case of Khettaras and Qanats, they have a slope of 10% to 13%, which improves the distribution, functioning, and flow of water.

According to the analysis, it is observed that Nasca and Marrakech share similarities in their soils, which allow water retention and texture, while Tabriz is characterized by a finer texture and a greater water retention capacity. This attribute potentially improves the stability and durability of the ancestral system by minimizing the effects of erosion and wear of the most eroded material.

According to the analysis, the pre-Inca culture used advanced construction techniques to address erosion caused by the flow of groundwater, using stable elements in both the vertical and horizontal planes through a lintel system that could withstand soil pressure. Additionally, they used materials such as river stones to ensure proper seating and improve adhesion.

Author Contributions: Methodology, D.E.; Investigation, J.V., M.A., A.G. and P.C.; Supervision, V.R. All authors have read and agreed to the published version of the manuscript.

Funding: This research received no external funding.

Data Availability Statement: All data are in the manuscript.

Conflicts of Interest: The authors declare no conflict of interest.

References

1. Programa Mundial de la UNESCO de Evaluación de los Recursos Hídricos. Hechos y Cifras, el 3er Informe Sobre el Desarrollo de los Recursos Hídricos en el Mundo: El Agua en un Mundo en Constante Cambio. 2009. Available online: https://unesdoc.unesco.org/ark:/48223/pf0000374903_spa?posInSet=1&queryId=N-EXPLORE-6626a201-062c-44e7-ae89-c605c84b36fb (accessed on 21 May 2023).
2. Fundación Aquae. Datos Interesantes de la Distribución del Agua en la Tierra. 2014. Available online: https://www.fundacionaquae.org/principales-datos-del-agua-en-el-mundo/ (accessed on 23 June 2023).
3. Beraaouz, M.; Abioui, M.; Hssaisoune, M.; Martínez-Frías, J. Khettaras in the Tafilalet Oasis (Morocco): Contribution to the promotion of tourism and sustainable development. *Built Herit.* **2022**, *6*, 24. [CrossRef]
4. Cesar Minga, J.; Elorza, F.J.; Rodriguez, R.; Iglesias, A.; Esenarro, D. Assessment of Water Resources Pollution Associated with Mining Activities in the Parac Subbasin of the Rimac River. *Water* **2023**, *15*, 965. [CrossRef]
5. Hssaisoune, M.; Bouchaou, L.; Sifeddine, A.; Bouimetarhan, I.; Chehbouni, A. Moroccan Groundwater Resources and Evolution with Global Climate Changes. *Geociences* **2020**, *10*, 81. [CrossRef]
6. Ahmed, A.T. Water quality for irrigation and drinking water use of Aflaj in Oman. *Water Sci. Technol.* **2015**, *15*, 421–428. [CrossRef]
7. Peroni, L. The Khettara Water Management Ancient Techniques Promoted in Morocco. 2018. Available online: http://www.ideassonline.org/public/pdf/Khettaras-ENG.pdf (accessed on 12 June 2023).
8. Khaneiki, M.L. *Territorial Water Cooperation in the Central Plateau of Iran*; Springer: Dordrecht, The Netherlands, 2019. [CrossRef]
9. Ponce Vega, L.A. Puquios, Qanats y Manantiales: Gestion del Agua en el Perú Antiguo. *Agric. Soc. Desarrollo* **2015**, *12*, 279. Available online: https://www.scielo.org.mx/scielo.php?script=sci_arttext&pid=S1870-54722015000300002 (accessed on 15 June 2023). [CrossRef]
10. Barnes, M.; Fleming, D. Filtration-Gallery Irrigation in the Spanish New World. *Lat. Am. Antiq.* **1991**, *2*, 48–68. [CrossRef]
11. Schreiber, K.; Lancho, J. *Irrigation and Society in the Peruvian Desert: The Puquios of Nasca*; Lexington Books: Lanham, MD, USA, 2003; Available online: https://rowman.com/ISBN/9780739106419/Irrigation-and-Society-in-the-Peruvian-Desert-The-Puquios-of-Nasca (accessed on 26 July 2023).

12. Schreiber, K.; Lancho, J. Aguas en el Desierto: Los Puquios de Nasca. Fundación Editorial, Pontificia Universidad Católica del Perú. Lima. 2006. Available online: https://repositorio.pucp.edu.pe/index/handle/123456789/181450 (accessed on 13 May 2023).
13. El Faiz, M.; Ruf, T. An Introduction to the Khettara in Morocco: Two Contrasting Cases. In *Water and Sustainability in Arid Regions*; Springer: Dordrecht, The Netherlands, 2010. [CrossRef]
14. Sanaan Bensi, N. The Qanat System: A Reflection on the Heritage of the Extraction of Hidden Waters. In *Adaptive Strategies for Water Heritage*; Hein, C., Ed.; Springer: Cham, Switzerland, 2020. [CrossRef]
15. Chakhchar, A. Vulnerability and Adaptation Strategies to Climate Change on Water Resources and Agriculture in Morocco: Focus on Marrakech-Tensift-Al Haouz Region. 2018. Available online: https://www.researchgate.net/publication/323368838_VULNERABILITY_AND_ADAPTATION_STRATEGIES_TO_CLIMATE_CHANGE_ON_WATER_RESOURCES_AND_AGRICULTURE_IN_MOROCCO_FOCUS_ON_MARRAKECH-TENSIFT-AL_HAOUZ_REGION (accessed on 14 July 2022).
16. Negro, S.; Amorós, S.; Fuentes, M.D.C.; Palma, R.C. Los Acueductos con Galerías Filtrantes de Nasca y su Gestión Patrimonia. 2019. Available online: https://cicopperu.files.wordpress.com/2021/09/galerias-filtrantes-nasca-sandra-negro.pdf (accessed on 23 November 2022).
17. English, P.W. The Origin and Spread of Qanats in the Old World. *Proc. Am. Philos. Soc.* **1968**, *112*, 170–181. Available online: https://www.ircwash.org/sites/default/files/English-1968-Origin.pdf (accessed on 25 June 2022).
18. El Faiz, M. *Marrakech: Patrimoine en Péril*; Acte Sud/Eddif.: Arles, France, 2002.
19. Kabiri, L. Gestion des ressources en eau dans les oasis de sud marocain: Cas de Ferkla (Tinjdad, Errachidia, Maroc). Tema 3. Gestion durable des ressources naturelles de l'espace oasien. In Proceedings of the Simposio Internacional: Desarrollo Agricole Durable des Systèmes Oasiens, Erfoud, Morocco, 8–10 March 2005.
20. Critchley, W.R.S.; Reij, C.; Wilcocks, T.J. Indigenous Soil and Water Conservation: A Review of the State of Knowledge and Prospects for Building on Traditions. *Land Degrad. Dev.* **1994**, *5*, 293–314. [CrossRef]
21. Gómez, A.; Esenarro, D.; Martínez, P.; Vílchez, S.; Raymundo, V. Cálculo Térmico para la Implementación de Muros Verdes como Aislantes Térmicos en las Fachadas Este y Oeste en las Áreas Adyacentes de la Facultad de Ciencias Biológicas, Universidad Ricardo Palma (URP) en Lima, Perú 2023. *Edificaciones* **2023**, *13*, 2301. [CrossRef]
22. Bouchaou, L.; Tagma, T.; Boutaleb, S.; Hssaisoune, M.; El Morjani, Z.E.A. Climate Change and Its Impacts on Groundwater Resources in Morocco: The Case of the Souss-Massa Basin. In *Climate Change Effects on Groundwater Resources*; Taylor and Francis: Abington, UK, 2011; Volume 2007, pp. 129–144. Available online: https://www.tandfonline.com/doi/full/10.1080/02626667.2021.1982137 (accessed on 11 May 2023).
23. Gleeson, T.; VanderSteen, J.; Sophocleous, M.A.; Taniguchi, M.; Alley, W.M.; Allen, D.M.; Zhou, Y. Groundwater Sustainability Strategies. *Nat. Geosci.* **2010**, *3*, 378–379. [CrossRef]
24. Wada, Y.; Van Beek, L.P.H.; Van Kempen, C.M.; Reckman, J.W.T.M.; Vasak, S.; Bierkens, M.F.P. Global Depletion of Groundwater Resources. *Geophys. Res. Lett.* **2010**, *37*, 1–5. [CrossRef]
25. Gleeson, T.; Richter, B. How much groundwater can we pump and protect environmental flows through time? Presumptive standards for conjunctive management of aquifers and rivers. *River Res. Appl.* **2018**, *34*, 83–92. [CrossRef]
26. Gobezie, W.J.; Teferi, E.; Dile, Y.T.; Bayabil, H.K.; Ayele, G.T.; Ebrahim, G.Y. Modeling Surface Water–Groundwater Interactions: Evidence from Borkena Catchment, Awash River Basin, Ethiopia. *Hydrology* **2023**, *10*, 42. [CrossRef]
27. Khaleghi, N.; Kovacs, F. Rehabilitation strategies for Tehran University Qanat in the frame of sustainable development. *Int. J. Architect. Eng. Urban Plan* **2019**, *29*, 223–231.
28. Sophocleous, M. Interactions between groundwater and surface water: The state of the science. *Hydrogeol. J.* **2002**, *10*, 52–67. [CrossRef]
29. Landsat y Copernicus. Nasca, Peru. (Satelital Image). Google Earth. 2015. Available online: https://earth.google.com/web/search/ciudad+de+nasca/@-14.82632895,-74.9103416,640.79433675a,766.11375989d,35y,0h,0t,0r/data=CnAaRhJACiUweDkxMTQxZThkY2ZkYWFhMjc6MHhkYzQ2NmZiNjMyNWU0Yzg0KhdBY3VlZHVjdG9zIGRlIENhbnRhbGxvYxgCIAEiJgokCQfbmfwcpz9AEcrm6DtjjT9AGf4II-p3ch_AIR4-OO2_OSDA (accessed on 11 May 2023).
30. Landsat y Copernicus. Marrakech, Morroco. (Satelital Image). Google Earth. 2015. Available online: https://earth.google.com/web/search/mechour+kasbah/@31.60260203,-7.98728899,481.7647163a,17224.52085367d,35y,0h,0t,0r/data=CnkaTxJJCiQweGRhZmVmYjBmZGJmMThhZjoweGUyNzkzZDQ3ODg0OGY2Y2EZojN6f82ZP0Ah_G8lOzbyH8AqD21lY2hvdWFyIGthc2JhaBgCIAEiJgokCcIrWXWAuD9AEeHSXrGViT9AGV27CZ3kBB_AIeRvebv_bCDA (accessed on 5 January 2023).
31. Landsat y Copernicus. Tabriz, Iran. (Satelital Image). Google Earth. 2015. Available online: https://earth.google.com/web/@37.99820862,46.39003121,1705.06272979a,10114.13474792d,30y,-0h,0t,0r (accessed on 8 January 2023).
32. Rostworowski, M. Sistema Hidráulicos de los Señoríos Costeros Prehispánicos. Capítulo 5 de: Ensayos de Historia Andina II. Nasca Pampas, género, Hechicería. Obras Completas VI. 2006. Available online: https://www.scielo.org.mx/scielo.php?script=sci_nlinks&ref=441950&pid=S1870-54722015000300002000014&lng=es (accessed on 14 June 2023).
33. Ccaracoocha PET. Afianzamiento Hídrico de la sub Cuenca del rio Nazca en las sub Cuencas de los Ríos Rangramayo, Toro Rumi y Tranca: Elaboración de Estudio Hidrológico para la Acreditación de la Disponibilidad Superficial del PIP 222081 (Formato Anexo N° 6). 2016. Available online: https://repositorio.ana.gob.pe/handle/20.500.12543/5358 (accessed on 30 March 2023).
34. Vos, J.; Marshall, A. Conquering the desert: Drip irrigation in the Chavimochic systems in Peru. In *Drip Irrigation for Agriculture*; Routledge: London, UK, 2017. [CrossRef]
35. Clarkson, P.B.; Dorn, R.I. Nuevas Fechas Cronométricas para los Puquios de Nasca, Perú. *Antigüedad Latinoam.* **1995**, *6*, 56–69. [CrossRef]

36. Benqlilou, H.; Bensaid, S. Protection and performance of the ancestral water supply system 'Khettara'as a sustainable alternative for arid regions. *Water Sci. Technol. Water Supply* **2013**, *13*, 1452–1462. [CrossRef]
37. Khardi, Y.; Lacombe, G.; Kuper, M.; Taky, A.; Bouarfa, S.; Hammani, A. Pomper ou disparaître: Le dilemme du renforcement des khettaras par le pompage solaire dans les oasis du Maroc. *Cah. Agric.* **2023**, *32*, 1. [CrossRef]
38. El-Ouahmani, N.; Chahouri, A.; Zekhnini, A.; Azim, K.; Choukr-Allah, R.; Yacoubi, B. Effects of irrigation with municipal treated wastewater on soil's heavy metals accumulation and turf leaves under drip and sprinkler systems (Case study: Agadir, Southern Morocco). *Int. J. Recycl. Org. Waste Agric.* **2021**, *22*, 309–317. [CrossRef]
39. Zaina, F.; Branduini, P.; Zavvari, F. Applying ICOMOS-IFLA Principles for the Conservation, Management and Reuse of a Historical Hydraulic System: The No-Ras Qanat in North-Western Iran. *Heritage* **2022**, *5*, 3165–3187. [CrossRef]
40. Branduini, P.; Federico, Z.; Zavvari, F.; Mazloumi, Y.N. Qanats as Endangered Traditional Hydraulic Heritage. An integrated Methodology for Documenting, Restoring and Resuing an Ancient Iranian Qanat. In Proceedings of the Joint International Event 9th ARQUEOLÓGICA 2.0 & 3rd GEORES, Valencia, Spain, 26–28 April 2021. [CrossRef]
41. Save the Tigris Campaign. *Damming the Kurdistan Region of Iraq: Structural Gaps in the KRG dam Construction Policies;* Save the Tigris: Amsterdam, The Netherlands, 2020; 86p.
42. Neissi, L.; Albaji, M.; Nasab, S.B. Combination of GIS and AHP for site selection of pressurized irrigation systems in the Izeh plain, Iran. *Agric. Water Manag.* **2020**, *231*, 106004. [CrossRef]
43. Al-Kashani, A. *Historia de Al-Jaito;* Mahin Mohbeli: Teherán, Irán, 1968; Available online: https://scholar.google.com/scholar_lookup?title=History+of+Al-Jaito.&author=Al-Kashani,+A.&publication_year=1968 (accessed on 13 March 2023).
44. Esenarro, D.; Vásquez, P.; Morales, W.; Raymundo, V. Centro de Interpretación para la Revalorización de la Flora y Fauna en Cusco, Perú. *Edificios* **2023**, *13*, 2345. [CrossRef]
45. Esenarro, D.; Chicche, P.; Chichipe, V.; Vilchez, A.; Cobeñas, P.; Raymundo, V. Bioclimatic Criteria for a Guest House in the District of Canta—Lima. In Proceedings of the 2022 11th International Conference on Power Science and Engineering (ICPSE), Eskisehir, Turkey, 23–25 September 2022; pp. 1–9. [CrossRef]
46. Clifton, C.; Evans, R.; Hayes, S.; Hirji, R.; Puz, G.; Pizzaro, C. *Water and Climate Change: Impacts on Groundwater Resources and Adaptation Options;* Water Working Notes; World Bank: Washington, DC, USA, 2010.
47. Panwar, S.; Chakrapani, G.J. Climate Change and Its Impact on Groundwater Resources. *Curr. Sci.* **2013**, *105*, 37–46.
48. Dragoni, W.; Sukhija, B.S. Climate change and groundwater: A short review. *Clim. Chang. Groundw. Geol. Soc. Lond. Spec. Publ.* **2008**, *288*, 1–12. [CrossRef]
49. Weatherspark. Compare the Climate and Weather in Tabriz, Nasca, and Marrakesh. 2023. Available online: https://weatherspark.com/compare/y/104056~23228~32742/Comparison-of-the-Average-Weather-in-Tabriz-Nazca-and-Marrakesh (accessed on 23 June 2023).
50. Bolaños Chavarría, S.; Betancur Vargas, T. Estado del Arte sobre el Cambio Climático y las Aguas Subterráneas. Ejemplos en Colombia. *Rev. Polit.* **2018**, *14*, 52–64. Available online: https://revistas.elpoli.edu.co/index.php/pol/article/view/984 (accessed on 25 June 2023). [CrossRef]
51. Meteoblue. Simulated Historical Weather and Climate Data for Hacienda Cantalloc. 2023. Available online: https://www.meteoblue.com/en/weather/historyclimate/climatemodelled/hacienda-cantalloc_peru_3945738 (accessed on 15 July 2023).
52. Máximo, A. Acueductos de Nasca "La Joya de la Obra Hidráulica de la Cultura Preinca". Sdsu.edu. 2016. Available online: https://ponce.sdsu.edu/acueductos_nazca_crispin_gomez.pdf (accessed on 15 May 2023).
53. Meteoblue. Simulated Historical Climate & Weather Data for 31.6°N 8°W. 2023. Available online: https://www.meteoblue.com/en/weather/historyclimate/climatemodelled/31.602N-7.995E482_Africa%2FCasablanca (accessed on 15 July 2023).
54. Meteoblue. Simulated Historical Climate & Weather Data for 38°N 46.39°E. 2023. Available online: https://www.meteoblue.com/en/weather/historyclimate/climatemodelled/37.998N46.389E1708_Asia%2FTehran (accessed on 15 July 2023).
55. Onebuilding.org. Available online: https://climate.onebuilding.org/WMO_Region_3_South_America/PER_Peru/index.html (accessed on 15 July 2023).
56. Weather-Atlas. Weather Forecast and Temperature for Today Marrakesh, Morocco. 2023. Available online: https://www.weather-atlas.com/en/morocco/marrakesh (accessed on 20 July 2023).
57. Weather-Atlas. Climate and Monthly Weather Forecast Tabriz, Iran. 2023. Available online: https://www.weather-atlas.com/en/iran/tabriz-climate (accessed on 20 July 2023).
58. Owor, M.; Taylor, R.G.; Tindimugaya, C.; Mwesigwa, D. Rainfall intensity and groundwater recharge: Empirical evidence from the Upper Nile Basin. *Environ. Res. Lett.* **2009**, *4*, 35009. [CrossRef]
59. Taylor, R.G.; Scanlon, B.; Döll, P.; Rodell, M.; van Beek, R.; Wada, Y.; Longuevergne, L.; Leblanc, M.; Famiglietti, J.S.; Edmunds, M.; et al. Ground water and climate change. *Nat. Clim. Chang.* **2013**, *3*, 322–329. [CrossRef]
60. International Association of Hydrogeologists and British Geological Survey, Strategic Overview Series: Global Change and Groundwater. Available online: https://iah.org/wp-content/uploads/2016/07/IAH-Global-Change-Groundwater-14-June-2016.pdf (accessed on 10 February 2023).
61. Döll, P.; Flörke, M. Global-scale estimation of diffuse groundwater recharge. *Syst. Res.* **2008**, *12*, 863–885.
62. Corporación de Reconstrucción y Desarrollo de Ica (CRYDI) Tahal Consulting Engineers. Proyecto Nazca: Estudio para el Desarrollo de las Aguas Subterráneas y la Agricultura. 1966. Volume 1. Available online: https://repositorio.ana.gob.pe/handle/20.500.12543/4286 (accessed on 15 July 2023).

63. Babault, J.; Van den Driessche, J.; Teixell, A. *Longitudinal to Transverse Drainage Network Evolution in the High Atlas (Morocco): The Role of Tectonics*; Tectonics, American Geophysical Union (AGU): Hoboken, NJ, USA, 2012; Volume 31, p. TC4020. Available online: https://core.ac.uk/download/pdf/48218187.pdf (accessed on 23 March 2023).
64. Ait Mlouk, M.; Algouti, A.; Algouti, A.; Ourhzif, Z. Evaluación de la erosion de riberas de ríos en regions semiáridas mediante datos de teledetección y SIG: Caso del río Rdat, Marrakech, Morroco. *Estud. Geol.* **2018**, *74*, e081. [CrossRef]
65. el Mehdi Saidi, M.; Bouloumou, Y.; Ed-Daoudi, S.; Aresmouk, M.E.H. Les Crues de l'Oued Issil en Amont de Marrakech (Maroc), un Risque Naturel Recurrent The Floods of The Wadi Issil Upstream of Marrakesh (Morocco), a Recurring Natural Hazard. *Eur. Sci. J.* **2013**, *9*, 189–208. Available online: https://eujournal.org/index.php/esj/article/view/1681 (accessed on 15 June 2023).
66. Molle, F.; Tanouti, O. The reappropriation of water in overexploited basins: The Case of the Tensift Basin (Morocco). *Etudes Rurales* **2013**, *192*, 79–96. Available online: https://www.researchgate.net/publication/283795410_The_reappropriation_of_water_in_overexploited_basins_The_Case_of_the_Tensift_Basin_Morocco (accessed on 15 June 2023).
67. Asadollahfardi, G.; Taklify, A.; Ghanbari, A. Application of artificial neural network to predict TDS in talkheh rud river. *J. Irrig. Drain. Eng.* **2012**, *138*, 363–370. [CrossRef]
68. Cultura.pe. Formulario de Presentación. *Acueductos de Nasca.* 2019. Available online: https://patrimoniomundial.cultura.pe/sites/default/files/li/pdf/1.%20Acueductos%20de%20Nasca%20-%20Esp_reduce.pdf (accessed on 18 July 2023).
69. Schreiber, K.; Lancho Rojas, J. The puquios of Nasca. *Lat. Am. Antiq.* **1995**, *6*, 229–254. [CrossRef]
70. Regal Alberti, B. Los Acueductos Precolombinos de Nasca. 1943. Available online: http://repositorio.pucp.edu.pe/index/handle/123456789/53471 (accessed on 15 February 2023).
71. Herterich, J.G.; Griffiths, I.M. A mathematical model of the erosion process in a channel bend. *Tribol. Int.* **2021**, *163*, 107175. [CrossRef]
72. Usda.gov. Soil Survey Manual Soil Science Division Staff. 1977. Available online: https://www.nrcs.usda.gov/sites/default/files/2022-09/The-Soil-Survey-Manual.pdf (accessed on 18 March 2023).
73. Gob.pe. Prevención del Fenómeno del Niño. Available online: http://bvpad.indeci.gob.pe/doc/estudios_CS/Region_Ica/nasca/nasca.pdf (accessed on 17 March 2023).
74. Dewam, M.; Famouri, J. The Soils of Iran. 1964. Available online: https://edepot.wur.nl/486029 (accessed on 13 June 2023).
75. Marañés Corbacho, A.; Sánchez Garrido, J.A.; de Haro Lozano, S.; Sánchez Gómez, S.T.; del Moral Torres, F. *Soil Analysis. Methodology and Interpretation*; Universidad de Almería, Servicio de Publicaciones: Almería, Spain, 1998; p. 184.
76. González-Naranjo, V.; Leal, M.; Lillo, J.; De Bustamante, I.; del Pino Palacios-Díaz, M. Guía de Caracterización Edáfica para Actividades de Regeneración de Aguas Residuals en Usos Ambientales. 2012; pp. 10–11. Available online: https://www.researchgate.net/publication/259621127_guia_de_caracterizacion_edafica_para_actividades_de_regeneracion_de_aguas_residuales_en_usos_ambientales (accessed on 13 March 2023).
77. Beresford-Jones, D.G.; Arce, T.; Whaley, O.Q.; Chepstow-Lusty, A.J. The Role of Prosopis in Ecological and Landscape Change in the Samaca Basin, Lower Ica Valley, South Coast Peru from the Early Horizon to the Late Intermediate Period. *Lat. Am. Antiq.* **2009**, *20*, 303–332. [CrossRef]
78. Fitzwilliam, S. The Living Khettaras of Southern Morocco: A Traditional Water Harvesting Technology on the Brink. Available online: https://www.academia.edu/5069877/The_Living_Khett%C4%81ras_of_Southern_Morocco_A_Traditional_Water_Harvesting_Technology_on_the_Brink (accessed on 13 March 2023).
79. Cartwright, M. "Acueducto". Ancient History Encyclopedia. 2012. Available online: https://www.ancient.eu/aqueduct/ (accessed on 18 May 2023).
80. MATEE/DAT-UNDP. Programme de lutte contre la désertification et lutte contre la pauvreté par la sauvegarde et la valorisation des oasis. Composante Tafilalet. In *Plaquette de Présentation du Projet*; Ministry of Spatial Planning, Water and the Environment: Rabat, Morocco, 2006; 55p.
81. El Khoumsi, W.; Hammani, A.; Kuper, M.; Bouaziz, A. La durabilité du système oasien face à la détérioration des ressources en eaux souterraines: Cas de la palmeraie de Tafilalet. *Rev. Marocaine Sci. Agron. Vét.* **2017**, *5*, 41–51.
82. Montes-Hernández, R.; Romero-Contreras, A.T.; Solís-Morelos, C.; Rivera-Herrejón, M.G.; Zamorano-Camiro, S. Las galerías filtrantes del Alto Lerma: Usos y manejos sociales. *Econ. Soc. Territ.* **2011**, *11*, 455–485. Available online: https://www.scielo.org.mx/scielo.php?script=sci_arttext&pid=S1405-84212011000200007 (accessed on 25 March 2023). [CrossRef]

water

MDPI

Article

Spatiotemporal Economic Analysis of Corn and Wheat Production in the Texas High Plains

Aminun Naher [1], Lal K. Almas [1,*], Bridget Guerrero [1] and Sania Shaheen [1,2]

1 Paul Engler College of Agriculture and Natural Sciences, West Texas A&M University, Canyon, TX 79016, USA; bonoecon@gmail.com (A.N.); bguerrero@wtamu.edu (B.G.); saniaashaheen@gmail.com (S.S.)
2 International Institute of Islamic Economics (IIIE), Islamic International University, Islamabad 44000, Pakistan
* Correspondence: lalmas@wtamu.edu

Abstract: The aim of this study is to visualize the historical changes in wheat and corn cropping patterns in the Texas High Plains from the perspective of geographical concentration and spatial autocorrelation. Historical county-level agricultural census data were collected from the United States Department of Agriculture and the National Agricultural Statistics Service from 1978 to 2017. Exploratory data analysis techniques were employed to examine the geographical concentration and spatial dependence of crop production among nearby locations. The results of temporal changes indicate that the harvested acres of corn and wheat tended to decrease throughout the study period. Total and irrigated harvested corn and wheat acreages were concentrated in a smaller number of counties over time while wheat production was mainly concentrated in the northern part of the region. The Moran's I test statistic for total and irrigated areas of cropland suggest that there was spatial dependence among the neighboring counties in crop production in this region. In summary, there was a spatiotemporal change in cropping patterns in the Texas High Plains over the study period. Based on the results of the spatiotemporal changes in cropping patterns in the Texas High Plains, policy makers should promote and support non-irrigated varieties of crops in order to decrease the dependence on irrigation water from the Ogallala Aquifer.

Keywords: cropping pattern; Moran's I test statistic; Ogallala Aquifer; spatiotemporal change; Texas High Plains

1. Introduction

The Texas High Plains is one of the most extensive agricultural areas in the United States. This region covers about 37,676 square miles and is comprises 39 counties. Agriculture is one of the major economic drivers in this region, which is a hub for substantial agriculture production in the state of Texas. In 2017, there were 29,360,229 acres of cropland in Texas of which 6,595,607 acres (78%) were contained in the 39 counties of the Texas High Plains [1].

The history of agriculture involves the human-induced spatial movement of crop production. The agricultural industry in this region faces unique agronomic, environmental, and economic challenges due to extreme weather conditions and water scarcity. The climate is semi-arid, and climatic changes are leading to reduced regional rainfall and increased crop water demand. Therefore, irrigated agriculture production in this region faces great threat [2]. Due to low rainfall rates, most of the agricultural producers in the region rely on the Ogallala Aquifer for irrigation water [3]. The Ogallala Aquifer was formed by ancient runoff from the Rocky Mountains. This aquifer was first discovered by the United States Geological Survey (USGS) in the 1890s. After World War II, farmers used large-scale irrigation technologies to extract ground water from the Ogallala Aquifer [4].

The use of groundwater has been increasing for irrigation. However, according to the USGS, after 1974, the water tables have declined significantly from pre-development levels [5]. In 1978, the total harvested area of irrigated cropland was 4,393,257 acres in the Texas High Plains [6]. However, in 2017, this area had decreased to only 2,940,888 acres [1]. The main irrigated crops grown in this area are corn, cotton, sorghum, and wheat; cotton, sorghum, and wheat can be grown in both irrigated and non-irrigated areas. In 2017, the total irrigated acres harvested for corn, cotton, sorghum, and wheat were 519,029, 1,517,214, 98,708, and 236,879 acres, respectively. The availability of irrigated water and the temperate climatic conditions have made this area suitable for crop production. From 2012 to 2017, the Texas High Plains ranked third and fifth among other states in the U.S. for cotton and sorghum production, respectively. For the same period, the Texas High Plains also ranked 15th for corn and wheat production nationwide [7].

1.1. Spatial and Temporal Analysis in Agriculture

Understanding the spatial and temporal changes in the production of major crops in a specific area is important for effective, evidence-based agricultural and economic policies. Historical changes in geographical distribution and concentration of livestock production were examined in the United States. The results indicated that the greatest change in geographical concentration was in laying hen egg production and pullet inventory over the study period. The results also suggested that geographical concentrations in the other livestock industry sectors were not as high as those in broiler production [8]. Another study examined the historical change in the spatial movement of plants and animals from 1879 to 2007. The results of the study indicate that corn production in the United States increased dramatically during the twentieth century. The results also revealed that environmental, biological, and spatial changes play a vital role in crop production. There are also other factors such as soil type, elevation, rainfall, pests and disease, sunlight, and temperature that limit agricultural productivity [9].

Another study investigated the relationship between producers' crop price expectations and groundwater pumping decisions. County-level data were collected from Northwest Kansas Groundwater Management District 4 (GMD4) and monthly precipitation data were collected from the PRISM Climatic Group. Kansas monthly cash price data were used to construct expected crop prices from 1997 to 2016. This study focused on the five most common irrigated crops (namely alfalfa, corn, sorghum grain, soybean, and wheat). The estimated results suggested that producers of northwest Kansas adjust the quantity of groundwater pumped in response to changes in precipitation for various irrigated crops. However, there was no statistically significant relationship found between crop price expectations and groundwater pumping decisions [10].

1.2. Exploratory Spatial Data Analysis

The Geographical Information System (GIS) is a very powerful tool for visualizing spatial patterns. GIS-based maps can be created to explore historical patterns of crops, urbanization trends, land use or cover changes, and water use in industry. The change in production decisions or land use can be easily visualized from these maps. Several studies that have applied GIS-based approaches for exploratory spatial data analysis are reviewed in this section.

Reference [11] proposed a method that allows long-term mapping of cropping patterns using time-series crop maps. Crop maps were derived from supervised classification of remote sensing data. This study applied GIS overlay analysis operations to derive the spatial and temporal relationships between crops. The results of the study show that the application of the method to the study area revealed a large variability in cropping patterns. Guerrero et al. (2019) focused on the impact of dairy industry expansion on water use, crop composition, and the local economy. Data on dairy cow inventory and annual irrigated crop acres were collected from the Federal Milk Marketing Order and Farm Service Agency, respectively. The Wilcoxon test and SAS PROC NPERWAY methods were employed to determine the significant difference between the number of acres cultivated at the beginning of 2000 and the most recent data for 2015. Moran's I statistics were presented to visualize the spatial autocorrelation between neighboring dairy industries in the Texas High Plains. The total irrigated area decreased by 17.8 percent from 2000 to 2015, indicating a trade-off between increased irrigation requirements due to dairy feed demand and the overall irrigation demand in the region. Moran's I statistics suggested that the spatial autocorrelation of dairy inventory by county in the study region experienced a positive increase from 2000 to 2015. The authors of [12] examined the urbanization trends of Hebei Province in China using the Geographical Information System (GIS) and remote sensing. The objective of the study was to explore the temporal and spatial characteristics of urban expansion and to examine land cover changes due to urbanization between 1987 and 2001. To achieve these objectives, multi-annual socio-economic statistics and two types of satellite multi-spatial images were collected from 1934 to 2001. GIS software (MapInfo5.0) was used to create maps of the urban area of Shijiazhuang City in different historical periods. The results indicated that the urban area of Shijiazhuang City expanded by 96% from 1934 to 2001. However, the annual growth rate varied significantly in different periods, and the fastest expansion stage was from 1981 to 2001. The results from the landscape change due to high-speed urbanization show that urban regions have increased sharply while agricultural land has decreased significantly.

1.3. Cropping Patterns in the Texas High Plains

Agriculture in the Texas High Plains is different from that in other areas of the United States. The Texas High Plains are a semi-arid region, and irrigation is vital to this region. The depletion of groundwater sources is a growing concern for crop production in this region. This section reviews studies related to cropping patterns in the Texas High Plains. The authors of [4,13] focused on the historical change in groundwater availability mainly from the Ogallala Aquifer, the short and long-term effects of agriculture's adaption to water resources, and the threat of drought. Data were collected from the census of agriculture and the United States geological survey, and the baseline model was used to assess the adaptation of groundwater in agricultural production. A placebo test was used to explore the local spillover effects throughout the nearby counties of the Ogallala Aquifer. Groundwater reduces the negative impact of drought on water-intensive crops. The results of the study indicated that from 1970 to 1997, irrigation increased by 11 percent for those counties that lie above the Ogallala Aquifer. Since corn is a water-intensive crop, the irrigated area of corn decreased throughout this time period in the study region. The authors of [14] examined the production levels and management practices of corn producers in the Texas High Plains with reduced or limited levels of irrigation. Corn has a high evapotranspiration (ET) demand (both daily and seasonally) in the Texas High Plains [15,16]. Although corn yield varied from year to year, there has been a clear linear upward trend from 1975 to 2015. Management practices are more important than breeding when water exists in limited conditions. Irrigation management is the most effective way to sustain high crop productivity. The authors stated that breeding for drought tolerance in corn is a major goal to improve yield stability under drought conditions. The results show that newly developed drought-tolerant corn hybrids provide yield benefits of 10–15 percent under limited (reduced) irrigation water levels. The results of the study suggest that management practices for irrigated corn in the Texas High Plains require proper management, hybrid selection, a high seeding rate, and planting date planning to achieve higher yields. Subsequently, the reference [17] analyzed that irrigated agricultural production in eight states relies on water from the Ogallala Aquifer (Southern South Dakota, Southeast Wyoming, Eastern Colorado, Nebraska, Western Kansas, Eastern New Mexico, Northwest Oklahoma, and Northwest Texas). Crop production data and weather information were collected from 1960 to 2007 for 205 counties from the abovementioned eight states. A regression technique was used to estimate the irrigation elasticity (IR), which is related to the ratio of county dry matter yields to county share of irrigated area. The results of the study showed that in 2007, most of the irrigated agricultural production was produced in Nebraska which was worth around USD two billion. Nebraska covers 36% of the total study area and accounts for 69%

of the total volume of water. Moreover, the results explain that Nebraska benefited most from water withdrawn for irrigation compared to Kansas and Texas.

Based on previous literature, very few studies have examined the spatiotemporal patterns of crop production in the Texas High Plains. Therefore, this research adds to the existing literature by analyzing county-level, time-series data on crop production. In particular, several exploratory data analysis techniques were employed to examine and visualize changes in spatiotemporal patterns of crop production in this region over the past 40 years.

Few studies have examined the spatiotemporal patterns of agricultural production in the Texas High Plains. In addition, most of these studies on the evolution of regional economic activity in this region have focused mainly on the impacts of one-time major events. To help regional producers and the public make better and more informed decisions, it is essential to have visuals that communicate complex information about the spatiotemporal dynamics of changes in regional economic activity as a simple value. Therefore, this study examines the historical changes in spatiotemporal patterns of crop production in the Texas High Plains from the perspective of geographical concentration and spatial autocorrelation. Specifically, this study attempts to answer the following research questions: (i) Does the pattern of geographical concentration of corn and wheat crop production show spatial trends and if so, do those trends change over time? (ii) Is there any spatial dependence in the production of major crops (corn and wheat) across the counties of the Texas High Plains?

From this research, regional producers and the public will benefit from the ability to access and visualize the spatial and temporal patterns of regional crop production activity information. Further, considering the spatiotemporal change of cropping patterns in the Texas High Plains, the results of this study may provide information to policy makers about where conversion to non-irrigated varieties may occur first as dependence on irrigation water from Ogallala Aquifer is reduced.

2. Materials and Methods

This section explains the study area, methods, and data sources for this study.

Methodology: This study examined the spatial and temporal changes in cropland acreage in the Texas High Plains from the perspective of geographical concentration and spatial dependence. First, to examine how much (or little) the acreage of cropland in the Texas High Plains has changed over time, maps were generated for each variable of interest. To create maps, categories were created which were dependent on the range of values. Therefore, the maps can be used to identify counties with an extremely large (or small) number of acres. Second, to analyze the overall change in geographical concentration, the Gini coefficient, one of the most commonly used measures of geographical concentration of industries, was calculated [9,18–21]. In this study, the geographical concentration of cropland acreage refers to the relative share of harvested acreage contributed by each county. In particular, the Gini coefficient was calculated as:

$$G = (\sum_{i=1}^{n}(2i - n - 1)x_i)/n^2\mu \quad (1)$$

where x is the number of acres harvested, n is the total number of counties, i is the rank of values in ascending order, and μ is the mean value of x. The Gini coefficient takes a value between zero and one. A value of zero means that each county harvests the same number of acres, while a value of one indicates that all of the production is concentrated in a single county.

Third, to examine the spatiotemporal changes in geographical concentration, size distributions were calculated for all variables of interest. To do so, the counties were first ranked in descending order based on the number of acres. The cumulative distribution of acres harvested was then generated. The number of counties with 25, 50, and 75 percent of total acres harvested was then determined. A map-based visual representation of geographical concentration for each census year was then created. Additionally, to investigate spatiotemporal changes in irrigation decisions, a quantile map was generated using data on irrigated acreage as a percent of total acreage for each variable of interest and each census year.

Finally, to examine the spatial autocorrelation or dependence of crop choices across the study region, the Moran's I statistic was calculated [22]. Spatial autocorrelation is characterized by a correlation among nearby locations. Specifically, the Moran's I statistic measures how one county's spatial information content is similar to that of the surrounding counties. In this study, the statistical test was performed to test whether the relative proportion of each harvested crop was randomly distributed across counties. In this study, the relative proportion of each crop was computed as the ratio of the number of acres harvested for each crop to the total number of acres of cropland. The test was conducted using both total acreage and irrigated acreage figures. The Moran's I statistic ranges from −1 to 1. A value of −1 indicates perfect dispersion and perfect clustering of dissimilar values; a value of 0 indicates that there is no autocorrelation among the neighboring counties; and a value of 1 means perfect clustering of similar values. In other words, a higher value of Moran's I indicates that the observations are clustered near other high values relative to lower values [23]. The *p*-value of Moran's I index determines whether the null hypothesis of no spatial autocorrelation can be rejected.

Data Sources

Study Area: This study analyzes temporal changes in spatial patterns of crop production at the country level in the Texas High Plains. The study area includes the following 39 counties in the Northern and Southern High Plains Texas Agricultural Statistics Service (TASS) districts: Andrews, Armstrong, Bailey, Briscoe, Carson, Castro, Cochran, Crosby, Dallam, Dawson, Deaf Smith, Floyd, Gaines, Glasscock, Gray, Hale, Hansford, Hartley, Hemphill, Hockley, Howard, Hutchinson, Lamb, Lipscomb, Lubbock, Lynn, Martin, Midland, Moore, Ochiltree, Oldham, Parmer, Potter, Randall, Roberts, Sherman, Swisher, Terry, and Yoakum (Figure 1). The region is comprised mostly of agricultural land, with nearly 11.4 million acres of cropland in 2017 [24].

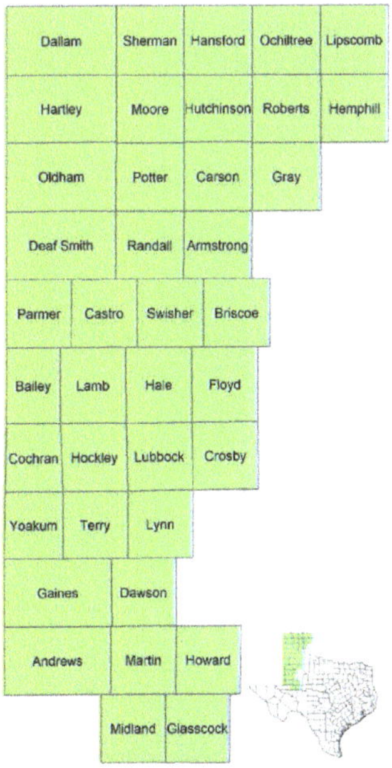

Figure 1. The 39-county area of the Northern and Southern High Plains Texas Agricultural Statistics Service (TASS) districts and their location in the state of Texas.

The climate is semi-arid with low rainfall. Annual precipitation varies widely across these 39 counties with long-term averages of 14.2–23.7 inches [25]. Due to low precipitation, crop production in the area is highly dependent on irrigation water from the Ogallala Aquifer. Given the depletion of the Ogallala Aquifer, it is important for policy purposes to examine historical changes in spatiotemporal patterns of crop production in the region. The Northern and Southern High Plains Texas Agricultural Statistics Service (TASS) districts and their location in the state of Texas is shown in Figure 1.

Agricultural Census Data. Historical county-level agricultural census data were collected for the years 1978, 1982, 1987, 1992, 1997, 2002, 2007, 2012, and 2017 from the United States Department of Agriculture (USDA) and National Agricultural Statistics Service (NASS). Two crops, corn and wheat, were selected for this study because these are the most prominent crops in the Texas High Plains and census data were available for these crops throughout the study period. The variables included in this study are total and irrigated harvested cropland area and number of farms for each selected crop. Census data were chosen for analysis because it presents a nearly complete, county-level enumeration of crop production data in the U.S. making it possible to examine spatial variations between countries over multiple time periods [10].

3. Results

3.1. Corn Harvested for Grain

The total and irrigated harvested corn grain acres are presented in Figure 2. The total harvested corn acres decreased by approximately 9% from 799,000 acres in 1978 to 726,000 acres in 2017. From 1978 to 1987, both the total and irrigated harvested corn acreages declined fast and reached a minimum of 409,000 and 397,000, respectively. Overall, except for the years 2012 to 2017, both total and irrigated acres of corn harvested for grain changed in the same direction. This suggests a decline in the irrigated percentage of the region's corn acreage.

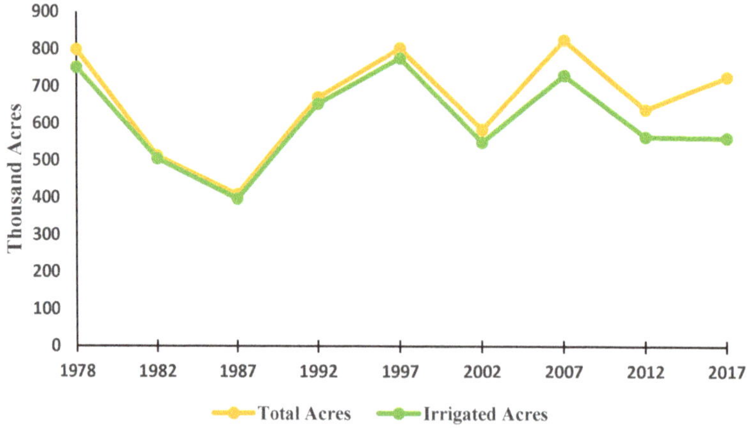

Figure 2. Total and irrigated corn harvested area for grain (in thousand acres) in the Texas High Plains, 1978–2017.

The maps for irrigated and non-irrigated acres of corn harvested for grain by county from 1978 to 2017 are presented in Figure 3. Parmer County had the largest corn acreage (more than 150,000 of harvested acres) in 1978 and 1982. Nevertheless, its total harvested corn acres gradually decreased over time. In 1987 and 1992, the counties with the largest share of corn acreage were Parmer, Hale, Castro, and Dallam. These counties had at least 50,000 and 75,000 acres of their cropland acres planted with corn in 1987 and 1992, respectively. From 1997 to 2012, Dallam had become the county with the largest corn acreage. In 2017, however, Sherman had become the county with the largest corn acreage.

A similar trend was found when focusing solely on the irrigated acres of corn harvested for grain. Specifically, Parmer County had the largest corn acreage from 1978 to 1992. From 1997 to 2012, Dallam County had the largest share of irrigated corn acreage, whereas Sherman took first place in 2017.

This study also estimated the Gini coefficients for total and irrigated acres of corn for grain for all census years. Table 1 explains the estimated coefficients of Gini for total and irrigated acres of corn for grain. The Table 1 results report that for total corn acreage, the Gini coefficient ranged between 0.636 and 0.790. For irrigated corn acreage, the Gini coefficient varied between 0.683 and 0.797. The high value of the Gini coefficients indicates that both total and irrigated corn acreages were consistently concentrated in a smaller number of counties over time.

Table 1. Gini coefficients of total and irrigated corn.

Year	Total	Irrigated
1978	0.756	0.773
1982	0.790	0.797
1987	0.749	0.753
1992	0.717	0.723
1997	0.715	0.724
2002	0.782	0.786
2007	0.742	0.746
2012	0.738	0.741
2017	0.636	0.683

Notes: Source: authors' own calculations. This table reports the estimated Gini coefficients for total and irrigated corn acres harvested for grain, 1978–2017.

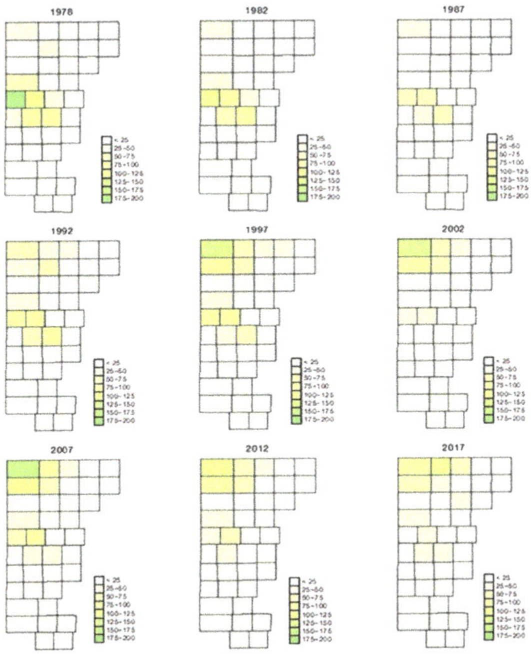

Figure 3. Irrigated and non-irrigated corn harvested area for grain (in thousand acres) by county, 1978–2017.

The geographical concentration of total corn acreage is presented in Figure 4. Overall, a total of two to three counties covered a quarter of total acres of corn harvested for grain: Parmer and Castro for the census years 1978 and 1982; Parmer and Hale for the census year 1987; Parmer, Hale, and Castro for the census year 1992; Dallam and Castro for the census year 1997; Dallam and Hartley for the census years 2002, 2007, and 2012; and Sherman and Dallam for the census year 2017. The combined land area in the top two or three counties covered less than 10 percent of the total land area. Overall, there was a small change in the geographical concentration of total corn acreage. A similar trend was observed when considering the irrigated corn acreage. These results are expected as most of the corn acreage was irrigated.

Figure 4. Concentration of irrigated and non-irrigated corn harvested area for grain (in thousand acres) by county, 1978–2017.

3.2. Wheat Harvested for Grain

The region's total and irrigated acres of wheat harvested for grain during the census years 1978 and 2017 are depicted in Figure 5. The total acres had been clearly much more volatile than the irrigated acres. The total acres decreased from 1.4 million acres in 1978 to 1.1 million acres in 2017, whereas the irrigated acres decreased from 595,775 acres in 1978 to only 250,723 in 2017.

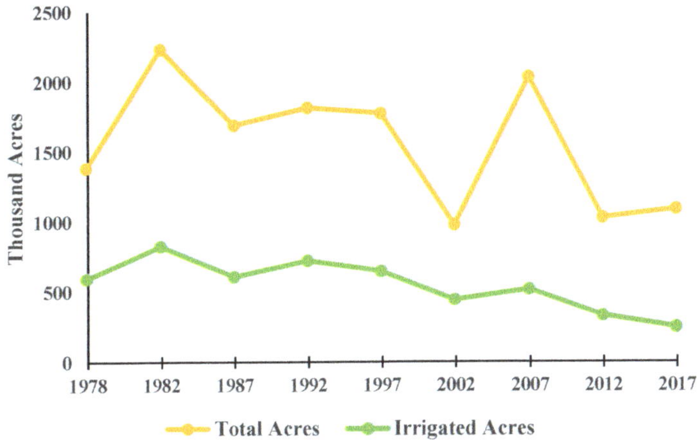

Figure 5. Total and irrigated wheat harvested area for grain (in thousand acres) in the Texas High Plains, 1978–2017.

Next, the maps of total and irrigated acres of wheat harvested for grain are illustrated in Figure 6. Overall, wheat production was concentrated in the northern part of the region. Focusing on total wheat acreage, Ochiltree had the largest number of acres in most census years. During the census year 1992, Ochiltree was considered wheat-dense (having more than 150,000 acres of wheat harvested). In the most recent census year, Deaf Smith was the county with the largest wheat acreage in the area. Similar patterns were observed when considering irrigated wheat acreage.

This study also estimated the Gini coefficients for total and irrigated acres of wheat harvested for grain for the census period 1978–2017. The estimated Gini coefficients for total and irrigated acres of wheat harvested for grain are reported in Table 2.

Table 2. Estimated Gini coefficients for total and irrigated wheat harvested for grain.

Year	Total	Irrigated
1978	0.542	0.647
1982	0.485	0.586
1987	0.527	0.624
1992	0.522	0.632
1997	0.535	0.643
2002	0.488	0.619
2007	0.535	0.622
2012	0.539	0.570
2017	0.554	0.616

Notes: Source: authors' own calculations. Estimated Gini coefficients for total and irrigated wheat harvested for grain acres, 1978–2017.

The estimated Gini coefficients for total and irrigated acres of wheat harvested for grain are reported in Table 2. For total wheat acreage, the Gini coefficient value ranged between 0.488 and 0.554. For irrigated wheat acreage, the Gini coefficient varied between 0.570 and 0.647. Similar to corn crops, the results suggest that wheat acreage was concentrated in a small number of counties. As the Gini coefficient values for the case of irrigated acres were higher than those for the case of total acres, irrigated wheat acreage was more concentrated in fewer counties than total (irrigated and non-irrigated) wheat acreage.

Further, spatial changes in the geographical concentration of total and irrigated acres of wheat harvested for grain are shown in Figure 7.

Figure 6. Irrigated and non-irrigated wheat harvested area for grain (in thousand acres) by county, 1978–2017.

As can be seen from Figure 7, the fewest number of counties with 25 percent of total wheat acres were two to four counties, depending on the census year. In 1978, the top three counties were Ochiltree, Deaf Smith, and Hansford. During the census years 1982 and 2017, six counties covered more than 25 percent of the harvested wheat acreage: Ochiltree, Deaf Smith, Hansford, Randall, Dallam, and Sherman. These counties, however, covered less than 15 percent of the total land area. Similar patterns were observed when focusing on the case of irrigated wheat acreage. However, these counties covered less than 15 percent of the total land area. Overall, wheat acreage was concentrated in the northern part of the region.

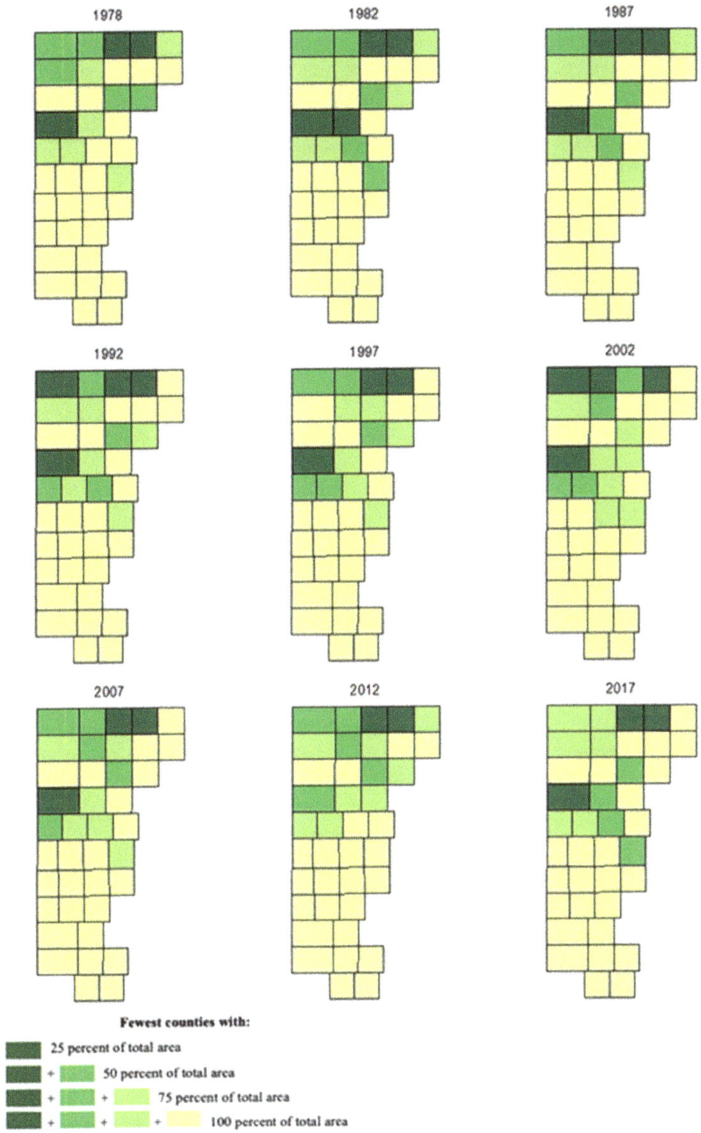

Figure 7. Concentration of irrigated and non-irrigated wheat harvested area for grain (in thousand acres) by county, 1978–2017.

3.3. Spatial Autocorrelation

To assess spatial autocorrelation or dependence in crop choices among neighboring counties, Moran's I statistics were calculated to test whether the relative proportion of each crop harvested was randomly distributed across counties. The test results for total acreage (irrigated and non-irrigated) are reported in Table 3. Overall, the Moran's I statistics were positive and statistically significant at the 5 percent significance level, indicating the existence of spatial dependence. The test statistic ranged from 0.421 to 0.670, when considering both irrigated and non-irrigated cropland area and from 0.448 to 0.690, when considering only irrigated cropland area. For wheat, it is clear that the Moran's I statistics declined over time when considering the proportion of wheat relative to the total cropland area. In fact, the null hypothesis of no spatial autocorrelation cannot be rejected for the census year 2017. Nevertheless, when considering only irrigated cropland, the Moran's I statistics were consistently high for the case of wheat, indicating strong autocorrelation across neighboring counties. Overall, the spatial autocorrelation results are in line with the exploratory data analysis results presented in the previous sections.

Table 3. Estimated Moran's I statistics for total and irrigated acreage.

	Total		Irrigated	
Year	Corn	Wheat	Corn	Wheat
1978	0.551 *	0.917 *	0.503 *	0.904 *
1982	0.486 *	0.753 *	0.448 *	0.907 *
1987	0.518 *	0.835 *	0.491 *	0.920 *
1992	0.590 *	0.922 *	0.543 *	0.919 *
1997	0.625 *	0.676 *	0.599 *	0.903 *
2002	0.670 *	0.518 *	0.690 *	0.824 *
2007	0.638 *	0.677 *	0.581 *	0.821 *
2012	0.653 *	0.170 *	0.605 *	0.754 *
2017	0.421 *	−0.005	0.621 *	0.744 *

Notes: Source: authors' own calculations. * Denotes a rejection of the null hypothesis of no spatial autocorrelation at the 5 percent significance level.

4. Discussion

The total harvested cropland acres of wheat and corn were more volatile than the irrigated harvested cropland acres. The crop maps for both total and irrigated harvested acres of cropland for all variables of interest show that counties in the center of the Texas High Plains have the largest share of harvested acres and visualize the change in county-level harvested cropland over time. Further, 88 percent of the total land area of the study region is situated over the Ogallala Aquifer. Moreover, the counties in southern region of the Texas High Plains have a lower saturated thickness than the northern part. Therefore, these areas are not as affected by declines in the Ogallala Aquifer. Some other factors such as weather variables and soil quality might have some correlations which warrant further investigation.

As the Gini coefficient values for the case of irrigated acres were higher than those for the case of total acres, irrigated wheat acreage was more concentrated in fewer counties than total (irrigated and non-irrigated) wheat acreage.

Moreover, for quantiles of size distribution, total (irrigated and non-irrigated) and irrigated harvested acres of cropland corn and wheat were ranked from largest to smallest and the minimum number of counties with 25 percent, 50 percent, and 75 percent of total acres harvested were identified at the county level throughout the study period. There was almost no change in the number of counties with 25 percent of the total and irrigated harvested cropland acres concentration over the study period whereas there was a small change in the geographical concentration of total corn acreage where only two to three counties covered a quarter of the total acres of corn harvested for grain. However, wheat production was concentrated in the northern part of the region. The saturated thickness of the Ogallala Aquifer in the northern part of the Texas High Plains is higher (up to 500 feet than the southern or center part (up to 200 feet) which might drive a high concentration of wheat production in the northern region. Spatiotemporal changes in the proportion of irrigated harvested areas relative to total harvested area for each crop were also examined. Since corn is mostly an irrigated crop, more than 80 percent of the total corn acreage in most were irrigated in the Texas High Plains.

For corn, the Moran's I statistics are relatively stable over time when considering both irrigated and non-irrigated cropland area. However, for wheat, it is clear that the Moran's I statistics have declined over time when considering the proportion of wheat relative to the total cropland area.

5. Conclusions

This study examined the temporal changes in county-level spatial patterns of corn and wheat crop production in the Texas High Plains. Historical agricultural census data on acres of corn harvested for grain with acres of wheat harvested for historical census years were considered in the analysis. Total (irrigated and non-irrigated) crop acres were analyzed. The study also analyzed the spatial and temporal changes in corn and wheat acres in the Texas High Plains from the standpoint of geographical concentration and spatial dependence. Maps were generated for each variable of interest in order to examine how much cropland acreage in the Texas High Plains has changed over time. The temporal analysis results show that during the study period, the total and irrigated harvested acres showed a downward trend for almost all the variables of interest from 1978 to 2017.

To analyze the overall change in geographical concentration and spatiotemporal changes, the Gini coefficient and the quantiles of size distributions, respectively, were computed for all variables of interest. The high values of the Gini coefficients indicate that both total and irrigated corn acreages were consistently concentrated in a smaller number of counties over time. Similar to corn crops, the wheat acreage was concentrated in a small number of counties. Overall, considering the spatiotemporal change in corn and wheat cropping patterns in the Texas High Plains, policy makers should promote and support the non-irrigated varieties of crops to reduce the dependence on irrigation water from the Ogallala Aquifer.

Author Contributions: This research work titled "The Spatiotemporal Economic Analysis of Wheat and Corn Production in the Texas High Plains" was carried out in collaboration among all the authors. A.N. developed main idea of this study and was responsible for the majority of the writing. She also collected and organized the data to estimate the model and interpret the results. S.S. wrote introduction, specified the model, and wrote the conclusion and recommendations. L.K.A. supervised the overall write up throughout and B.G. reviewed, edited, corrected, and formatted this manuscript. All authors have read and agreed to the published version of the manuscript.

Funding: This research was funded in part by the USDA-ARS Ogallala Aquifer Program.

Data Availability Statement: Not Applicable.

Acknowledgments: We acknowledge the technical support of West Texas A&M University for providing access to materials pertaining to this research work.

Conflicts of Interest: The authors declare no conflict of interest.

References

1. United States Department of Agriculture, National Agricultural Statistics Service. Census of Agriculture. 2017. Available online: https://www.nass.usda.gov/Publications/AgCensus/2017/Full_Report/Volume_1,_Chapter_1_State_Level/Texas/st48_1_0001_0001.pdf (accessed on 25 June 2020).
2. Thayer, A. Climate, Water, Water Markets, and Texas Agriculture: Three Essays. Doctoral Thesis, Texas A & M University, College Station, TX, USA, 2018. Available online: https://oaktrust.library.tamu.edu/bitstream/handle/1969.1/174341/THAYER-DISSERTATION-2018.pdf?sequence=1&isAllowed=y.https://oaktrust.library.tamu.edu/bitstream/handle/1969.1/174341/THAYER-DISSERTATION-2018.pdf?sequence=1&isAllowed=y (accessed on 25 February 2021).
3. McCullough, L.R. Water Use and the Regional Economic Impact of the Cotton Industry on the Southern Ogallala Aquifer. Master's Thesis, West Texas A&M University, Canyon, TX, USA, 2016. Available online: https://wtamu-ir.tdl.org/items/1e375e64-4f63-4104-bd50-a7e33f896e5e (accessed on 25 February 2021).
4. Hornbeck, R.; Keskin, P. The historically evolving Impact of the Ogallala Aquifer: Agricultural adaptation to groundwater and drought. *Am. Econ. J. Appl. Econ.* **2014**, *6*, 190–219. [CrossRef]

5. McGuire, L.V.; Johnson, R.M.; Schieffer, L.R.; Stanton, S.J.; Sebree, K.S.; Verstraeten, M.I. *Water in Storage and Approaches to Groundwater Management, High Plains Aquifer 2000*; U.S. Geological Survey Circular: Reston, VI, USA, 2003; Circular 1243. Available online: https://pubs.usgs.gov/circ/2003/circ1243/ (accessed on 8 October 2023).
6. United States Department of Agriculture, National Agricultural Statistics Service 1978. Census of Agriculture. Available online: http://libusda05.serverfarm.cornell.edu/usda/AgCensusImages/1978/01/43/181/Table-01.pdf.http://libusda05.serverfarm.cornell.edu/usda/AgCensusImages/1978/01/43/181/Table-01.pdf (accessed on 20 June 2020).
7. Benavidez, J.; Guerrero, B.; Dudensing, R.; Jones, D.; Reynolds, S. The Impact of Agribusiness in the High Plains Trade Area. Agriculture Council of Amarillo Chamber of Commerce. 2019. Available online: https://amarillo.tamu.edu/files/2019/12/Impact-of-AgriBusiness.pdf (accessed on 25 February 2021).
8. McBride, D.W. Change in U.S. Livestock Production, 1969–1992. Rural Economy Decision, Economic Research Services, U.S. Department of Agriculture.Agricultural Economic Report (AER754). 1997. Available online: https://www.ers.usda.gov/publications/pub-details/?pubid=40795 (accessed on 8 October 2023).
9. Beddow, J.M.; Pardey, P.G. Moving matters: The effect of location on crop production. *J. Econ. Hist.* **2015**, *75*, 219–249. [CrossRef]
10. Sukcharoen, K.; Golden, B.; Vestal, M.; Guerrero, B. Do crop price expectation matter? An analysis of groundwater pumping decisions in western Kansas. *Agric. Water Manag.* **2020**, *231*, 106021. [CrossRef]
11. Martinez- Casasnovas, A.J.; Martin-Montero, A.; Casterad, A.M. Mapping multi-year cropping patterns in small irrigation districts from time- series analysis of Landsat TM images. *Eur. J. Agron.* **2005**, *23*, 159–169. [CrossRef]
12. Xiao, J.; Shen, Y.; Ge, J.; Tateishi, R.; Tang, C.; Liang, C.; Huang, Z. Evaluating urban expansion and land use change in Shijiazhuang, China, by using GIS and remote sensing. *Landsc. Urban Plan.* **2006**, *15*, 69–80. [CrossRef]
13. Hornbeck, R. The Enduring Impact of the American Dust Bowl: Short- and Long-Run Adjustments to Environmental Catastrophe. *Am. Econ. Rev.* **2012**, *102*, 1477–1507. [CrossRef]
14. Xue, Q.; Marek, H.T.; Xu, W.; Bell, J. Irrigated corn production and management in the Texas High Plains. *J. Contemp. Water Res. Educ.* **2017**, *162*, 31–41. [CrossRef]
15. Howell, T.; Yazar, A.; Schneider, A.; Dusek, D.; Copeland, K. Yield and water use efficiency of corn in response to LEPA irrigation. *Trans. Am. Soc. Agric. Eng.* **1995**, *38*, 1737–1747. [CrossRef]
16. Howell, T.A.; Evett, S.R.; Tolk, J.A.; Schneider, A.D.; Steiner, J.L. Evapotranspiration of corn Southern High Plains. In *Evapotranspiration and Irrigation Scheduling, Proceedings of the International Conference, San Antonio, TX, USA, 3–6 November 1996*; Camp, C.R., Sadler, E.J., Yoder, R.E., Eds.; ASAE: St. Joseph, MI, USA, 1996; pp. 158–166.
17. Garcia, F.; Fulginiti, L.; Perrin, R. What is the Use Value of Irrigation Water from the High Plains Aquifer? *Am. J. Agr. Econ.* **2019**, *101*, 455–466. [CrossRef]
18. Krugman, P. Increasing returns and economic geography. *J. Political Econ.* **1991**, *99*, 483–499. [CrossRef]
19. Haggett, P.; Cliff, A.D.; Frey, A. Locational analysis in human geography. *J. Econ. Hum. Geogr.* **1977**, *68*, 363–367.
20. Cullis, J.; Van Koppen, B. *Applying the Gini Coefficient to Measure Inequality of Water Use in the Olifants River Water Management Area, South Africa*; Research Report 113; International Water Management Institute (IWMI): Colombo, Sri Lanka, 2007.
21. Liu, H.; Jia, Y.; Niu, C.; Gan, Y. Spatial pattern analysis of regional water use profile based on the Gini coefficient and location quotient. *Am. Water Resour. Assoc.* **2019**, *55*, 1349–1366. [CrossRef]
22. Moran, P. A Test for the Serial Independence of Residuals. *Biometrika* **1950**, *37*, 178–181. [CrossRef]
23. Peng, J.; Liu, Y.; Liu, Z.; Yang, Y. Mapping spatial non-stationarity of human-natural factors associated with agricultural landscape multifunctionality in Beijing–Tianjin–Hebei region, China. *Agric. Ecosyst. Environ.* **2017**, *246*, 221–233. [CrossRef]
24. United States Department of Agriculture, National Agricultural Statistics Service. U.S. Census of Agriculture for 2017. USDA National Agricultural Statistical Service. 2019. Available online: https://www.nass.usda.gov/Publications/AgCensus/2017/index.php (accessed on 24 September 2020).
25. PRISM Climate Group. PRISM Data Explorer. Oregon State University, 2020. Available online: https://prism.oregonstate.edu/explorer/ (accessed on 24 September 2020).

water

MDPI

Article

Effect of Deficit Irrigation on Growth Parameters of the *Salvia splendens* L. Plant

Işık Sezen [1,*], Sevda Yağanoğlu [2], Elif Akpınar Külekçi [1] and Ayşe Karahan [3]

1 Department of Landscape Architecture, Faculty of Architecture and Design, Atatürk University, Erzurum 25240, Turkey; eakpinar@atauni.edu.tr
2 Atatürk University Herbal Production Application and Research Center, Erzurum 25240, Turkey; sevdayaganoglu@atauni.edu.tr
3 Department of Landscape Architecture, Atatürk University Graduate School of Natural and Applied Sciences, Erzurum 25240, Turkey; ayseozkur70@hotmail.com
* Correspondence: isiksezen@atauni.edu.tr; Tel.: +90-4422316133

Abstract: This study aims to investigate alterations in the developmental parameters of *Salvia splendens* L., a commonly utilized seasonal flower associated with excessive water consumption in urban green spaces, through the implementation of deficit irrigation practices. Four distinct irrigation treatments, which entailed maintaining the evaporation pot's water level at 100% (control), 75%, 50%, and 25% of the pot's water-holding capacity, were established. This study scrutinized 18 growth parameters to assess the impact of varying water application levels. The findings of this research revealed that *Salvia splendens* L. plants exhibited more substantial improvements in 17 out of 18 assessed parameters when subjected to 75% water application (representing a 25% reduction in water supply) in comparison to 100% water application (with no reduction). Notably, the only parameter negatively affected by reduced water availability in *Salvia splendens* L. was the diameter of the flowers. Thus, it is recommended to reduce water application by 25% when cultivating *Salvia splendens* L. in urban areas. Such a measure is expected to yield substantial water conservation benefits in urban landscaping. Consequently, it is advisable to promote the frequent utilization of *Salvia splendens* L. plants in urban green spaces, given their robust development even under conditions of water scarcity.

Keywords: urban landscape; deficit irrigation; seasonal flower; water consumption; *Salvia splendens* L.

1. Introduction

The escalation of global population, rapid urbanization, and the intensified density of agricultural and industrial regions have precipitated heightened competition for freshwater resources. This burgeoning water scarcity issue is particularly critical in urban areas, where landscaping plays a pivotal role [1,2]. Consequently, there arises a compelling imperative to prioritize water management within the realm of landscaping, emphasizing the dual goals of water conservation and enhanced irrigation efficiency [2–4].

In light of the escalating impact of drought stress on plant viability due to irrigation restrictions, the selection of drought-resistant plant species has assumed increasing significance in the pursuit of sustainable landscape development [2]. Presently, the judicious use of water resources has become more paramount than ever, attributing to the limitation of water resources and the staggering consumption of 65–80% of available water for irrigation purposes [5,6]. Notably, the substantial water consumption in open green spaces, primarily aimed at sustaining plant life, underscores the imperative need for the judicious utilization of water resources in landscaping endeavors [7].

The effective management of water resources has attained global prominence and holds equal relevance in the context of Turkey. Consequently, innovative approaches aimed at economically harnessing water resources for irrigating ornamental plants, especially within landscaping domains, have come to the forefront. Given that Turkey allocates a significant portion of its usable water resources, approximately 75%, for irrigation, comprehensive efforts are warranted to effectively plan existing water resources and promote their judicious use. The implementation of regulated irrigation practices during specific periods for ornamental plants known for their high water consumption in landscaping is an essential strategy for achieving water conservation [6,8].

Regulated irrigation, defined as the controlled allocation of water to plants, has gained extensive global recognition and application [6,8–11]. Regulated irrigation signifies a methodology applied at prescribed levels and under specific protocols to ensure that it does not imperil plant well-being. Its primary objective is to achieve water savings while maintaining adequate irrigation. In the realm of regulated irrigation, water utilization efficiency is augmented, and productivity is minimally affected. A pivotal characteristic of regulated irrigation is the provision of a consistent water volume to plants, consequently enabling a higher yield per unit area [8,10].

The pivotal step in the quest for water-efficient landscaping lies in the planning and design phases. Prior to selecting plant species, an in-depth assessment of the physical and environmental conditions of the area is necessary to address existing challenges. Subsequently, a strategic evaluation of the purpose of the area, plant species selection, and classification of these species based on their water consumption rates is imperative [7,12,13]. In the context of landscaping applications, a zoning strategy should be adopted where plants are categorized according to their water consumption levels, facilitating the informed placement of these plants based on their specific irrigation requirements [13,14].

While a plethora of research exists on regulated irrigation, most of these studies have predominantly focused on field and garden plants, such as *Gossypium hirsutum* L. [15], *Zea mays* L. [16], *Beta vulgaris* sp. [17], *Malus domestica* Borkh. [18], *Solanum tuberosum* L. [19], *Helianthus annuus* L. [20], and *Olea europaea* L. [21]. Remarkably, Demirel et al. [22] have noted the insufficiency of studies focusing on the application of water restrictions to ornamental plants. Consequently, efforts have been directed toward exploring the influence of regulated irrigation on seasonal ornamental plants, including Zinnia elegans [23], *Petunia violacea* [24], *Impatiens walleriana* L. [25], *Cyclamen persicum* Mill. [26], *Primula* sp. [27], *Tagates erecta* L. [8], *Chrysanthemum morifolium* R. [28], and *Pelargonium domesticum* [29], with regards to their developmental patterns.

Gül et al. [30] have underscored the high water consumption associated with seasonal flowers commonly used in urban areas [31]. In an effort to mitigate water consumption for irrigation purposes in urban zones, recommendations have been made to curtail the cultivation of seasonal flowers [32]. Nonetheless, it is acknowledged that seasonal flowers are indispensable components of the urban landscape [33], contributing to the aesthetic enhancement of various areas such as parks, gardens, residences, woodlands, medians, intersections, squares, balconies, flower pots, roofs, and modular flower beds. Seasonal flowers bring forth a spectrum of colors and vitality to the urban landscape, ameliorating monotony, serving as space-fillers, and allowing for annual variations, all at a low cost. They boast prolonged flowering periods and are easily cultivated in greenhouses.

In a study assessing the impact of regulated irrigation on the developmental parameters of *Tagates erecta*, Sezen et al. [8] observed that a 25% reduction in water had an insignificant impact on the number of flowers, chlorophyll content, plant height, leaf area, and root dry weight. In essence, there was no significant disparity between full irrigation and a 25% reduction in water application. Therefore, a 25% reduction in water usage holds the promise of cost savings. Uçar and Kazaz [28], in their research on chrysanthemums, expounded on the significant influence of different irrigation practices on leaf area and the number of flowers. Furthermore, they noted that different irrigation strategies significantly affected the number of flowers in the *Chrysanthemum morifolium* R. plant. In their exploration of the effect of different irrigation levels on the vegetative characteristics of *Pelargonium domesticum*, Doğan et al. [29] discerned that water scarcity expedited flowering.

The research hypothesis is rooted in the supposition that *Salvia splendens* L., a frequently employed seasonal flower in urban settings, receives an excess of water. This study seeks to ascertain that the application of controlled water restrictions at specific rates will not detrimentally affect plant growth parameters, thereby yielding water savings. The primary objective of this research is to scrutinize the impact of regulated irrigation systems on the developmental patterns of *Salvia splendens* L., a member of the Labiatae family, which holds an indispensable role in enhancing the seasonality of urban landscapes but has been subjected to superfluous water provision due to inadequate recognition of its actual water needs.

2. Materials and Methods

The research conducted in 2022 was situated within the confines of the Atatürk University Plant Production Application and Research Center Greenhouses. *Salvia splendens* L., a plant material renowned for its burgeoning commercial value in recent years, was selected as the subject of the study. Notably, the germination period for *Salvia splendens* L. seeds typically occurs between December and April, although in regions characterized by colder climates, sowing is recommended between March and April. On 1 March 2022, *Salvia splendens* L. F1 hybrid seeds, boasting a remarkable 90% germination rate, were sown. The growth medium utilized for planting the seeds was a blend of peat, fiber, and vegetable soil, comprising 65% peat and 35% soil content and characterized by an organic matter ratio of 75%. The pH value of this medium ranged from 5 to 7.

Salvia splendens L. seeds commenced germination approximately 9 days post-sowing, with significant germination occurring after 15 days. Once the plants attained a suitable size, they were transplanted into 5 L pots. Following the guidelines provided by [34], the potting medium was composed of 2 parts loamy soil, 1 part peat, and 1 part sand, developed by the John Innes Horticultural Institute, a renowned source of superior growing mediums used for garden flowers. This experimental soil, formed by mixing washed and sieved river sand and peat in proportions of 50%, 25%, and 25%, respectively, was meticulously prepared to create the potting medium.

For the irrigation application, once the plants were transplanted into the pots, the soil was brought to field capacity (FC) for all subjects, and the irrigation levels were subsequently administered. With the exception of the control group, all other groups were subjected to irrigation at rates of 75%, 50%, and 25%, respectively. The initiation of both the control and irrigation groups was heralded by diversionary water applications.

The irrigation interval was determined based on the evaporation amount recorded from the evaporation pot until it reached a range of 5–10 mm. Consequently, irrigation was executed upon reaching the anticipated evaporation levels. The total evaporation amount was applied equally across the entire pot surface, with the entire evaporation quantity designated for the control group and 75%, 50%, and 25% allocated to the other groups.

The amount of water allocated to the control group was computed in accordance with Equation: $I = kp \times Ep \times A$

I: Irrigation water (liter × pot− − 1)
kp: Evaporation vessel coefficient (1 for control, 0.75–0.50–0.25 for other applications)
Ep: Total evaporation read from the reduced evaporation pan (mm)
A: Pot area (m^2)

On 30 March 2022, the *Salvia splendens* L. plants were transplanted into 5 L pots, and six replications were conducted for each irrigation application. This resulted in six replications for each of the 25%, 50%, 75%, and 100% irrigation levels, totaling 24 pots of *Salvia splendens* L. plants. After potting, the plants were thoroughly watered to reach field capacity.

Subsequently, to monitor evaporation, a 60 cm diameter, 25 cm deep, and 15 cm high wooden base support unit made from 2 mm thick gray board was employed. Additionally, a custom water-filled evaporation measuring device, crafted specifically for this study by Atatürk University Central Workshops, was utilized. The measurement of evaporated water was achieved using a millimeter-scale wooden ruler. Furthermore, temperature and humidity measurements encompassing average, maximum, and minimum values were conducted employing a temperature-humidity measuring instrument.

Throughout the experiment, ambient temperature was maintained at a constant range of 25–30 °C, and irrigation was administered based on environmental evaporation rates. The study entailed the creation of four distinct irrigation scenarios, with each determined as a percentage of usable water holding capacity compared to the pot volume. Prior to the experiment, the pot capacity (field capacity) and usable water holding capacity (UWHC) of each pot were meticulously determined [21]. The pots were filled to their field capacity upon initial planting. The study continued by adjusting water application in subsequent stages based on the calculated evaporation, with the control group receiving 100% water and the other groups receiving 75%, 50%, and 25%. The intervals between irrigation applications were adapted depending on the evaporation levels.

It is noteworthy that the water used in this research was sourced from Atatürk University and was confirmed to be suitable for irrigation. Following the irrigation application, various growth parameters of the plants, such as the number of flowers, flower stem thickness, flower diameter, flower height, flower fresh weight, flower dry weight, leaf chlorophyll value, leaf area, root length, and root fresh weight, were meticulously observed and documented.

Cumulative evaporation levels were measured in millimeters using a modified (reduced) evaporation pot located within the greenhouse, and the water amount was presented in milliliters relative to the pot volume. The cumulative quantities of irrigation water applied to the flower seedlings, both in the control group and the other irrigation groups, over the course of the 55-day experimental period spanning from the seedling planting date (30 March 2022) to the final irrigation application (23 May 2022), were recorded in milliliters.

In this study, irrigation procedures commenced on 30 March 2022, with all pots brought to field capacity, and the research was successfully concluded on 23 May 2022. Over the duration of the trial period, a total of 319 mm of evaporation occurred from the evaporation pan. As a result, the daily average evaporation rate was calculated at 4.89 mm/day. In each irrigation group, 269 mm of irrigation water was allocated to the control group, whereas the other groups received 67.25 mm, 134.5 mm, and 201.75 mm at 25%, 50%, and 75% irrigation levels, respectively. The specific values pertaining to irrigation water and evaporation amounts applied during the research can be found in Table 1.

Table 1. Irrigation water in research subjects (mm).

Date	Evaporation Amount (mm)	25%	50%	75%	100%
30 March 2022	5	1.25	2.5	3.75	5
31 March 2022	5	1.25	2.5	3.75	5
1 April 2022	5	1.25	2.5	3.75	5
4 April 2022	9	2.25	4.5	6.75	9
5 April 2022	6	1.5	3	4.5	6
6 April 2022	5	1.25	2.5	3.75	5
8 April 2022	10	2.5	5	7.5	10
11 April 2022	15	3.75	7.5	11.25	15
12 April 2022	6	1.5	3	4.5	6
13 April 2022	6	1.5	3	4.5	6
15 April 2022	12	3	6	9	12
18 April 2022	20	5	10	15	20
20 April 2022	10	2.5	5	7.5	10
22 April 2022	14	3.5	7	10.5	14
25 April 2022	16	4	8	12	16
27 April 2022	11	2.75	5.5	8.25	11

Table 1. *Cont.*

Date	Evaporation Amount (mm)	25%	50%	75%	100%
29 April 2022	14	3.5	7	10.5	14
2 May 2022	9	2.25	4.5	6.75	9
4 May 2022	6	1.5	3	4.5	6
6 May 2022	3	0.75	1.5	2.25	3
9 May 2022	9	2.25	4.5	6.75	9
11 May 2022	5	1.25	2.5	3.75	5
13 May 2022	10	2.5	5	7.5	10
16 May 2022	21	5.25	10.5	15.75	21
18 May 2022	11	2.75	5.5	8.25	11
20 May 2022	11	2.75	5.5	8.25	11
23 May 2022	15	3.75	7.5	11.25	15
Total	269	67.25	134.5	201.75	269

In the realm of seasonal floriculture, irrigation practices were enacted upon the attainment of a daily evaporation rate ranging between 5 and 10 mm within the confines of the greenhouse. As indicated in Table 1, this condition resulted in varying irrigation intervals, spanning from 1 to 3 days. The pinnacle of evaporation during the research occurred on 16 May 2022, registering an imposing rate of 21 mm/day for the control group. Conversely, the nadir of evaporation was noted on 6 May 2022, with a minimal rate of 3 mm/day for the control group.

Throughout the course of this study, the nexus between the recorded temperature, relative humidity, and evaporation levels was explored. This exploration was conducted between 30 March 2022, and 23 May 2022. It was observed that temperature and humidity levels remained constant in the region until April 20th. Subsequently, the upsurge in relative humidity, transpiring between 27 April 2022, and 15 May 2022, exhibited a clear correlation with temperature fluctuations. Notably, evaporation rates remained consistent during this period. The intricate interplay between evaporation, temperature, and humidity is illustrated in Figure 1.

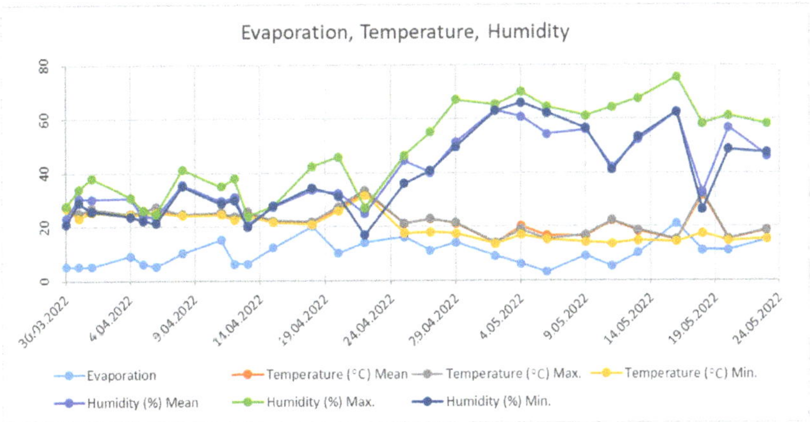

Figure 1. The relationship between evaporation, temperature, and humidity.

The method flow chart is given in Figure 2.

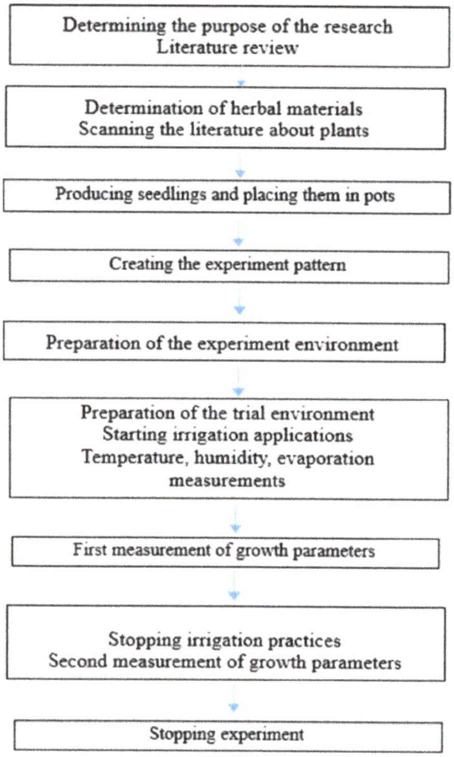

Figure 2. Method flow chart.

3. Results and Discussion

The assessment of plant development involved the measurement of plant height, plant diameter, flower stem thickness, flower diameter, and flower height, which were performed using a digital caliper. Additionally, leaf chlorophyll values were quantified using a SPAD-502 chlorophyll meter on 29 April 2022, marking the 30th day following the initiation of water restriction on the seedlings. Furthermore, the count of new buds was meticulously determined.

Subsequently, on the 43rd day since the commencement of the water deficit conditions, a comparative analysis was conducted to scrutinize the plant growth of the replications that most aptly represented the distinct irrigation applications of *Salvia splendens* L. plants, encompassing 25%, 50%, 75%, and 100% irrigation levels. It is noteworthy that this particular section, while not obligatory, may be appended to the manuscript if the ensuing discussion entails intricacies or lengthiness beyond the ordinary scope.

The salient differences in the developmental trajectories of *Salvia splendens* L. plants in response to varying irrigation applications are visually depicted in Figure 3.

On the 55th day of the water scarcity period, a discernible phenomenon was recorded, whereby the *Salvia splendens* L. plants subjected to a 25% irrigation regimen exhibited signs of desiccation. Consequently, the experiment was promptly terminated, thereby discontinuing the water restriction regimen imposed on the *Salvia splendens* L. plants.

Figure 3. Growth difference in the *Salvia splendens* L. plant according to irrigation applications.

In Table 2, the trial schedule of the *Salvia splendens* L. plant is given.

Table 2. Trial schedule in *Salvia splendens* L. plant.

Seed Planting	Planting Seedlings in Pots	Measurement 1	Measurement 2	Dry Weight Measurement
1 March 2022	30 March 2022	29 April 2022	23 May 2022	9 June 2022
	30th day of sowing	60th day of sowing	84th day of seed sowing	Flowers, vegetative parts, and roots of the plant were kept dry for 17 days.
	The beginning of the water shortage	30th day of water restriction application to seedlings	55th day of water restriction application to seedlings	
		The date on which flowering occurs in most applications	The date of establishment of all recurrences in the application, given 25% water	

In the course of this experimental study, six repetitions of *Salvia splendens* L. plant applications with a 25% irrigation rate were conducted. In four of these repetitions, there was a complete absence of flowering, while one repetition yielded no instances of flowering, and in another repetition, the plant succumbed to its environmental conditions before reaching full maturation. Conversely, within all instances of plant applications featuring a 50% irrigation rate, flowering manifested on the 30th day subsequent to the initiation of the water scarcity period. In a consistent manner, across all six repetitions of applications with a 75% irrigation rate, flowering occurred on average after 62 days, commencing 56 days following the initial seed sowing. These findings substantiate that the provision of a 25% irrigation rate to *Salvia splendens* L. plants leads to a marked reduction in the flowering rate.

Moreover, it is noteworthy that within the group subjected to 75% irrigation, flowering occurred six days prior to the cohort exposed to 50% irrigation and four days earlier than the set receiving 100% irrigation. Although flowering was observed in all repetitions within the 50%, 75%, and 100% irrigation groups, it is evident that the 75% irrigation application, which corresponds to a 25% water restriction, demonstrated the most expedient onset of flowering. In Table S1, changes in growth values of *The excerpt of Table S1 has been added to the main text.* L according to irrigation practices and multiple comparison tests are given.

The impact of water restrictions on *Salvia splendens* L. plants was examined. On the 30th day of this restriction, a discernible disparity in the number of flowers was noted, with the application of 75% water (equivalent to a 25% water reduction) yielding the highest number of flowers in the first rank, followed by the 50% application in the second rank, and the 100% application in the third rank, while the 25% application exhibited a complete absence of flowering. Subsequent measurements conducted on the 55th day of the water shortage, which marked the conclusion of the experiment due to the desiccation of the 25% application group, corroborated these findings. Once again, the 75% water application ranked highest in terms of flower yield, followed by the 100% application in the second rank, the 50% application in the third rank, and the 25% application in the fourth rank. These outcomes underscore the superior effectiveness of the 75% water application (constituting a 25% reduction in water availability) in stimulating flower production. On the 30th and 55th days of this restriction, the highest number of flowers was found in the application where 75% water was applied, while the values in the other three water applications were quite low. This observation is aligned with the findings of Doğan et al. [24], who similarly noted that water scarcity had an accelerating effect on flowering in their investigation assessing the impact of various irrigation levels on the vegetative characteristics of *Pelargonium domesticum*.

Moreover, measurements conducted on the 30th day of the water shortage indicated that the flower stem thickness was notably greater in the 75% water application, with the respective rankings as follows: 75% in the first rank, 50% in the second rank, 100% in the third rank, and 25% in the fourth rank. The measurements taken on the 55th day of the water shortage, in concurrence with the termination of the experiment due to the desiccation of the 25% application group, again affirmed the prominence of the 75% water application in terms of flower stem thickness. Specifically, the rankings in this instance were as follows: 75% in the first rank, 100% in the second rank, 50% in the third rank, and 25% in the fourth rank.

In the investigation concerning water restriction applied to *Salvia splendens* L. plants, the examination of various parameters revealed noteworthy findings. On the 30th day of water scarcity, it was observed that the flower diameter was greater in the group receiving 75% water in comparison to the other treatments. The respective flower diameters were ranked as follows: 75% in the first position, 50% in the second position, 100% in the third position, and 25% in the fourth position.

Subsequent measurements on the 55th day of the water scarcity period, which marked the termination of the experiment due to the desiccation of the 25% water application group, disclosed a different pattern. In this case, the highest flower diameter was found in the 100% water treatment group, followed by the 75% treatment in the second position, the 50% treatment in the third position, and the 25% treatment in the fourth position.

In her scholarly investigation of *Chrysanthemum morifolium* R., Turan [35] observed that augmenting the water supply exhibited a favorable influence on both the yield and quality attributes. In a parallel context, on the 30th day of water restriction, the measurements revealed that flower height was significantly greater in the 75% water application group when compared to other treatments, with respective averages of 75% in the first position, 100% in the second position, 50% in the third position, and 25% in the fourth position.

Likewise, the measurement on the 55th day of the water scarcity period confirmed this trend, as the 75% water application group once again demonstrated the highest flower height, followed by the 100% treatment in the second position, the 50% treatment in the third position, and the 25% treatment in the fourth position.

In the realm of vegetative characteristics, measurements conducted on both the 30th and 55th days of the water scarcity period revealed that the plant height and diameter were superior in the 75% water application group (representing a 25% reduction in water supply) compared to other treatments.

The results of the water restriction study further indicated that the wet weight of vegetative parts on the 55th day was highest in the 75% water application group, followed by the 100% treatment, the 50% treatment, and the 25% treatment. Moreover, after the drying period following the 55th day of water restriction (23 May 2022), the measurements of dry weight showed that the 75% water application group yielded the highest dry weight for vegetative parts, followed by the 100% treatment, the 50% treatment, and the 25% treatment.

The leaf chlorophyll value of the *Salvia splendens* plant was measured on the 30th day of water curtailment, and the average chlorophyll value was higher in the application with 25% water compared to the application with 75% and 100% water, while it was lower than the applications with 50% water. The highest chlorophyll value was seen in the first rank with 50%, in the second rank with 25%, in the third rank with 75%, and in the fourth rank with 100%. Chlorophyll values were checked for the second time on the 55th day of the water shortage, and measurements could not be made because the plants were dry in 25% applications, while an increase in chlorophyll value was observed in 50% applications and a decrease in 75% and 100% applications. In the measurements made on the 55th day of the water shortage, the highest value in the average chlorophyll value was observed in the first order of 50%, in the second line of 75%, and in the third line of 100%. In the measurement made on the 55th day of the water shortage, an increase in the average chlorophyll value was observed in the 50% application and a decrease in the 75% and 100% applications. According to these results, in the *Salvia splendens* plant, with too much water restriction application, the chlorophyll value decreases to almost zero, while not applying any reduction (100% water application, 0% water reduction) also reduces the chlorophyll value. Adding water between 50 and 75% increases the chlorophyll value. As a matter of fact, Tüfenkçi [36], in his study carried out to determine the effect of different levels of water restriction applications on some yield parameters of the quinoa plant, stated that the application of 50% less irrigation water compared to the application with 100% irrigation did not statistically lead to losses in yield. Kırnak and Doğan [37], in their study to determine the effect of restricted irrigation on some quantitative parameters in the melon plant, stated that the effect of different irrigation water levels on chlorophyll under each irrigation system was found to be significant at the $p < 0.05$ level.

Furthermore, the number of plant shoots was directly influenced by the level of water scarcity, with an increase in water restrictions resulting in a decrease in the number of shoots. This outcome was in alignment with previous research by Curti et al. [38] and Tufenkçi [36], which noted that plants experienced reduced vegetative growth and shoot numbers under conditions of water stress.

Regarding leaf area, measurements were not feasible for the 25% water application due to leaf loss. Nevertheless, it was observed that the leaf area was largest in the 75% water application group, followed by the 100% treatment and the 50% treatment.

Additionally, on the 55th day of water scarcity, the number of new buds was highest in the 75% water application group, followed by the 100% treatment, the 50% treatment, and the 25% treatment.

The study also revealed that root length and root wet weight were greater in the 75% water application group compared to other treatments, indicating the efficacy of the 25% water restriction in promoting root development in *Salvia splendens* L. plants.

The application of the Duncan Multiple Comparison test has provided insight into the variations among irrigation treatments concerning the growth parameters of the *Salvia splendens* L. plant.

On the 30th day of the water scarcity period, the statistical analysis revealed significant differences among all groups with respect to the number of flowers ($p < 0.05$). However, in the measurements conducted on the 55th day of water shortage, the disparities in the number of flowers were not statistically significant between the 25%, 50%, and 100% irrigation applications, whereas a significant difference ($p < 0.05$) persisted between the 75% irrigation application and the remaining groups. Notably, the 75% irrigation application, representing a 25% reduction in water supply, yielded a higher number of flowers. It is pertinent to mention that Uçar and Kazaz [26] similarly reported significant variations in the number of flowers in *Chrysanthemum morifolium* R. as a consequence of distinct irrigation practices.

Moreover, the analysis of flower stem thickness demonstrated no significant differences among the irrigation applications in the measurements made on the 30th day. Conversely, on the 55th day, the variances in flower stem thickness between the irrigation applications were highly significant ($p < 0.01$).

Likewise, the measurements on the 30th day indicated no substantial distinctions in flower diameter among the irrigation applications. However, on the 55th day, while the differences between the 75% and 100% irrigation applications and flower diameter were deemed insignificant, a highly significant difference ($p < 0.01$) persisted between these applications and the rest.

According to the Duncan Multiple Comparison test, on the 30th day, the distinctions in flower height among the irrigation applications were insignificant. On the same day, a significant difference ($p < 0.05$) was observed between the 75% irrigation application and the other groups. Yet, on the 54th day, the variance between the 75% and 100% irrigation applications and flower height was considered insignificant, whereas it remained highly significant ($p < 0.01$) when comparing these applications with the other groups. In a similar vein, Sezen et al. [8] reported that plant height exhibited no significant alterations in a 75% water application (equivalent to a 25% water reduction), and a 50% water limitation did not adversely affect plant height.

Conversely, on the 30th day, there were no significant distinctions in plant diameter among the irrigation applications. However, on the 55th day, while the differences between the 50%, 75%, and 100% irrigation applications and plant diameter were considered insignificant, a highly significant difference ($p < 0.01$) was noted between the 25% application and the other groups. Doğan et al. [29] also reported that different irrigation practices did not yield significant disparities in plant diameter values.

The leaf area measurements demonstrated insignificant differences among the 50%, 75%, and 100% irrigation applications. This finding aligns with the results of a study conducted by Sezen et al. [8], who found that the variance in leaf area between a 75% water application and a 100% water application was statistically insignificant.

The number of buds showed no significant distinction between the irrigation applications.

However, in terms of the dry weight of the vegetative parts, there were significant differences ($p < 0.05$) among all applications.

According to the Duncan Multiple Comparison test, leaf area measurements made on the 55th day revealed no significant differences between the irrigation applications.

Furthermore, on the 55th day, the differences in the number of plant shoots between the 50%, 75%, and 100% irrigation applications were deemed insignificant. However, a highly significant difference ($p < 0.01$) was observed between the 25% application and the other groups.

Root length, root fresh weight, and root dry weight displayed no significant differences among the 50%, 75%, and 100% irrigation applications. Conversely, the variance between the 25% application and the other groups was considered significant ($p < 0.05$). This observation resonates with the findings of Ekinci and Başbağ [39], who suggested that water limitation fosters root development. In accordance with Kaçar [40], the accelerated development of the main root in the cotton plant under drought conditions serves as an indicator of the plant's ability to reach soil moisture.

In summary, the Duncan Multiple Comparison test has elucidated the significant and insignificant variations in growth parameters resulting from distinct irrigation practices applied to the *Salvia splendens* L. plant. These findings provide valuable insights into the plant's response to water restriction and its impact on various growth aspects.

Effective use of water resources has become increasingly important all over the world and in Turkey. For this reason, new approaches aiming at the economical consumption of water used in irrigating ornamental plants, especially in landscape areas, have come to the fore. One of these approaches is that in Turkey, where a large proportion of usable water resources, such as 75%, is used for irrigation purposes, many intermediate steps need to be taken for the planning of existing water resources and effective water use. Deficit irrigation approaches should be developed in certain periods for ornamental plants that consume a lot of water and are used in landscape areas. With deficit irrigation approaches, water savings should be ensured [6,8].

Deficit irrigation is an irrigation approach applied at certain levels and processes, provided that it does not endanger the plant. Its main purpose is to save water by making adequate use of irrigation. In deficit irrigation, water use efficiency is increased and productivity is minimally affected. The most important feature of deficit irrigation is to obtain more income per unit area by giving the same amount of water to the plant and irrigating more area with the increased water [6,8,10].

In summary, the findings underscore the nuanced effects of varying water application rates on multiple aspects of *Salvia splendens* L. plants' growth and development. The relationship between these parameters and the overall water application rates is graphically presented in Figure 4.

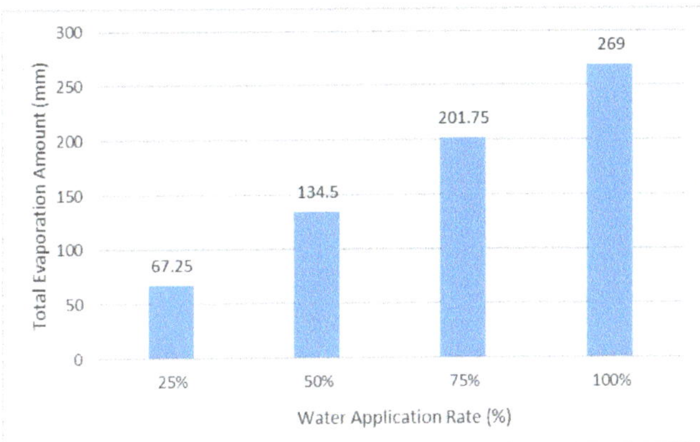

Figure 4. The relationship between the total evaporation amount and water application rates.

The respective values for flower stem thickness (30th), flower diameter (30th), and flower length (30th) within each irrigation treatment are depicted in Figure 5.

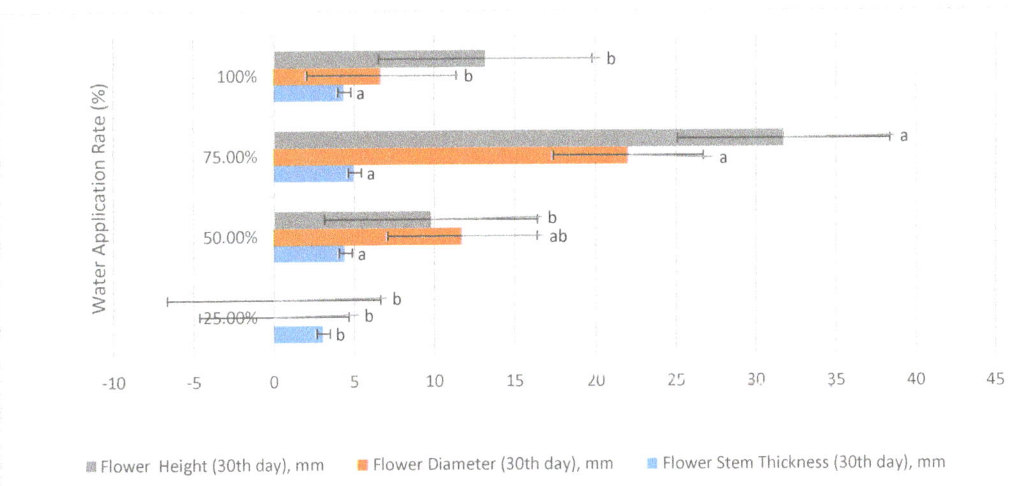

Figure 5. Flower height (30th day), flower diameter (30th day) and flower stem thickness (30th day) values for each irrigation application (Letters a, b indicate statistical difference).

The number of flowers (30 days) and number of flowers (55 days) values for each irrigation application are given in Figure 6.

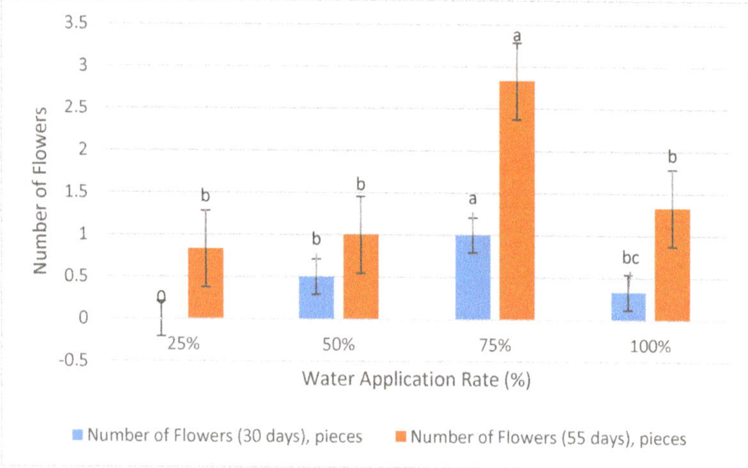

Figure 6. Number of flowers (30 days) and number of flowers (55 days) values for each irrigation application (Letters a, b, c indicate statistical difference).

Flower stem thickness (55th), flower diameter (55th), and flower length (55th) values for each irrigation application are given in Figure 7.

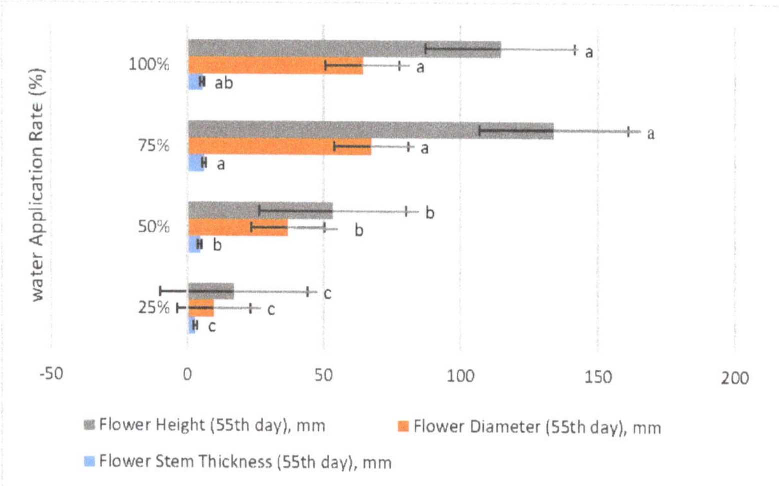

Figure 7. Flower height (55th day), flower diameter (55th day) and flower stem thickness (55th day) values for each irrigation application (Letters a, b, c indicate statistical difference).

Flower dry weight and flower fresh weight values for each irrigation application are given in Figure 8.

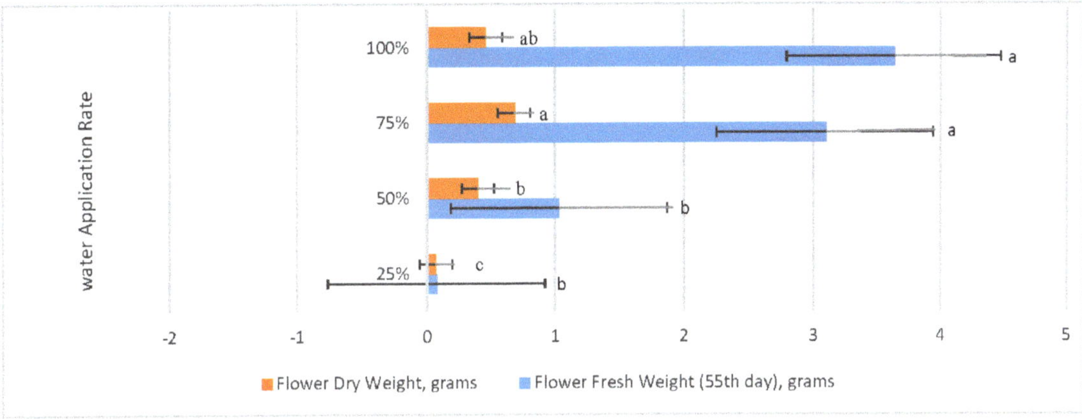

Figure 8. Flower dry weight and flower fresh weight values for each irrigation application (Letters a, b, c indicate statistical difference).

Leaf chlorophyll value (55th day) and leaf chlorophyll value (30th day) values for each irrigation application are given in Figure 9.

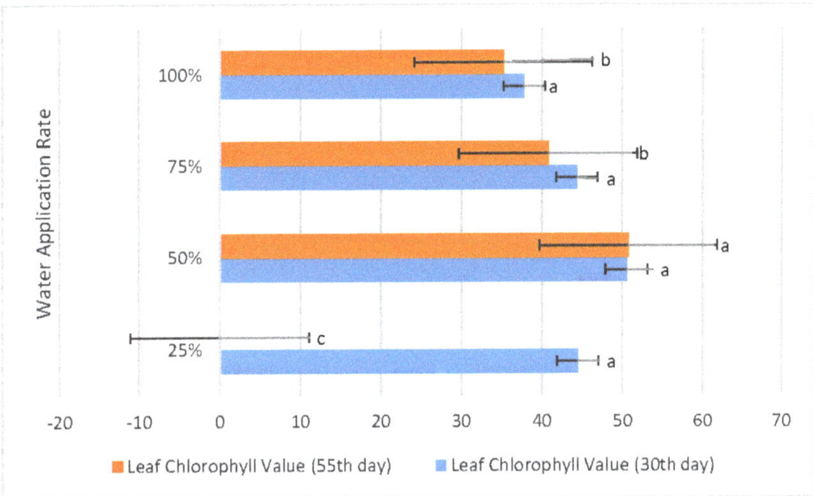

Figure 9. Leaf chlorophyll value (55th day) and leaf chlorophyll value (30th day) values for each irrigation application (Letters a, b, c indicate statistical difference).

The figures illustrating the plant height values on the 30th and 55th days for each irrigation treatment are provided in Figure 10.

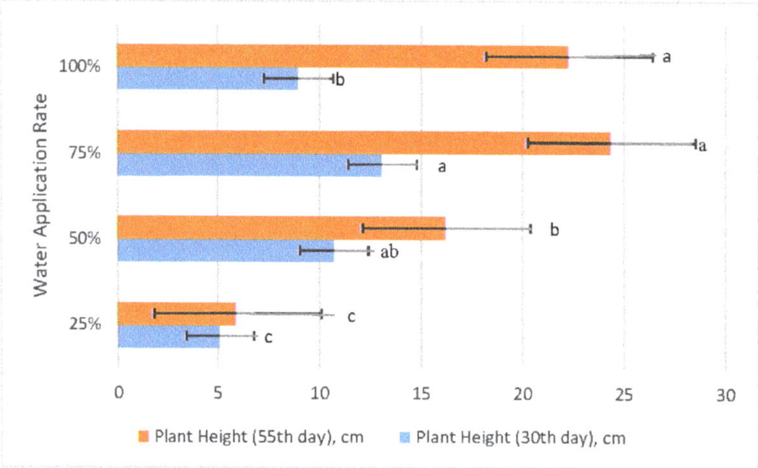

Figure 10. Plant height (55th day) and plant height (30th day) values for each irrigation application (Letters a, b, c indicate statistical difference).

Plant diameter (55th day) and plant diameter (30th day) values for each irrigation application are given in Figure 11.

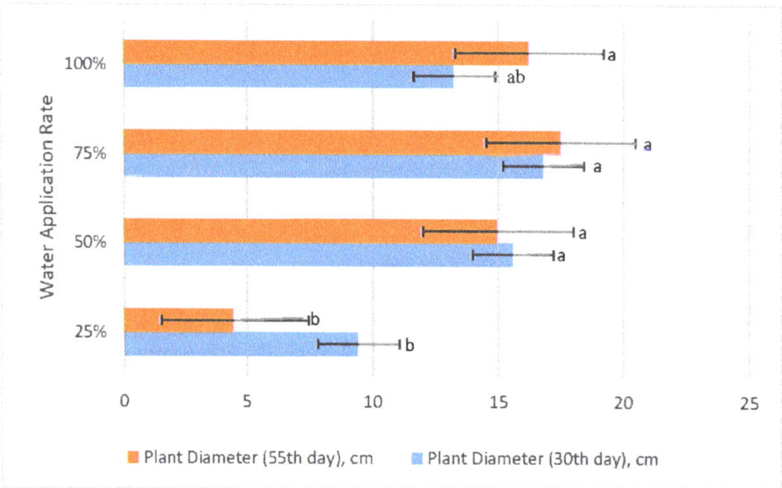

Figure 11. Plant diameter (55th day) and plant diameter (30th day) values for each irrigation application (Letters a, b indicate statistical difference).

The vegetative part dry weight and vegetative part wet weight values for each irrigation application are given in Figure 12.

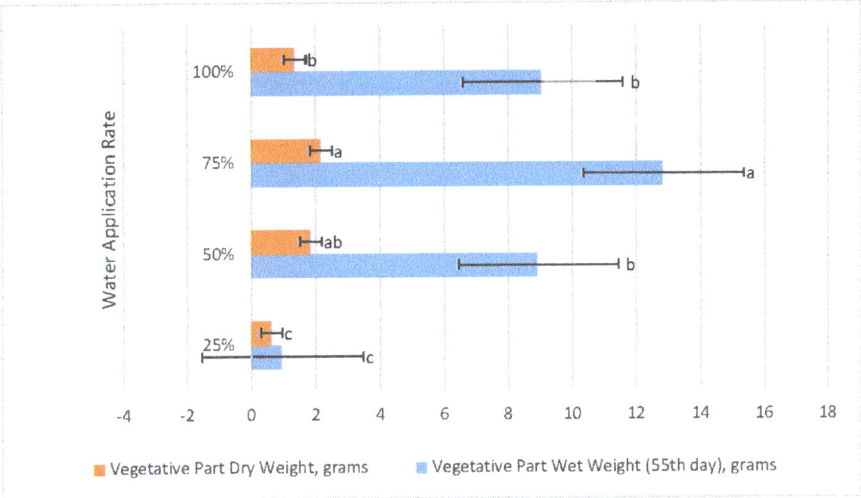

Figure 12. Vegetative part dry weight and vegetative part wet weight values for each irrigation application (Letters a, b, c indicate statistical difference).

Root dry weight and root fresh weight values for each irrigation application are given in Figure 13.

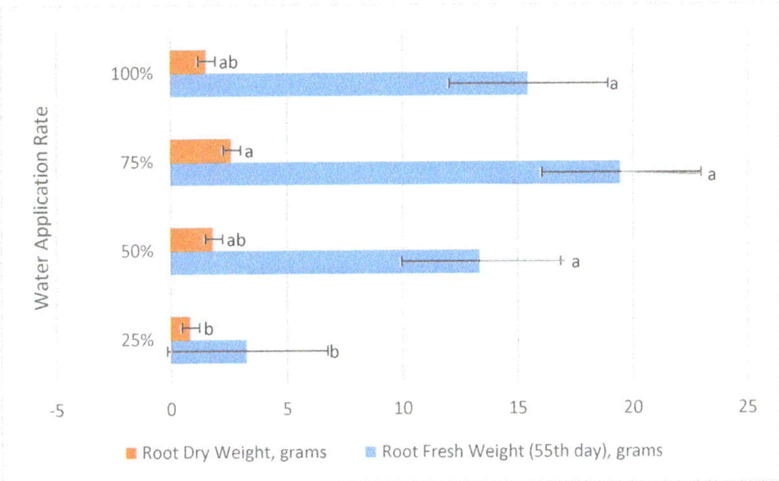

Figure 13. Root dry weight and root fresh weight values for each irrigation application (Letters a, b indicate statistical difference).

4. Conclusions

Seasonal flowers, a commonly utilized element in urban landscaping, are well-recognized for their substantial water consumption. Among these seasonal flowers, *Salvia splendens* L. stands out as the most prevalent choice. It is noteworthy that the existing literature indicates a scarcity of comprehensive studies addressing water restriction practices in ornamental plants. This study was conducted to address this research gap, with the aim of investigating the impacts of diverse water application regimes across 18 parameters related to plant growth.

Within the growth parameters, the difference between 75% irrigation and 100% irrigation in the parameters of number of flowers (30th days), number of flowers (55th days), flower stem thickness (30th days), flower stem thickness (55th day), flower diameter (30th day), flower length (30th day), flower length (55th day), flower dry weight, plant height (30th day), plant diameter (30th day), vegetative part wet weight (55th day), vegetative part dry weight, leaf area, root length (55th day), and root dry weight was statistically significant. It was observed that the values of growth parameters were higher with 75% irrigation. Although the difference between 75% irrigation and 100% irrigation in other parameters is not statistically significant, it is seen that the growth parameters in 75% irrigation have higher values than the development parameters in 100% irrigation. As a result, it was observed that 75% irrigation (25% water reduction) positively affected the growth parameters of the Salvia splendes plant. Especially in *Salvia splendens* L., the factors negatively affected by water scarcity were the fresh weight of the flower (day 55) and the number of plant shoots (day 55). Notably, the only aspect negatively affected by the water shortage in *Salvia splendens* L. was the development of flower diameter. Consequently, the results underscore the favorable influence of a 75% water application, presenting a 25% water reduction, on overall plant growth in *Salvia splendens* L., outperforming the results of a 100% water application with no water reduction.

In light of these findings, it is recommended to implement a 25% reduction in water when incorporating *Salvia splendens* L. in urban landscaping. This strategic approach not only fosters significant water conservation but also encourages the widespread adoption of *Salvia splendens* L. in urban environments, contributing to water-saving initiatives.

Supplementary Materials: The following supporting information can be downloaded at: https://www.mdpi.com/article/10.3390/w15234187/s1, Table S1: Changes in growth values according to irrigation applications in *Salvia splendens* L. andmultiple comparison test.

Author Contributions: I.S., A.K. and S.Y. were responsible for composing the main manuscript text; S.Y. conducted statistical analysis; and I.S., S.Y., E.A.K. and A.K. conducted the greenhouse studies as part of the research. All authors have read and agreed to the published version of the manuscript.

Funding: This research was supported by Atatürk University Scientific Research Projects Coordination Unit as a Basic Research Project with the code FBA-2021-9142.

Data Availability Statement: Data are contained within the article and supplementary materials.

Acknowledgments: We would like to thank Atatürk Uni-versity Scientific Research Projects Coordination Unit for their support. The authors express their gratitude to the Atatürk University Scientific Research Projects Coordination Unit for their invaluable support.

Conflicts of Interest: The authors declare no conflict of interest.

References

1. Lea-Cox, J.D.; Ross, D.S. A review of the federal clean water act and the Maryland water quality improvement act: The rationale for developing a water and nutrient planning process for container nursery and greenhouse operations. *J. Environ. Hort.* **2001**, *19*, 226–229. [CrossRef]
2. Cai, X.; Starman, T.; Niu, G.; Hall, C.; Lombardini, L. Response of Selected Garden Roses to Drought Stress. *Hortscience* **2012**, *47*, 1050–1055. [CrossRef]
3. Nicolas, E.; Ferrandez, T.; Rubio, J.S.; Alarcon, J.J.; Sanchez-Blanco, M.J. Annual water status, development, and flowering patterns for Rosmarinus officinalis plants under different irrigation conditions. *Hortscience* **2008**, *43*, 1580–1585. [CrossRef]
4. Niu, G.; Rodriguez, D.S.; Cabrera, R.; McKenney, C.; Mackay, W. Determining water use and crop coefficients of five woody landscape plants. *J. Environ. Hort.* **2006**, *24*, 160–165. [CrossRef]
5. Evsahibioğlu, N.A.; Akhüzüm, T.; Çakmak, B. Water Management, Water use strategies and transboundary waters, Turkish Agricultural Engineering VII. *Tech. Congr.* **2010**, *1*, 119–134.
6. Bayramoğlu, E.; Ertek, A.; Demirel, Ö. Approach The Deficit Irrigation in Landscape Architecture for Water Conservation. *İnönü Univ. J. Art Des.* **2013**, *3*, 45–53.
7. Barış, M.E. Arid Landscape. *Sci. Tech. J.* **2007**, *478*, 22–27.

8. Sezen, I.; Yağanoğlu, S.; Akpınar Külekçi, E.; Karahan, A. *Effect of Restricted Irrigation on Developmental Parameters of Tagates erecta Plant. The Increasing Need for Adaptation to a Rapidly Changing Environment: The Challenges of Agriculture*; Polat, T., Okant, M., Eds.; İKSAD Publishing House: Ankara, Turkey, 2023; pp. 55–76.
9. Trimmer, W.L. Applying partial irrigation in Pakistan. *J. Irrig. Drain. Eng.* **1990**, *116*, 342–353. [CrossRef]
10. English, M.J.; Musich, J.T.; Murty, V.V.N. Deficit irrigation. In *Management of Farm Irrigation Systems*; Hoffman, G.J., Howell, T.A., Soloman, K.H., Eds.; ASAE: St Joseph, ML, USA, 1990.
11. Jurriens, M.; Wester, P. *Protective Irrigation in India*; 1994 Annual Report; International Institute for Land Reclamation and Improvement: Wagenmgen, The Netherlands, 1994.
12. Çetinkale Demirkan, G.; Akat, H. Water Efficient Landscape Arrangements Approaches in Arid Regions 'Xeriscape'. In Proceedings of the 3RD ASM International Congress of Agriculture and Environment, Antalya, Turkey, 16–18 November 2017.
13. Sezen, I.; Esringü, A.; Yardımcı, K. Water Efficient Use for Sustainability of Water Resources in Urban Areas: Xeriscape. *Kent Akad.* **2018**, *11*, 474–485.
14. Çorbacı, Ö.M.; Özyavuz, M.; Yazgan, M.E. Water-wise in Landscape Architecture: Xeriscape. *J. Agric. Sci. Res.* **2011**, *4*, 25–31.
15. Ektiren, Y.; Değirmenci, H. Effect of Deficit Irrigation Applications on Plant Leaf Nutrition Elements of Cotton (*Gossypium hirsutum* L.). *KSU J. Agric. Nat.* **2018**, *21*, 691–698. [CrossRef]
16. Gençoğlan, C.; Yazar, A. The Effects of Deficit Irrigations on Corn Yield and Water Use Efficiency. *Turk. J. Agric. For.* **1999**, *23*, 233–241.
17. Süheri, S.; Topak, R.; Yavuz, D. The Effects of Different Irrigation Regimes on Yield and Water Use Efficiency of Sugar Bee. *Selcuk. J. Agric. Food Sci.* **2007**, *21*, 37–45.
18. Emre, M.; Küçükyumuk, C.; Yıldız, H. The effect of periodic restricted irrigation practices on income in apple production. *Derim* **2016**, *33*, 77–92. [CrossRef]
19. Ayas, S. Water-Yield Relationship of Limited Irrigated Potatoes. Ph.D. Thesis, Uludağ University Institute of Science and Technology, Bursa, Türkiye, 2007.
20. Gençoğlan, C.; Gençoğlan, C.; Akbay, C.; Uçan, K. Deficit Irrigation Analysis in Sunflower (*Helianthus annuus* L.). *KSU J. Sci. Eng.* **2005**, *8*, 138–144.
21. Camoglu, G. The effects of water stress on evapotranspiration and leaf temperatures of two olive (*Olea europaea* L.) cultivars. *Zemdirb.-Agric.* **2013**, *100*, 91–98. [CrossRef]
22. Demirel, K.; Camoğlu, G.; Akcal, A.; Genç, L.; Nar, H. Investigation of the Effects of Different Irrigation Levels on Plant Properties of Zinnia. In Proceedings of the 1st International, 14th National Congress on Agricultural Structures and Irrigation, Antalya, Turkey, 26–28 September 2018.
23. Riaz, A.; Martin, P.; Riaz, S.; Younis, A.; Hameed, M. Comparative performance of two zinnia (*Zinnia elegans*) cultivars under high-temperature and limited water conditions in Pakistan. In Proceedings of the I International Trials Conference: Assessment of Ornamental Plants, Wisley, UK, 4–7 July 2011; Volume 980, pp. 69–78.
24. Shams, J.; Najafi, P.; Etemadi, N.; Shams, A. Effects of irrigation levels on morphology and physiology characteristics of Petunia violacea. *Res. Crops* **2012**, *13*, 307–310.
25. Andersson, N.E. The influence of water stress and air velocity on growth of Impatiens walleriana and Petunia x hybrid. *Sci. Hortic.* **2011**, *128*, 146–151. [CrossRef]
26. Demirel, K.; Çatıkkaş, G.R.; Kesebir, B.; Çamoğlu, G.; Nar, H. Determination of Changes in Physiological and Morphological Properties of Cyclamen in Different Water Stress Levels. *J. Agric. Fac. Bursa Uludag Univ.* **2020**, *34*, 55–70.
27. Demirel, K.; Yıldırım, D.; Ayanoğlu, Z.; Albayrak, F.; Kuşak, İ.; Ersoy, Ç.; Budak, N.B.; Nar, H.; Çamoğlu, G.; Akçal, A.; et al. Investigation of the Effects of Water Stress on Morphological and Physiological Properties of Primula. *COMU J. Agric. Fac.* **2020**, *8*, 347–358.
28. Uçar, Y.; Kazaz, S. Effects of Different Irrigation Schedulings on Quality of Chrysanthemum. *J. Agric. Sci.* **2016**, *22*, 385–397.
29. Doğan, S.; Demirel, K.; Çamoğlu, G.; Nar, H.; Akçal, A. Determination of the Effects of Different Irrigation Levels on the Plant Characteristics of Regal Geranium. *Lapseki Vocat. Sch. J. Appl. Res.* **2020**, *1*, 1–15.
30. Gül, A.; Özçelik, H.; Uzun, Ö.F. The Characteristics and Adaptation of Some Native Groundcover Plants in Isparta Province. Isparta Province. *Süleyman Demirel Univ. J. Nat. Appl. Sci.* **2012**, *16*, 133–145.
31. Yazıcı, N.; Dönmez, Ş.; Kuş Şahin, C. Evaluation in terms of Xeric Landscape Design of Some Plants Used in Landscape Designs in Isparta. *Kastamonu Univ. J. For. Fac.* **2014**, *14*, 199–208.
32. Çetin, N.; Mansuroğlu, S. Determination of Plant Species can be used in Xeriscape design under Mediterranean conditions: The Sample of Antalya/Konyaaltı. *J. Agric. Fac. Ege Univ.* **2018**, *55*, 11–18.
33. Alp, Ş.; Onat, İ.; Kubik, M. *Ornamental Plants Production Techniques and Care Requirements*; Istanbul Metropolitan Municipality Parks and Gardens Green Areas Department Publications: Ankara, Türkiye, 2011; 156p.
34. Oral, N.; Açıkgöz, E. *Garden Flowers*; Çevre Publications, 1991; 171p.
35. Turan, A. The Effect of Different Irrigation Interval and Water Amount on Yield and Quality Parameters of Chrysanthemum (*Chrysanthemum morifolium* Ramat). Master's Thesis, Süleyman Demirel University Graduate School of Applied and Natural Sciences, Isparta, Turkey, 2013.
36. Tufenkci, S. The Effect of Different Levels of Water Deficit on Some Yield Parameters of Quinoa. *Eregli J. Agric. Sci.* **2021**, *1*, 27–37.

37. Kırnak, H.; Doğan, E. The Effects of Deficit Irrigation on Some Quantitative Parameters of Muskmelon with Subsurface and Surface Drip Irrigation Systems. *J. Agric. Fac. Gaziosmanpasa Univ.* **2017**, *34*, 80–86.
38. Curti, R.N.; Andrade, A.J.; Bramardi, S.; Velásquez, B.; Daniel Bertero, H. Ecogeographic structure of phenotypic diversity in cultivated populations of quinoa from Northwest Argentina. *Ann. Appl. Biol.* **2012**, *160*, 114–125. [CrossRef]
39. Ekinci, R.; Başbağ, S. Determination of Effects of Deficit Irrigation on Some Morphological Properties of Cotton (*G. hirsutum* L.). *Yüzüncü Yıl Univ. J. Agric. Sci.* **2019**, *29*, 792–800.
40. Kaçar, M.M. Examination of the Change of Water Stress Index in Cotton Plant of Different Water and Fertilizer Systems. Master's Thesis, Çukurova University Institute of Science, Adana, Turkey, 2007.

MDPI AG
Grosspeteranlage 5
4052 Basel
Switzerland
Tel.: +41 61 683 77 34

Water Editorial Office
E-mail: water@mdpi.com
www.mdpi.com/journal/water

www.ingramcontent.com/pod-product-compliance
Lightning Source LLC
Chambersburg PA
CBHW061609210926
51850CB00009B/2264
* 9 7 8 3 7 2 5 8 6 6 5 4 0 *